Lecture Notes in Computer Science 1008

Edited by G. Goos, J. Hartmanis and J. van Leeuwen

Advisory Board: W. Brauer   D. Gries   J. Stoer

Springer
Berlin
Heidelberg
New York
Barcelona
Budapest
Hong Kong
London
Milan
Paris
Santa Clara
Singapore
Tokyo

Bart Preneel (Ed.)

Fast Software Encryption

Second International Workshop
Leuven, Belgium, December 14-16, 1994
Proceedings

Springer

Series Editors

Gerhard Goos
Universität Karlsruhe
Vincenz-Priessnitz-Straße 3, D-76128 Karlsruhe, Germany

Juris Hartmanis
Department of Computer Science, Cornell University
4130 Upson Hall, Ithaca, NY 14853, USA

Jan van Leeuwen
Department of Computer Science, Utrecht University
Padualaan 14, 3584 CH Utrecht, The Netherlands

Volume Editor

Bart Preneel
Department Elektrotechniek-ESAT, Katholieke Universiteit Leuven
Kardinaal Mercierlaan 94, B-3001 Heverlee, Belgium

Cataloging-in-Publication data applied for

Die Deutsche Bibliothek - CIP-Einheitsaufnahme

Fast software encryption : second international workshop,
Leuven, Belgium, December 14 - 16, 1994 ; proceedings / Bart
Preneel (ed.). - Berlin ; Heidelberg ; New York ; Barcelona ;
Budapest ; Hong Kong ; London ; Milan ; Paris ; Tokyo :
Springer, 1995
  (Lecture notes in computer science ; Vol. 1008)
  ISBN 3-540-60590-8
NE: Preneel, Bart [Hrsg.]; GT

CR Subject Classification (1991): E.3, F.2.1, E.4, G.2.1, G.4

ISBN 3-540-60590-8 Springer-Verlag Berlin Heidelberg New York

Typesetting: Camera-ready by author
SPIN 10487173   06/3142 – 5 4 3 2 1 0   Printed on acid-free paper

CONTENTS

Session 3: Stream Ciphers–Cryptanalysis
Chair: Cunsheng Ding

Session 4: Block Ciphers—Differential Cryptanalysis
Chair: Eli Biham

Session 5: Block Ciphers–Linear Cryptanalysis
Chair: Bart Preneel

Session 6: Odds and Ends
Chair: Ross Anderson

Session 7: New Algorithms and Protocols
Chair: Eli Biham

Session 8: Recent Results
Chair: Bart Preneel

Author Index

Fast Software Encryption

Katholieke Universiteit Leuven

Leuven, Belgium, December 14–16, 1994

Organizing committee:

Bart Preneel (ESAT, Katholieke Universiteit Leuven) – Chair
Ross Anderson (University Computer Laboratory, Cambridge)
Eli Biham (Technion, Haifa)
Cunsheng Ding (University of Turku, Finland)
Dieter Gollmann (Royal Holloway College, University of London)
James Massey (Swiss Federal Institute of Technology, Zürich)

In cooperation with:

The International Association for Cryptologic Research (IACR)

Sponsored by:

Europay International
Microsoft
Uti-maco Belgium

Introduction

B. Preneel

Katholieke Universiteit Leuven
ESAT-COSIC
K. Mercierlaan 94, B-3001 Heverlee, Belgium

bart.preneel@esat.kuleuven.ac.be

1   An Overview

This volume contains the proceedings of the second workshop on fast software encryption, which was held at the Katholieke Universiteit Leuven in Belgium from the 14th to the 16th of December 1994. It followed a workshop held the previous year at Cambridge, whose proceedings were published as Lecture Notes in Computer Science, Vol. 809. The workshop was organized in cooperation with the International Association for Cryptologic Research (IACR).

The goal of this series of workshops is to advance our understanding of software algorithms for two cryptographic primitives which require very high speeds, namely encryption algorithms and hash functions. The reason for the interest in software is clear. With the proliferation of personal computers and workstations, everyone will have sufficient computing power on his desk or in his briefcase. On the other hand, performance of networks keeps increasing, and applications such as multimedia make use of this increased bandwidth. However, it is quite unlikely that most users will spend a significant amount of money on hardware solutions to secure their applications. Good software solutions require fast algorithms; unfortunately, for performance reasons many applications use weak solutions such as exoring with a constant string or ciphers which are easy to break.

As long as no fast provably secure solutions are available, we will have to live with ad hoc solutions. Designing secure ad hoc solutions is probably not difficult, but developing solutions which are both fast and secure is a hard problem. Progress in this direction is made by the combined effort of making new proposals on the one hand, and trying to find weaknesses in these proposals on the other hand. In parallel, the generalization of specific attacks leads to the development of a theory which supports both the design and analysis activity. Although some powerful tools have been developed, such as correlation attacks, linear attacks, and differential attacks, we are still very far from a theory on how to design ciphers.

At the workshop, about 30 research contributions were presented on this topic. Almost all authors have found the time to produce a final version, which has been incorporated into this volume after a brief review. As could be expected, fewer proposals for new ciphers appeared than in the first edition. However, this volume contains six proposals, as well as new results on the security of these new proposals.

For the stream cipher papers, we have the following results: U. Baum and S. Blackburn introduce a generalization of clock control, W.G. Chambers studies random permutations and their application to ciphers, and C. Ding presents a thorough study of cyclotomic sequences. Cryptanalysis is done by R. Anderson, who proposes a new algorithm for cryptanalysis of nonlinear filter generators and breaks the FISH algorithm proposed at last year's workshop and by E. Biham and P.C. Kocher, who demonstrate a practical attack on the PKZIP encryption algorithm. J.Dj. Golić explains how linear cryptanalysis can be extended to stream ciphers. A. Klapper discusses feedback with carry shift registers over certain rings of algebraic integers, and D.J.C. MacKay shows how free energy minimization can be used in the optimization of fast correlation attacks on stream ciphers.

Concrete proposals for new stream ciphers can be found in the contributions of Ross Anderson (PIKE), W.G. Chambers, and J.Dj. Golić.

Three new block ciphers are proposed by R.L. Rivest (RC5), M. Blaze and B. Schneier (McGuffin) and D.J. Wheeler and R.M. Needham (TEA). L.R. Knudsen develops extensions of differential attacks. J.L. Massey discusses differential cryptanalysis of SAFER-K64 (presented at last year's workshop) and presents a variant with a 128-bit key (SAFER-K128). V. Rijmen and B. Preneel show how differential attacks to DES can be extended to hash functions based on DES. The same authors also cryptanalyze the McGuffin algorithm of M. Blaze and B. Schneier. B.S. Kaliski Jr. and M.J.B. Robshaw study the linear cryptanalysis of FEAL using multiple linear relations. U. Blocher and M. Dichtl discuss a problem which occurs in linear cryptanalysis of the DES when more than one S-box is active per round. S. Vaudenay investigates multipermutations and shows how the absence of this property can be used to cryptanalyze MD4 and a variant of SAFER-K64.

Supporting theory is discussed by H. Dobbertin (bent functions), X. Lai (additive and linear structures), J. Daemen (correlation matrices), K. Nyberg (S-boxes with controllable linearity and differential uniformity) and L. O'Connor (linear approximation tables).

S. Lucks proposes a new NP-hard problem on which hash functions and stream ciphers can be based. M. Roe presents methods to reverse engineer a key escrow device. W. Penzhorn proposes a fast algorithm for homophonic coding based on arithmetic coding. M. Roe gives a list of benchmarks for encryption algorithm and hash functions.

No paper was submitted to the proceedings for the following two presentations: P. Landrock: "Do we need two theories?" (on the relation between block ciphers and hash functions) and J.-J. Quisquater: "Printed documents, images and sounds as messages: new open problems for cryptographic functions."

The next edition of the workshop will be held in Cambridge, U.K. from February 21-23, 1996 under chairmanship of Dieter Gollmann (Royal Holloway, University of London). It will be part of the 6-month research programme on "Computer Security, Cryptology and Coding Theory" at the Isaac Newton Institute for Mathematical Sciences in Cambridge.

2   Discussion Sessions

In this section we try to summarize the issues that came up in the discussions between presentations and in the scheduled discussion sessions.

2.1   Stream Ciphers Versus Block Ciphers

Currently there exist a number of block cipher algorithms in the open literature, including CAST, DES, FEAL, GOST 28147, IDEA, LOKI, LUCIFER, RC5, SAFER, TEA, WAKE, etc. DES and Lucifer were designed in the seventies, FEAL in the late eighties, but all the other proposals are less than five years old. The design principles employed in these algorithms serve as examples for the development of new block ciphers and also provide targets for the development of cryptanalytic attacks. System developers can select one of these for a particular application.

However, the situation for stream ciphers is entirely different, since many types of stream ciphers have been proposed and analyzed, but few standardized and published algorithms are available. The SEAL stream cipher described in the proceedings of the 1993 Cambridge workshop is a notable exception. This can in part be explained by the fact that stream ciphers are found mostly in hardware implementations, which often use proprietary algorithms. Especially in environments where error propagation and synchronization are important, specific stream ciphers are much more efficient than the Output FeedBack mode (OFB) or the Cipher FeedBack mode (CFB) of a block cipher.

The current situation may be described as follows: for block ciphers, we have little theory, but many algorithms, while for stream ciphers, we have much theory, but very few completely specified algorithms. It should be noted however that recent developments in cryptanalysis (such as the discovery of differential and linear cryptanalysis) have stimulated the development of theoretical work on block ciphers; these proceedings form a clear witness to this.

It was recognized that a definite need exists for good stream cipher algorithms. More specifically, we need concrete designs for stream ciphers based on existing or new theoretical work in the area, and which may be identified by "catchy" names, as in the case of block ciphers. If these algorithms are placed in the public domain, they could serve as good design examples and also as targets for the development of effective cryptanalytic attacks. This in turn would lead to the establishment of proven design principles for stream ciphers. Also, more stream ciphers would be used in software applications.

2.2   Criteria for New Algorithms

One very important observation is that there exists a definite need for fast encryption algorithms and hash functions, designed specifically for software implementation. Older algorithms, like DES, are largely based on bit-oriented operations; therefore straightforward implementations are very slow on general-purpose computers. It should be noted however that this has not stopped certain groups from developing very fast DES implementations, which are only two to three times slower than most new block ciphers (only WAKE is an order of magnitude faster than all the others). Another important consideration is that encryption algorithms implemented in software appear to be less vulnerable to export control restrictions than hardware devices. It was also remarked that cryptographic algorithms are for the time being completely absent in all benchmarks for computers or compilers.

Designers of software algorithms should aim to exploit the parallelism offered by byte/word oriented operations, since shift/rotate operations and integer addition/exor are efficiently performed by most modern computers. Multiplication is also an obvious candidate, but it is slow compared to the other operations (multiplication of two 32-bit registers requires 40 cycles on a 80486 processor). For the design of stream ciphers this would imply the development/extension of the theory of linear feedback shift-registers to the general case of byte/word operations.

It was agreed that look-up tables are a useful building block, but in order to obtain a high speed, the tables should fit into the computer's cache memory. A popular size for the tables seems to be 8 Kbyte. An alternative is to use data-dependent operations, following the example set by RC5.

A challenge was put forth (by E. Biham) for anyone to develop an encryption algorithm based on *bent functions*. In spite of the current development of the theory of bent functions, few cryptographic applications have been presented thus far. In response to this it was mentioned that C. Adams and S. Tavares have proposed the CAST block cipher, which uses S-boxes with 8 input bits and 32 output bits, where the individual output bits are bent functions.

It was concluded that for optimal performance under various circumstances, different algorithms will be required: there is a need for very compact algorithms (for example for smart cards), high speed algorithms (for hardware and for general purpose processors), fast self-synchronizing stream ciphers (for communications applications), and algorithms which lend themselves to implementation in a FPGA (Field Programmable Gate Array).

2.3   Provable Security

It was suggested (by J.L. Massey) that we should continue to pursue the development of a theory of provable security. Ideally, we would like to state the equivalence of Shannon's Channel Coding Theorem for cryptography. This would guarantee the existence of provably secure cryptographic systems and provide us with the confidence of pursuing an attainable goal. This pursuit would probably warrant a fresh look at complexity theory. From a cryptographic viewpoint it is important to distinguish between P/NP-complexity and Boolean complexity, which is expressed for example, in terms of gate count. This subject has already received some attention in the past and is in dire need of further exploitation.

2.4   Hash Functions and Secrecy

Experience has shown that hash functions are less susceptible to export control restrictions than cryptographic algorithms dedicated to secrecy. It was suggested to consider the development of special modes of hash functions, which are capable of providing secrecy. It was remarked that this requires different properties from the hash functions as those for which hash functions have been designed and evaluated. Also, the use of stream ciphers for authentication should receive further attention.

Acknowledgment

I would like to thank the participants for their contributions, and for providing the papers in LaTeX format, which has enabled me to edit this volume in electronic format. I also would like to thank the organizing committee:

- Ross Anderson (University Computer Laboratory, Cambridge)
- Eli Biham (Technion, Haifa)
- Cunsheng Ding (University of Turku, Finland)
- Dieter Gollmann (Royal Holloway College, University of London)
- James L. Massey (Swiss Federal Institute of Technology, Zürich)

for the hard work and for the fine collaboration, and Lars R. Knudsen and Kaisa Nyberg for helping with the review of a paper. I am grateful to Walter Penzhorn, who made available his notes on the discussion sessions. Special thanks go to Rita De Wolf and Vincent Rijmen for assisting with the local organization, and to our sponsors Europay International, Microsoft, and Uti-maco Belgium for their generous support.

Clock-controlled pseudorandom generators on finite groups*

Ulrich Baum[1] and Simon Blackburn[2]

[1] Institut für Informatik V, Universität Bonn,
Römerstraße 164, 53117 Bonn, Germany.
uli@leon.cs.uni-bonn.de
[2] Department of Mathematics, Royal Holloway,
University of London, Egham, Surrey TW20 0EX, UK.
uhah058@vax.rhbnc.ac.uk

Abstract. As a generalisation of clock-controlled shift registers, we consider a class of key-stream generators where a clocking sequence is used to control a "pseudorandom" walk on a finite group.

1 Introduction

Cascades of clock-controlled shift registers have been extensively studied as candidates for secure key-stream generators [2]. It is possible to regard any clock-controlled shift register used in this scheme as a finite state machine whose states are the elements of a finite cyclic group of order equal to the period of the register (see Section 2 for details). In this paper, we generalise this construction to the case where an arbitrary finite group is used as state space of each component of the cascade. Our motivation for this generalisation is twofold. Firstly since many finite groups can be implemented efficiently in both hardware and software, it is realistic to hope that some of the new generators constructed will be useful in practice. Secondly, the broader perspective gained should improve our understanding of the strengths and weaknesses of the generators currently in use.

The rest of this paper is organised as follows. Section 2 contains a description of our generalisation of a clock-controlled shift register. Sections 3 and 4 discuss the design criteria for such generators when used in a cascade. Next, we present some simple examples of our construction and discuss the statistical properties and security of their output sequences. Finally, we set out our conclusions and suggest areas for further research.

---

* This research was supported in part by Deutscher Akademischer Austauschdienst and the British Council under the British-German Academic Research Collaboration programme. Simon Blackburn is supported by E.P.S.R.C. Research Grant GR/H23719

2   The basic setup

A *(clock-controlled) group generator* consists of the following three components:

(1) A *control generator* C producing a periodic binary clocking sequence $(c_i)_{i \geq 0}$ of least period γ.
(2) A *finite group* $G = \langle g_0, g_1 \rangle$ generated by two elements g_0 and g_1.
(3) An *output function* $f : G \rightarrow GF(2)$.

The control generator may be any finite state machine which produces a binary output, e.g., an LFSR or a cascade of clock-controlled group generators. This allows us to build cascades from very simple group generators.

The clocking sequence is used to control a walk on the group G as follows. In each step, we move from $g \in G$ to gg_0 if the next bit of the clocking sequence is 0 or to gg_1 otherwise. Starting with the identity of G, this defines the *state sequence* $(q_i)_{i \geq 0}$ of our generator:

$$q_{-1} := 1$$

$$q_i := q_{i-1}g_{c_i} \quad (i \geq 0)$$

Hence $q_i = g_{c_0} \cdot g_{c_1} \cdots g_{c_i}$. In each step, the generator computes an output bit s_i by applying f to the current state:

$$s_i := f(q_i)$$

We note that clock-controlled LFSRs are a special case of our construction: Let $G = \langle g \rangle$ be cyclic of order $2^l - 1$ and choose a pair $(g_0 = g^{e_0}, g_1 = g^{e_1})$ generating G. For fixed $0 \neq \beta, \alpha \in GF(2^l)$, where α is a primitive element of $GF(2^l)$, define the output function f by

$$f(g^e) := Tr(\beta \alpha^e).$$

The resulting generator is a clock-controlled LFSR that is stepped by e_0 or e_1 steps according to the clocking sequence. In particular, setting $g_0 = 1$ yields the well-known stop-and-go generator.

We will end this section by giving a simple example of a non-abelian group generator: Let $G = S_3 = \langle (123), (12) \rangle$ be the symmetric group of order 6 and define the output function f by

$$\{(1), (23), (12)\} \mapsto 0, \qquad \{(123), (132), (13)\} \mapsto 1.$$

Clocked by the m-sequence $0111010\ldots$ of period 7, we obtain the following pseudorandom sequence of least period 21:

$$101000111110110000110\ldots$$

This sequence has periodic linear complexity 18 and a good run length distribution.

3   The state sequence

It seems hard to prove general properties of the output sequence (s_i) produced by a clock-controlled group generator since these strongly depend on the chosen output function f. In this section, we are going to analyse the *state sequence* (q_i) of a clock-controlled group generator, which does not depend on the output function f.

PERIOD. First, we determine the least period ρ of the state sequence (q_i). Recall that γ denotes the least period of the clocking sequence.

Lemma 1. $\rho = \gamma \cdot \text{ord}(q)$, where $q := q_{\gamma-1} = g_{c_0} \cdot g_{c_1} \cdot \ldots \cdot g_{c_{\gamma-1}}$.

Proof. Since the clocking sequence (c_i) can be reconstructed from the state sequence (q_i), ρ equals the least period of the sequence of pairs (c_i, q_i), which is easy to determine: Obviously, it is a multiple of γ. By definition of q, we have $q_{k\gamma-1} = q^k$ for $k \in N$. Hence the sequence (c_i, q_i) has minimal period $l\gamma$, where $l = \min\{k \mid q^k = 1\} = \text{ord}(q)$.

To apply this lemma, we have to find the order of q. In general, there seems to be no better way than explicitly computing q from one cycle of the clocking sequence. However, this is not necessary in some special cases: Modulo its derived subgroup G', G is a two-generator abelian group. The *coset* of G' containing q is uniquely determined by the number of zeroes and ones in one cycle of the clocking sequence. Hence the minimum order of an element of this coset is a lower bound on the $\text{ord}(q)$. See Sections 5 and 6 for examples.

STATES REACHED. Next, we ask how many times each state occurs in one period of (q_i). In particular, we would like to know whether the sequence contains all possible states. Our intuition is that for good statistical properties of the output sequence, the state sequence should hit every element of G about equally often.

From the definition of q, it is clear that $q_{i+\gamma} = qq_i$ for all i. It follows that if we write a full period of the state sequence (q_i) as $(e \times \gamma)$-matrix , where $e := \text{ord}(q)$, we obtain

$$(q_{i\gamma+j})_{0\leq i<e, 0\leq j<\gamma} = \begin{pmatrix} q_0 & q_1 & \cdots q \\ qq_0 & qq_1 & \cdots q^2 \\ q^2 q_0 & q^2 q_1 & \cdots q^3 \\ \vdots & \vdots & \vdots & \vdots \\ q^{e-1}q_0 & q^{e-1}q_1 & \cdots 1 \end{pmatrix} .$$

Each column of this matrix contains a coset of the cyclic subgroup $U := \langle q \rangle$ of G. Hence we have the following.

Lemma 2. (a) The set $Q := \{q_i \mid i \geq 0\} \subseteq G$ of states reached by the generator is a union of cosets of the cyclic subgroup $U := \langle q \rangle$.
(b) All elements of the same coset occur equally often in (q_i).

(c) $Q = G$ *iff* $(q_0, q_1, \ldots, q_{\gamma-1})$ *contains a transversal of the cosets of* U *in* G.

(d) (q_i) *contains every element of* G *equally often iff* $(q_0, \ldots, q_{\gamma-1})$ *contains the same number of elements from each coset of* U *in* G.

In general, one will have to construct the states $(q_0, \ldots, q_{\gamma-1})$ to check if the condition in part (d) is satisfied. In some special cases however, this can be easily seen *a priori* from simple conditions on the clocking sequence, see the examples in Sections 5 and 6.

STATISTICAL PROPERTIES. One can expect the state sequence to have good statistical properties if the clocking sequence has reasonable statistics. To see why, suppose that our generator's clocking sequence consists of independent and identically distributed random bits. Then the state sequence corresponds to a *random walk* on G. By general Markov chain theory, the random walk converges exponentially fast to a distribution that periodically cycles through the uniform distributions on all cosets of the unique normal subgroup U of G of the form $U = \{g_0, g_1\}^k$. For details and further reference on random walks, see [1]. This indicates that the state sequence should still have good statistics if the generator is clocked by a reasonable pseudorandom sequence.

4   The output function

The choice of the output function f is crucial for our generator's performance. Of the $2^{|G|}$ possible output functions, we should choose one that

- is easy and fast to evaluate,
- guarantees high period and linear complexity of the output sequence,
- yields an output sequence with good statistical properties and
- disguises the group structure of the generator as much as possible.

For cryptographic applications, there should be a sufficient number of good output functions for a given generator so the choice of f can be used as (part of) a secret key.

Our ultimate goal is to find an efficient algorithm which when given a suitable group G and generators g_0, g_1, finds a set of good output functions. As this appears to be beyond our reach at the moment, we will discuss some properties that we consider desirable.

Firstly, we consider how to choose f such that the output sequence has large period σ. Obviously, σ divides the period $\rho = \gamma \mathrm{ord}(q)$ of the state sequence. How can we make sure that σ has the maximum possible value ρ? A counting argument shows that if the set Q of states reached by the state sequence is large enough, then $\sigma = \rho$ for most choices of f. However, this is not very helpful since we have to *construct* a suitable output function that has additional properties. Hence we will try to find (simple) conditions on f that guarantee the maximal possible period $\sigma = \rho$.

PROPERTY 1. The least common multiple of the least periods of all sequences $(f(hq^i))_{i \geq 1}$, $h \in Q$, should equal $\mathrm{ord}(q)$.

Since these sequences are all shifted decimations of (s_i) by γ, Property 1 is a necessary condition for $\sigma = \rho = \gamma\text{ord}(q)$ according to Lemma 6. Although it is not sufficient in general, there are some special cases where Property 1 guarantees $\sigma = \rho$:

Lemma 3. *Suppose that $\gamma = p^a$ and $\text{ord}(q) = p^b$ $(a, b \geq 1)$ are powers of the same prime p. Then Property 1 implies $\sigma = \rho$.*

Proof. As p is prime, $\sigma | \gamma\text{ord}(q) = p^{a+b}$ must be a power of p as well: $\sigma = p^c$. Since $b > 1$, it follows from Lemma 7 that $c > a$ and $b = c - a$.

PROPERTY 2. The function f should be nearly balanced (so $||f^{-1}(0)| - |f^{-1}(1)||$ should be small).

We want our output sequence to contain about same number of zeroes and ones. Property 2 guarantees this if we assume that the state sequence hits every element of G equally often.

   If the output sequence is to be cryptographically secure, it is desirable that deducing any information about the state sequence from the output sequence should be difficult. Thus f should disguise the group structure as much as possible.

PROPERTY 3. For any proper normal subgroup N of G, f should not be constant on every coset of N.

Otherwise, the generator is equivalent to one based on the smaller factor group G/N. In particular, f or its complement should not be group homomorphisms. In fact, if a close approximation to f fails Property 3, then f is also a poor choice because then the generator can be approximated by a smaller one.

   We now introduce a much stronger property, which is easily seen to imply Property 3.

PROPERTY 4. The function f does not correlate with multiplication in G:

$$\forall g \in G \setminus \{1\} : |\{x \in G \mid f(xg) = f(x)\}| = |G|/2.$$

To see why this requirement makes sense, suppose that for some $g \in G$, $f(x) = f(xg)$ for significantly more (or less) than half of all $x \in G$. Then for all positions $i < j$ with $q_j = q_i g$, s_j can be predicted by s_i (resp. its complement), with high probability. Such correlations could also be used to gain information about the state sequence from the output sequence. Hence for cryptographic strength of our generator, f should – at least approximately – have Property 4.

   Functions satisfying Property 4 have been studied in the case when G is an elementary abelian 2-group under the name of *bent functions* [5]. By analogy with this case, we say a function f is bent if it satisfies Property 4.

   Unfortunately, bent functions do not exist for all groups and are never completely balanced:

Theorem 4. *Let G be a group of order n. Suppose that f is a bent function on G and define $z := |f^{-1}(0)|$ to be the number of zeroes of f. Then n must be an even square and $z = (n \pm \sqrt{n})/2$.*

Proof. For any elements $x, y \in G$, define the integer $t(x, y)$ by

$$t(x, y) := \begin{cases} 1 & \text{if } f(x) = f(xy) \text{ and} \\ -1 & \text{if } f(x) \neq f(xy) \end{cases}.$$

We evaluate the sum $\sum_{x \in G} \sum_{y \in G} t(x, y)$ in two different ways: Firstly,

$$\sum_{x \in G} \sum_{y \in G} t(x, y) = \sum_{x \in G} t(x, 1) + \sum_{x \in G} \sum_{y \in G \setminus \{1\}} t(x, y) = n + 0 = n$$

since because f is bent, $\sum_{x \in G} t(x, y) = 0$ for any $y \neq 1$. Secondly,

$$\sum_{x \in G} \sum_{y \in G} t(x, y) = \sum_{x \in f^{-1}(0)} \sum_{y \in G} t(x, y) + \sum_{x \in f^{-1}(1)} \sum_{y \in G} t(x, y)$$
$$= z(2z - n) + (n - z)(n - 2z)$$
$$= (n - 2z)^2.$$

Thus $n = (n - 2z)^2$ and hence n is a square and $z = (n \pm \sqrt{n})/2$. The theorem follows, since clearly n must be even.

It is not clear whether bent functions exist for every group whose order is an even square, nor do we know how to find them efficiently. If G is an elementary abelian 2-group, efficient constructions of bent functions are well known, see [5]. For small examples of non-abelian groups, we have found bent functions by exhaustive search. However, this approach is not feasible unless G is very small, say $|G| \leq 25$. The fact that bent functions are not perfectly balanced does not seem to be a problem in practice since if a balanced function is desired, we can complement an appropriate number of bits to produce a balanced function which still has reasonable correlation properties. Even if no bent function exists for a given group, we can search for "approximately bent" functions.

Despite their good correlation properties, we are not sure that bent functions are the ideal choice for cryptographically secure output functions. Because the property of being bent is so strong, it may well force extra structure upon the function which can be exploited in an attack. Some experimental evidence of such phenomena is presented in Sections 5 and 6. However, the structures we found would only exhibit themselves over long segments of the sequence, so they might not be detectable in practical situations.

5 Example 1: The quaternion group generator

In this section, we will have a closer look at a group generator based on the group of quaternions.

THE GENERATOR. In the notation of our basic setup, choose

$$G := \langle i, j \mid i^2 = j^2 = ijij = -1 \rangle = \{1, -1, i, -i, j, -j, ij, -ij\},$$

the quaternion group of order 8 with generators $g_0 := i$ and $g_1 := j$.

STATE SEQUENCE. Modulo the commutator subgroup $G' = \{1, -1\}$, the generators commute and have order 2. Hence the coset of G' containing a state q_i is uniquely determined by the number of zeroes and ones in the first i bits of the clocking sequence. Since all elements of $G \setminus G'$ have order 4, we know that the state q reached after a full period of the clocking sequence has order 4 iff one cycle of (c_i) contains an odd number of zeroes or ones, or both. In this case, the state sequence has period $\sigma = 4\gamma$.

From now on, suppose that the clocking sequence has least period $\gamma = 2^k$ and odd weight. Then it is clear that $q = \pm ij$ has order 4, so the state sequence has period $\rho = 4\gamma = 2^{k+2}$.

We claim that the state sequence hits every element of G the same number of times in each cycle. To see this, look at the subgroup generated by q:

$$U = \langle q \rangle = \{1, -1, ij, -ij\}.$$

Since U has index 2 in G and does not contain the generators $g_0 = i$ and $g_q = j$, the state sequence alternates between elements of U and elements of the coset $iU = jU = G \setminus U$. Since γ is even, it follows by Lemma 2 that every group element is reached exactly $\gamma/2$ times in one cycle of the state sequence.

OUTPUT FUNCTION. Next, we have to choose a suitable output function f satisfying Properties 1–4 of Section 4. Since 8 is not a square, no bent functions on G exist according to Theorem 4, so we have looked for functions that have Properties 1–3 and are approximately bent. By exhaustive search using the group theory package GAP [3], we have found the following 8 best candidates for f that look very much alike.

	1	-1	j	$-j$	i	$-i$	ij	$-ij$
f_1	1	0	1	1	0	0	1	0
f_2	0	1	1	1	0	0	1	0
f_3	1	0	0	0	1	1	1	0
f_4	0	1	0	0	1	1	1	0
f_5	1	0	1	1	0	0	0	1
f_6	0	1	1	1	0	0	0	1
f_7	1	0	0	0	1	1	0	1
f_8	0	1	0	0	1	1	0	1

If we use one of these output functions, the resulting binary output sequence of our generator has least period $\sigma = \rho = 4\gamma$ by Lemma 3. Since this is a power of 2, the linear complexity of the sequence is at least half the period according to Lemma 8.

IMPLEMENTATION. The quaternion group generator is very easy to implement in hard- or software: If we represent the elements of G by three bits according to

$$(a, b, c) \longmapsto (-1)^a i^b j^c,$$

multiplication by i means $(a, b, c) \mapsto (a \oplus b \oplus c, \bar{b}, c)$ while multiplication by j amounts to $(a, b, c) \mapsto (a \oplus b, b, \bar{c})$. The output function can be implemented by a lookup table or a few boolean gates. In this implementation, we can expect our generator to be about as fast as a 3-bit LFSR.

CASCADES. Since one quaternion group generator is too small to be useful in practice, we are interested in cascading them. If we can make sure that the clocking input of each generator in the cascade has odd weight, we know that the period and linear complexity of our sequence is increased by a factor of 4 in each stage. Unfortunately, the output of our generator has even weight since the output function has weight 4 and every group element is reached the same number of times. A simple (but cryptographically questionable) way to overcome this difficulty is to change one output bit in each cycle of each stage.

When building cascades of LFSR, it is common to add the input of each stage to its output. Our experiments indicate that that the sequences obtained from cascades of quaternion group generators have better statistical properties if we do this as well.

For cryptographic applications, a key of three bits per stage can be used to choose one of the 8 output functions $f_1, \ldots, f_8$ in each stage.

EXPERIMENTS. We have implemented a cascade of quaternion group generators in software using the group theory system GAP [3]. Each stage uses the output function f_1 from the table above and adds its input to its output. Between the stages, the first output bit is changed to guarantee that all clocking sequence in the cascade have odd weight. Clocking a four-stage cascade with the sequence $(1, 0, 0, 0)$, we obtain the following sequences:

(1011110001101001)

(1010111111011100110100000110011101111010100010011000010100110010)

(101000010000111111100001001100111001011100111100110001100000001
1100101110110000000010111101110000111001110001110110100010101010
0111010001011010101101000110011011000010011010011001001101010100
1001111011100101010111101000100101101100100100100011110111111111)

(1010010010111110111001001010101011001100000001010101001100010001
1010100000101011101111010000111110101100111111000011001111001001
1100001010001100100000101000100010101111001100100010000000110111
1100111000001000110110110011110011011111111001010010101000011101110
1100111101000001000011110101010101110110101011111111110110111010
0101001011010100010001111011000001010111000001101101110001110010
0110110100110111001011010011001101010001110111001100101011011000
0111000010110011011100101110001110110000001110100111110110001001

0111000111101011101100011111111110011001010100000000011001000100
1111110101111110111010000101101011110011010100101100110100011100
1001011111011001110101111101110110111111010011001110111010101100010
1001101101011101100011100110100110001010100111110000010110111011
1001101000010100010110100000000000100011111110101010100011101111
0000011110000001000100101110010100000010010100111000100100100111
0011100001100010011110000110011000000100100010011001111110001101
0010010111100110001100001001001000110101001000011010111001000100)

The output sequence of the fourth stage has period and linear complexity $4^5 = 1024$. The graph below shows that its linear complexity profile is nearly optimal, i.e. very close to the line $y = x/2$:

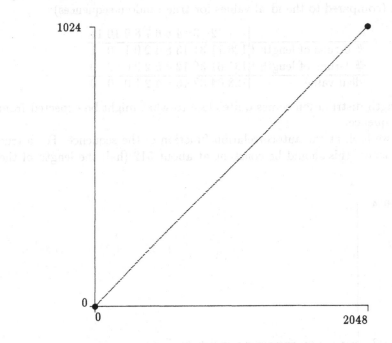

For all nonnegative integers k, the *k-error linear complexity* of a binary sequence is defined to be the smallest linear complexity that can be obtained by changing at most k bits of the sequence. The k-error linear complexity is an important indicator for the security of pseudorandom sequences: a sequence that has high linear complexity but is close to a sequence with low linear complexity, such as (0^n1), is cryptographically weak. See [6] for an algorithm which calculates the k-error complexity of a binary sequence of period 2^n. The k-error linear complexity of our sequence looks as follows:

k	k-error linear complexity
0	1024
1–254	514
255–256	513
257–510	257
511–512	1
≥ 513	0

Note that the linear complexity drops down to 514 if a single error is allowed. This happens because we changed one bit of the output sequence in order to give it odd weight. So in fact, the output sequence has linear complexity 514 which is quite close to the minimum value 512 from Lemma 8 and certainly far from the expected value of 1023 for a random sequence. This indicates that the sequence might have some hidden structure.

The sequence contains 511 zeroes and 512 ones with the following run length distribution (compared to the ideal values for true random sequences):

l	1	2	3	4	5	6	7	8	9	10	11
# 0-runs of length l	126	71	31	13	8	4	2	0	1	0	1
# 1-runs of length l	131	61	36	12	8	5	2	0	0	2	0
ideal value	128	64	32	16	8	4	2	1	0	0	0

This run length distribution comes quite close to what might be expected from a random sequence.

Finally, we look at the autocorrelation function of the sequence. For a true random sequence, this should be constant at about 512 (half the length of the sequence):

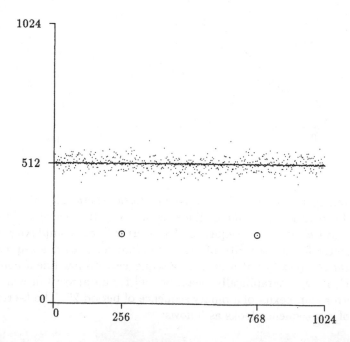

From the distinctive low values at shifts of 256 and 768, it can bee seen that the sequence has some global structure. In fact, we found the following symmetries: If the first half of the sequence is XORed to the second, we obtain (except for the first bit which was changed in the output) the alternating sequence $0101\ldots$. This problem is caused by the output function: A shift of $512 = 2\gamma$ in the state sequence is equivalent to multiplication by $q^2 = -1$. Since $f(-g)$ equals $\overline{f(g)}$ if $g \in \langle q \rangle$ and $f(g)$ otherwise, the sequence $(s_i \oplus s_{i+512})_{i \geq 0}$ alternates between 0 and 1.

CRYPTOGRAPHIC STRENGTH. Our sequence has sufficiently large period and linear complexity, a good local linear complexity profile and a good run length distribution. Unfortunately, it shows some strong correlations over large distances, and will thus be cryptographically weak if large portions of the sequence are used. However, the experiments indicate that the sequence is quite secure if only segments up to length 256 are used.

6   Example 2: The quaternion group of order 16

In a similar way, we can define a group generator using the quaternion group of order 16.

THE GENERATOR. We use the quaternion group of order 16 defined by

$$G := \langle h, k \mid k^4 = h^2, h^4 = 1, (h, k) = k^{-2} \rangle$$

and the generators $g_0 := h$ and $g_1 := k^3 h$.

STATE SEQUENCE. Modulo the derived subgroup $G' = \langle k^2, h^2 \rangle$, the two generators commute and have order 2. The group G contains four elements of order 8, which form the coset $g_0 g_1 G' = \{k, kh^2, k^3, k^3 h^2\}$. Hence q has order 8 iff one period of the clocking sequence contains an odd number of zeroes and an odd number of ones. In this case, the state sequence has least period 8γ.

The same argument as in Section 5 can be used to show that the state sequence hits every element of G the same number of times in one cycle.

OUTPUT FUNCTION. Again, we used GAP to search for a suitable output function. As the order of G is a square, bent functions on G may exist. By Theorem 4, such bent functions will not be balanced, but have either 6 or 10 zeroes. By exhaustive search, we have found 128 bent functions on G with 10 zeroes. The complements of these functions are the bent functions of G with 6 zeroes. Of these bent functions, we used the following as output function for the experiments:

1	h^2	h	h^3	$k^2 h$	$k^2 h^3$	$k^2 h^2$	k^2	k	kh^2	kh	kh^3	$k^3 h$	$k^3 h^3$	$k^3 h^2$	k^3
1	0	1	0	1	0	1	0	1	1	0	0	0	0	0	0

IMPLEMENTATION. There is an efficient four-bit representation of G similar to the one of the quaternion group of order 8 given in Section 5:

$$(a, b, c, d) \longmapsto k^{2a} h^{2b} g_0^c g_1^d.$$

The choice of 256 possible output functions will give us up to 8 bits of key for each stage of a cascade.

CASCADES. The comments in the previous section apply.

EXPERIMENTS. We have also implemented a cascade of these generators in GAP. Each stage uses the output function from the table above. As with the Q_8-generator, we add the input of each stage to its output and change the first output bit to make sure that all sequences in the cascade have odd weight. Clocking a three-stage cascade with the sequence $(1, 1, 1, 0, 0, 0, 0, 0)$, the output of the third stage is the following sequence of period and linear complexity $8^4 = 4096$.

```
( 1000100100011011010110111110101110000110000001000100100101110001010101011001011010110000011101100001010
  0010101101001110110100110100101001000101010101100011001001010010101100100111110100011101101101101110
  0101001101000010110110100100100001010010110100010011001110000001100110000010001100101011101001110
  0111011100000001100100000001010101111110000010111001000010000000101100100010010101010100011110101011
  1010111111100101111000100100100101011010101010110001100111110110001000110101111000010001110010010000
  0100101010010010010010011101101111010101101001001001000110011001110000010110011000010110011101110000000010
  10001011000000011011000000100110110000011010000101010100010001101000000011000010100011111110001111000011000
  00000001000100000001101001000101111011011000111111011110100101001011010001100011100011001100101100101010
  11111011000100111100000011100100010011011110111011010010010011110100110001000110001100011001110001010101010
  1001001011100110110010101110101010100100011001100111010011000011001011010000000001001010010110110000001
  10101101100010111111011101101010001000110001111100101011101010010010001000010101010100101100100110101111
  00000011001101111000111110000001110010001110001110001110010011100000010101001011011111000000110001101
  0100001001000011000100101011111001111010000011100011110010011110110100000110011001010000100101110010010
  001001100000010011100101110110011000011101011100000001000100000000010010010000010010001000011000010000110010
  001100000000110110100000010010110110111101100110110000100101010000010110010010010101001100001100001100000111
  01100111001011u11100100110100011100100011001100111000001101100000101001010010010100010001010101011100010
  1110010101011001100000100000100001101010001100100101001000010011000100100110011001011101101101011101101010100
  110010111010010010111110001110001110101010100011010010011011010011000100101011011111000010011010100110
  11010111011011010010110010010001001100110000001010110011001100011100000010001111000010001011011001011010001000100
  111010101101011011001001110000011100010000010000000111101000010110110101111100110101001101010001011111100
  0011100000010000000101110110110100101010100010010111101001010101101101010101101000011110010000010110010
  1110011001101000001110110000000100110010001010100101010101010000011101010010001011000011101011100001011
  11000101010000111000010111010010101001011100110000010111010110100110001000001000100100110001100100001110
  1011111010010111101011010010011011001110000110100001000110011001010110111011101010001110011001001001111
  101010001100011000001010010001000011110101010110011011100010110000100110010000010100011000000010100010101011
  110111100100110101010010010100111011001100010111010010110101001010001010100100000010000011010001000100110011
  110000100000000010010010010010010010100110011010010010001010011001100010001010100011000000111100111010000010001011
  01010011011011100000101011010010100000100111111011001000100111110100101011011111011011111000010110011011111
  110100100101100101011011100011101101111110110100100100100101010011001011111101101011111000101100101011111
  100011011111111100110110110001111111010010010010001010010000110011100101001000111110110100010001001000011
  001011111000011110100000100010010101011010110111100001100010011011101000110100110110101010101001101010011
  1111011000010011110010110110011110011000010001001010111000010010011100000100111011000011001110010010000111
  1000111100011111111110001000010110011011101000100010010110000011010000100110000101110011011000101001001000011
  1010100101011110100010101011100110011001101010101011100100001101101010110101101010100011000011110100000010
  0010011101000010000010111111110011000000010010100011011111101010101010101101011010100011001101100001100001110
  1111010001100111010011001001111110001111000011000000011010011100111100111110010100001010101010101010101010110
  0011110110001011001011011000001000011100010110110011010110110101111000001100100100110010011011010
  110101101100100010010011000100111110010110011110000101010011111101000110101101011000001110011100001110111 )
```

The linear complexity profile looks good as well:

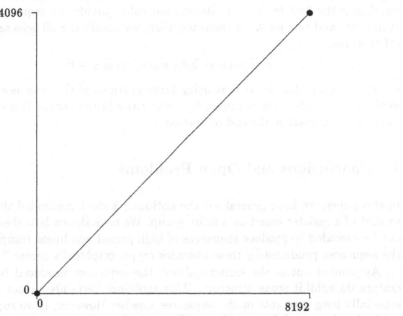

Its k-error linear complexity looks similar to the one of the sequence in Section 5:

k	k-error linear complexity
0	4096
1–1022	2561
1023–1534	1026
1535–1790	321
1791–1918	41
1919–1982	7
1983–1984	2
≥ 1985	0

Again, one might argue that the relatively low linear complexity of 2561 obtained when changing the first bit back to its original value hints at some global structure.

The sequence contains 2111 zeroes and 1985 ones with the following run length distribution, which is close enough to the ideal distribution for true random sequences:

l	1	2	3	4	5	6	7	8
# 0-runs of length l	522	269	114	61	33	12	8 2 1 1
# 1-runs of length l	486	267	137	66	32	15	11 4 3 3 0
ideal value	512	256	128	64	32	16	8 4 2 1 0

Finally, we have computed the autocorrelation function of the sequence. Over all shifts $\neq 0$, its minimum is 1920 and its maximum is 2222. All values being less than 9% off the ideal value of 2048, this seems acceptable.

CRYPTOGRAPHIC STRENGTH. So far, the test results indicate that we have found a good pseudorandom sequence. However, there are global correlations: When we change the first bit back to its original value, divide the sequence into four even parts and bitwise XOR them together, we obtain the all-zero sequence. In other words,

$$s_i \oplus s_{i+1024} \oplus s_{i+2048} \oplus s_{i+3072} = 0$$

for all i. Again, this shows that using large portions of the sequence should be avoided. This behaviour is caused by regularities in the output function similar to those discovered at the end of Section 5.

7   Conclusions and Open Problems

In this paper, we have generalised the notion of a clock controlled shift register to that of a register based on a finite group. We have shown how these registers can be cascaded to produce sequences of high period and linear complexity. Are the sequences produced by these cascades cryptographically secure ?

As pointed out in the sections above, the sequences produced by our generators do exhibit some structure. This structure certainly makes the use of especially long segments of the sequences unwise. However, if we suppose that only a portion of a period of the output sequence is used, the structures detailed in the previous sections do not seem to affect the security of the sequence since they involve terms of the output sequence that are very widely spaced. Since the local linear complexity profiles and run length distribution of the output sequences seem good, we have an indication that the sequences are secure when segments of reasonable length are used.

How can we maximise the assurance of security that our system gives ? The system depends on a careful choice of the output function of each generator in our cascade. We need to develop and expand the criteria given in Section 4. In particular, should we choose the output function to be bent? Maybe a function which only approximately satisfies the bent property but which performs better under other criteria is more appropriate. Further research – both experimental and theoretical – is needed on this matter. Is the form of the cascade we have used the most secure possible ? For example, we assume that the input to each stage is XORed with the output of the register. This operation seems to greatly improve the run length distribution of final output over that of a similar cascade with the XOR operation removed. Why is this, and can a different operation be introduced which increases the security of the output of the cascade ?

We believe that the concept of a bent function over a finite group is of interest in its own right, irrespective of its application in the situation outlined here. Do bent functions exist over any group whose order is an even square ? Certainly plenty of bent functions exist over the groups we have examined. Can large families of bent functions be constructed over certain families of groups ? The only constructions known so far apply only when the group is an elementary abelian 2-group.

In summary, the sequences produced by cascades of generators based on finite groups provide an interesting generalisation of the standard cascades of shift registers which can often be efficiently implemented. The output of such generators can have guaranteed minimum period and linear complexity. Experiments indicate that the sequences also have good linear complexity profiles and autocorrelation properties. However, further work is needed to establish that the sequences produced by cascades of this type are secure.

8   Appendix: well-known facts on period and linear complexity

For the convenience of the reader, this section contains a collection of well-known facts about period and linear complexity of binary sequences. These have been used in the proofs throughout this paper.

Lemma 5. *Let $(s_i)_{i \geq 0}$ be a periodic sequence of least period p. For $d, j \in N$, define the (shifted) d-decimation $(\sigma_i)_{i \geq 0}$ of (s_i) by $\sigma_i := s_{id+j}$. Then the following holds:*

(1) *The least period π of (σ_i) divides $p/(p,d)$.*
(2) *If $(p, d) = 1$ then $\pi = p$.*
(3) *For $(p, d) > 1$, π may be strictly less than $p/(p, d)$.*
(4) *If (s_i) has an irreducible minimal polynomial, then either $\pi = p/(p, d)$ or (σ_i) is the zero sequence.*

Proof. **(1)** For all i, we have

$$\sigma_{i+p/(p,d)} = s_{(i+p/(p,d))d+j} = s_{id+j+ \text{ lcm }(p,d)} = s_{id+j} = \sigma_i,$$

hence $\pi | p/(p, d)$.
   (2) For all i, we have

$$s_{(id+j)+\pi d} = s_{(i+\pi)d+j} = \sigma_{i+\pi} = \sigma_i = s_{id+j}.$$

As $(p, d) = 1$, $\{id + j \bmod p \mid i \geq 0\} = \{0, \ldots, p - 1\}$, hence $s_{i+\pi d} = s_i$ for all $i \geq 0$. It follows that $p|\pi d$. Since $(p, d) = 1$, we have $p|\pi$. Together with (1), our claim follows.
   (3) Example: Decimating the sequence 100010 by $d = 2$, we obtain 101 for $j = 0$ and 000 for $j = 1$.
   (4) [4, Ex. 9.5, p. 364].

Lemma 6. *Let $(s_i)_{i \geq 0}$ be a periodic sequence of least period md. For $j \geq 0$, let k_j denote the least period of the decimated sequence $(s_{id+j})_{i \geq 0}$. (Obviously, $k_j = k_{j \bmod d}$.) Then $m = lcm(k_0, \ldots, k_{d-1})$.*

Proof. Let $l := \text{lcm}(k_0, \ldots, k_{d-1}) = \text{lcm}(\{k_j\}_{j \geq 0})$. For all $j \geq 0$, we have $s_{ld+j} = s_j$ since l is a multiple of the period k_j. It follows that (s_i) is ld-periodic, hence $md \mid ld$ and $m \mid l$.
   On the other hand, $k_j \mid m$ for all j by Lemma 5(1), hence $l \mid m$.

Lemma 7. *Let $(s_i)_{i\geq 0}$ be a sequence of least period p^k. If we decimate this sequence by p^n, the least common multiple l of the least periods of all such decimations equals p^{k-n} for $k > n$ and 1 otherwise.*

Proof. Follows from the previous lemma and Lemma 5(1).

Lemma 8. *The linear complexity of a sequence $(s_i)_{i\geq 0}$ of least period p^n over a finite field of characteristic p is at least p^{n-1}. It equals p^n iff $s_0+s_1+\ldots+s_{p^n-1} \neq 0$.*

Proof. Since $x^{p^n} - 1 = (x-1)^{p^n}$ in characteristic p, the minimal polynomial of s has the form $(x-1)^l$ for some $l \leq p^n$. For $l < p^{n-1}$, it would divide $x^{p^{n-1}} - 1$, and the minimal period would be a divisor of p^{n-1}, contradicting our initial assumption. Hence $l \geq p^{n-1}$.

Let $s(x) := \sum_{i\geq 0} s_i x^i$ denote the generating function of s. Since $l \leq p^n$, we have $(x-1)^{p^n} s(x) = 0$. It follows that $l = p^n$ iff

$$(x-1)^{p^n-1}s(x) = \frac{x^{p^n} - 1}{x-1}s(x) = (x^{p^n-1} + x^{p^n-2} + \ldots + x + 1)s(x) \neq 0,$$

which proves the second claim.

References

1. P. DIACONIS, *Group Representations in Probability and Statistics*, Institute of Mathematical Statistics Lecture Notes – Monograph Series, 11, Hayward (CA), 1988.
2. D. GOLLMANN, W.G. CHAMBERS, *Clock-Controlled Shift Registers: A Review*, IEEE J-SAC 7/4 (1989), 525–533.
3. M. SCHÖNERT et al., *GAP: Groups, Algorithms and Programming*, RWTH Aachen, 1992.
4. R. LIDL, H. NIEDERREITER, *Introduction to finite fields and their applications*, Cambridge University Press, 1986.
5. R.A. RUEPPEL, *Analysis and Design of Stream Ciphers*, Springer, 1986.
6. M. STAMP, C.F. MARTIN, *An Algorithm for the k-Error Linear Complexity of Binary Sequences with Period 2^n*, IEEE Trans. Inform. Theory 39/4 (1993), 1398–1401.

On Random Mappings and Random Permutations

W. G. Chambers

Department of Electronic and Electrical Engineering,
King's College London, Strand, London WC2R 2LS, UK

w.chambers@kcl.ac.uk

1 Introduction

Much work has been done by many people, including the present author, to prove that certain classes of sequence-generator have guaranteed periods [1]. Here we examine what happens at the other extreme, with sequence generators which are finite-state machines modelled as having random next-state tables. The advantages are a lack of mathematical structure which might provide an entry for the cryptanalyst, and a huge choice of possibilities; the disadvantages are that there are no guarantees on anything, and as is well known there is a risk of getting a very short period.

Thus we consider a finite-state machine whose state is specified by an integer in the range $0, ...N - 1$, and which has a next-state function F which specifies the $n + 1$-th state s_{n+1} as $F(s_n)$ with s_n the n-th state. The output can also be regarded as a function of the state, but we are not so much concerned with this. Evidently the function F can be represented by a look-up table with addresses and entries in the range $0, ...N - 1$. Each function F corresponds to a state-diagram where the states correspond to N points $0, ...N - 1$, with point i (the predecessor) joined to point j (the successor) by an arrowed line if $F(i) = j$. Points lying on a loop or *cycle* are called cyclic points; a point i satisfying $F(i) = i$ is regarded as lying on a cycle of length 1. Every point has a unique successor, but some points have no predecessors and some have more than one predecessor. In general the state diagram will consist of a number of cycles, plus a number of directed trees rooted on cyclic points; in these trees the arrows are pointing to the root.

If the next-state function F is invertible, so that for every j in $0, ...N - 1$ there is an i satisfying $F(i) = j$, then the function will be called a permutation (an N-permutation), and the state-diagram will consist of a number of cycles without any trees rooted in them.

There are altogether N^N N-functions, and $N!$ N-permutations. Typically the state is represented by a number of bits, say n, so that we have $N = 2^n$. Normally n would be of the order of hundreds.

There is a considerable literature on this topic [2], [3], [4], [5], [6], [7].

2 Problems with Random Mappings

Random mappings in cryptography have attained a considerable notoriety, perhaps akin to the "Fool's Mate" in chess, as a trap for novices [2], [8]. Here are three reasons why random N-functions are best avoided:

1) On average, if we start from a specified state or point, we traverse of the order of $\sqrt{N}$ points before reaching a cyclic point, and the cycle we find ourselves on is typically of length of order $\sqrt{N}$.

2) A large number of starting points, of the order of $N/2$, are liable to end up on the same cycle.

3) It is not unlikely that the terminating cycle is a small fraction of $\sqrt{N}$. In fact the probability of finding a cycle of size $\beta\sqrt{N}$ is of the order of β, for small β.

Thus it can be shown that the probability of choosing at random an N-function with a tree exceeding ρN in size $(0.5 < \rho < 1)$ is asymptotically (for large N) equal to $\rho^{-1/2}-1$, so that we can find a tree of size $0.8N$ with probability 11.8%. It can also be shown [7] that the probability of finding a tree of size $> N/2$ rooted on a cycle of size $< \beta\sqrt{N}$ with $\beta \ll 1$ is about $\beta\sqrt{(2/\pi)}$.

On the other hand there are some points in favour of using general non-invertible mappings. We may be able to compensate for the "square-root" effect by doubling the number of bits specifying the state. Moreover the proof that a mapping is invertible is often constructive, telling us how to do it. This may be a source of cryptographic weakness.

3 Simple Properties of Random Permutations

The properties of random N-permutations are much more satisfactory than those of general random mappings. Thus we have the following elementary results:

1) The probability that two points lie on the same cycle of a randomly chosen N-permutation is $1/2$.

2) The probability of a given point lying on a cycle of length $\leq r$ (with $1 \leq r \leq N$) is simply r/N.

Thus although it is possible to choose a point on a 1-cycle the probability of this occurring is $1/N$, which is normally very tiny.

4 Simulating Cycle-lengths

In order to compare the cycle-structure of a given permutation with that of a typical random permutation it is useful to have a way of simulating the numbers and sizes of cycles in a random permutation. This can be done as follows: Suppose we are building up a permutation cycle by cycle and we have so far built up cycles containing all but R points. We now wish to find the probability that the next cycle to be formed has length l. From the points not yet included in the previous cycles we choose the unique point P whose numerical value is the least. Then we have to choose $l - 1$ further points not equal to P from the R points not included in the previous cycles. These are used to build up a string starting at P which is finally closed by the choice of P. The probability of obtaining a cycle of given length l in this way is $1/R$ if $l = 1$, and is

$$\frac{R-1}{R}\frac{R-2}{R-1}\cdots\frac{R-l+1}{R-l+2}\cdot\frac{1}{R-l+1} = \frac{1}{R}$$

if $l > 1$. Thus the probability is independent of l for $1 \leq l \leq R$, and this suggests the following Nim-like construction for simulating the numbers and the lengths of the cycles in a random N-permutation. We start with N objects. At any stage when there are R objects left $(R \geq 1)$ we choose at random an integer U in the range $1, \ldots, R$ with uniform probability. This gives the length of the next cycle. Then we decrease R by U and iterate unless R is now 0, when we stop. It is quite easy to simulate 10^8 permutations in this way.

5   Test Results with 2^{24} States

We carried out a test run on the generator

$$x_{n+3} = P[(ax_{n+2} + bx_{n+1} + x_n) \bmod 256], \tag{1}$$

where the 8-bit invertible mapping $P[.]$ was based on a method by Wheeler [9]. The x_n are 8-bit unsigned integers, and a and b are randomly chosen odd integers. This generator has 2^{24} states. We grouped the cycle-lengths l into ranges, Range-0 for $l = 1$, Range-1 for $l = 2$, Range-2 for $l = 3$ or 4, Range-3 for l from 5 to 8, up to Range-24 for l from $2^{23} + 1$ to 2^{24}. For each permutation we counted the number of cycle-lengths in each range. For the ranges corresponding to the shorter cycle lengths the distribution of counts is approximately Poisson, with a mean for any such range equal to the sum of the reciprocals of the lengths in the range. This was done for about 5000 permutations altogether, and the results scaled remarkably well with the results obtained by simulating the cycle-lengths of random permutations as described in the previous section. (See Table 1 for the counts found in this case.) Unstructured random permutations for P were also tried out, with similar results.

6   Generators Driven by a Periodic Input

One obvious way of reducing the risk of a short period is to drive the finite-state machine with a periodic input of period P from a sequence generator of guaranteed period. If P is very large, then nothing much can be said in general terms, but if P is very much less than the lengths of typical output sequences, then we have the following curious situation. The output can be regarded as the interleaving of P sequences generated by P finite-state machines with distinct structures. Thus if it is possible to de-interleave the sequences there is roughly a P-fold increase in the probability of one of them being found with a period

R	c=0	c=1	c=2	c=3	c=4	c=5	c=6	c=7	c=8	c=9	c=10
0	36792375	36781875	18396794	6129424	1533219	306707	51345	7226	919	109	7
1	60653544	30326549	7580746	1263713	158201	15826	1308	109	3	1	0
2	55810493	32542111	9496555	1847368	269100	31109	2994	239	29	2	0
3	53013429	33639358	10681920	2255302	359219	45440	4831	461	39	1	0
4	51529292	34170772	11323363	2501571	413340	54986	6042	582	51	1	0
5	50772158	34419894	11659964	2633092	447037	60395	6749	636	70	5	0
6	50391787	34533668	11836403	2704233	462232	63659	7261	682	70	5	0
7	50187758	34601249	11925662	2740190	471433	65429	7518	678	72	11	0
8	50096923	34628447	11968518	2756208	475991	65603	7474	775	56	5	0
9	50051307	34635363	11993991	2766419	477998	66537	7527	786	69	3	0
10	50020982	34647034	12006632	2769755	480499	66553	7788	685	69	3	0
11	50015585	34646548	12010145	2771860	480311	67100	7651	733	62	4	1
12	49996515	34663898	12010142	2773626	480899	66338	7779	736	64	3	0
13	49991107	34666006	12012180	2774882	480802	66399	7791	757	71	4	1
14	50001789	34657477	12006765	2776918	482266	66412	7567	732	69	5	0
15	49988613	34662906	12014393	2777989	481106	66386	7761	776	65	4	1
16	49997512	34656565	12012673	2777731	481101	65970	7633	741	69	3	2
17	49983626	34663103	12018489	2779363	480395	66495	7709	740	71	9	0
18	49985423	34657053	12019573	2779867	482147	67377	7735	736	82	6	1
19	49974912	34662769	12022805	2782080	481685	67147	7776	759	62	5	0
20	49969519	34664176	12024126	2784710	481899	66919	7770	798	73	10	0
21	49960804	34659303	12031792	2788560	483573	67427	7712	764	59	6	0
22	49968911	34645032	12027232	2836971	417030	101925	2899	0	0	0	0
23	53206024	25733028	19571408	1489540	0	0	0	0	0	0	0
24	30739151	69260849	0	0	0	0	0	0	0	0	0

Table 1. Table 1: Results from 10^8 simulated permutations on 2^{24} objects. The column labelled R gives the range-number (Section 5). A column labelled c=n gives the number of permutations with n cycles of lengths in the appropriate range. Thus the number 740 under c=7 with R=17 means that 740 permutations were found with 7 cycles of lengths in Range-17, that is of lengths from $2^{16} + 1$ to 2^{17}.

shorter than some stipulated value, in comparison with the original undriven machine. Thus suppose we have a 32-bit random finite-state machine; then there is a chance of roughly 2^{-16} of ending up in a 1-cycle. If this machine is driven in some way with a sequence of period $P = 2^{16}$ then the output will consist of P interleaved sequences, one of which may well have final period 1. Thus in the output there may be a value which is repeated every 2^{16} iterations.

On the other hand, the input period gives an almost certain guarantee on the output period, so that one way of allaying worries about short periods is to use an input sequence of very long period.

7   Generators that Modify their own Look-up Tables

Oddly enough, at the December 1993 workshop on "Fast Software Encryption" [10] there were no suggestions for generators that modify their own look-up tables as they run. There are certain practical objections to such a scheme of course, for instance if the generator needs to be restarted frequently or if random access to the key-stream is needed, but if one simply needs a long key-stream without any restarts such a method is feasible and of course it makes it harder for the cryptanalyst by presenting him as it were with a moving target.

If we have such a scheme then the internal description of the machine-state must include the look-up table. Now the point is the following: if the transformation from one state to the next is invertible, then we have a permutation for the next-state function, but otherwise we may be dealing with a general finite-state machine with its likelihood of much smaller loop sizes. Thus it would seem that the modification of the look-up tables should be carried out in a way that enables one to undo the transformation in a unique manner.

Thus as a simple example consider the following random-number generator proposed by Bob Jenkins and publicised on the Usenet service. The state variables are

a, b: Unsigned 32-bit integers
m[0]..m[255]: a lookup table of unsigned 32-bit integers
p: a counter cycling from 0 to 255

Initially p is set to zero and the other variables to random values. There are two internal unsigned 32-bit integer variables, x and y. Addition is carried out "modulo 2^{32}". We loop indefinitely on the following instructions:

```
x = m[p];
y = m[RS(x)]; /* RS(x) is a right-shift of x by 24, leaving an 8-bit result */
a = R(a); /* R() is a rotation left by 27 bits */
a = a + y; /* addition is mod 2^32 */
m[p] = a + b; /***/ /* see below */
output(b+y); /* put the value b+y on the output stream */
b = x;
p = (p + 1) mod 256; /* Step p */
```

The instruction marked /***/ evidently changes the lookup table. It is not hard to see that if we know the state variables at the end of this loop we could determine them at the start, including the value of the modified table-entry; thus we have what is in effect a permutation. The same thing would apply if we replaced /***/ by m[p] = m[p] + a + b, or m[p] = m[p] XOR (a + b), but not if we replaced /***/ by m[p] = m[p] + a, as we could not then deduce the initial value of b from the final values of the state variables.

8   A Cipher Claimed to Resemble A5

Two other encryption algorithms have recently been publicised on the Internet, without much theoretical backing. The first is "alleged RC4", which has similarities to the algorithm just described. Here the next state function is invertible.

The second (announced by Ross Anderson and Michael Roe) is purported to be very similar to the A5 cipher used in the GSM mobile telephone system [11]. It uses three binary linear feedback shift-registers with known (key-independent) primitive polynomials of degrees 19, 22 and 23 respectively. These registers are initially set in a key-dependent manner to non-zero values, and on each iteration they are stepped as follows: A control-bit is taken from a known position near the centre of each register. If two or three of the control bits are equal to 1, then the registers producing these bits are stepped. On the other hand if two or three of the control bits are 0, then the registers producing *these* bits are stepped. In effect the registers are mutually clock-controlled in a stop/go fashion, and it is easy to see that there is a probability of 3/4 that a register is stepped on any iteration of the algorithm. Thus the longest register would be expected to go through a complete cycle in roughly $P = (2^{23} - 1).4/3$ iterations.

Regarded as a finite-state machine the system has just under $2^{19+22+23} = 2^{64}$ states, and the next-state function is non-invertible. Thus we would expect to find eventual periodicities of the order of 2^{32}, after a precursor sequence of the same order of length. Instead, in a search for eventual periodicities the author found 237 cases (all distinct), and all of them had periods very close to small multiples of $P = (2^{23} - 1) \cdot 4/3$; moreover just over 40% of these cases had periods very close to P itself. (The precursor sequences were on the whole a little longer, of lengths something like $10P$ on average.) Evidently the shorter registers, one with a period very close to $P/16$ and the other with a period very close to $P/2$, are locking on to the period of the longest register, with respectively 16 cycles and 2 cycles for every cycle of the longest register. Further investigations have shown that this "lock-in" is a robust phenomenon, occurring independently of the choice of primitive polynomials, and even occurring if the three sequences of control bits are chosen as random periodic sequences with periods equal to numbers of the form $2^n - 1$. This topic is being pursued further.

The shortness of the period is probably not a hazard in normal use [11], where only a few hundred bits of output are required between key-changes, but perhaps one would be advised not to use this scheme as a random-number generator without further study.

9   Concluding Remarks

Encryption algorithms in which the look-up tables are continuously modified have the attractions of high speed and of making the analyst's task harder, but there may be a lingering doubt about the period, particularly when there is no significant theory available. The above discussion strongly suggests that the next-state function should be invertible, but one might like some further reassurance. The use of a "rekeyed" cipher is a good way of obtaining a guaranteed huge period. Here there is "driving" sequence generator D whose output is used to modify the state of a second generator G which provides the final output. The generator D has a known or lower-bounded huge value for its period. Thus a 32-bit cascade generator with a cascade of length 8, as suggested in [1], will have a period of $(2^{32} - 1)^8$. The speed of the driving generator D is not so critical, since the output generator G need not be rekeyed on every iteration. In this way we combine a generator D with fixed (but key-dependent) lookup tables with a generator G where the tables are continuously modified, and keep the advantages of both.

References

1. Chambers WG, "Two stream ciphers", *Lecture Notes in Computer Science*, **809**, 51-55 (1994)
2. Knuth D, *The art of computer programming, Vol 2: Seminumerical algorithms*, 2nd ed., Chapter 3 (Reading, Mass: Addison-Wesley) 1981
3. Kolchin VF, *Random Mappings*, (Optimization Software Inc., 1986), para 3.3
4. Flajolet P, Odlyzko AM, "Random mapping statistics", *Lecture Notes in Computer Science*, **434**, 329-354 (1990)
5. Arratia R, Tavare S, "The cycle structure of random permutations", *Annals of Probability*, **20**, 1567-1591 (1992)
6. Mutafchiev L, "Large trees in a random mapping pattern", *European Journal of Combinatorics*, **14**, 341-349 (1993)
7. DeLaurentis JM, "Components and cycles of a random function", *Lecture Notes in Computer Science*, **293**, 231-242 (1988)
8. Lewin R, *Complexity*, (London: J M Dent Ltd) 1993, p28
9. Wheeler D, "A bulk data-encryption algorithm", *Lecture Notes in Computer Science*, **809**, 127-134, (1994)
10. Anderson R (Ed.), "Fast Software Encryption", *Lecture Notes in Computer Science*, **809** (1994, Springer-Verlag)
11. Mouly M, and Pautet M-B, *The GSM System for Mobile Communications*, (1992: published by M. Mouly and Marie-B. Pautet, 49, rue Louise Bruneau, F-91120 Palaiseau, France), p249 and p481

Binary Cyclotomic Generators

Cunsheng Ding[*]

Department of Mathematics
University of Turku
Fin-20500 Turku, Finland

cding@ra.abo.fi

Abstract. In this paper a number of binary cyclotomic generators based on cyclotomy are described. A number of cryptographic properties of the generators are controlled. A general approach to control the linear complexity and its stability for periodic sequences over any field is shown. Two bridges between number theory and stream ciphers have been established, and the relations between the design and analysis of some stream ciphers and some number-theoretic problems are shown. A number of cryptographic ideas are pointed out.

1 Introduction

The word *cyclotomy* means "circle-division" and refers to the problem of dividing the circumference of the unit circle into a given number, n, of arcs of equal lengths [19]. Our interest in the theory of cyclotomy has stemmed from the rather remarkable fact that the cyclotomic numbers represent actually the difference property and the nonlinearity of some cryptographic functions from residue rings Z_p's to some abelian groups. The DSC and ADSC generators described in [11] are actually the cyclotomic generators of order 2. In this paper we describe some keystream generators based on the theory of cyclotomies modulo a prime p, a square of a prime, and the product of two distinct primes. These generators are all special natural sequence generators of Figure 1 [11], where a cryptographic function $f(x)$, which is a mapping from a residue ring Z_N to an abelian group $(G, +)$, applies to the register of the modulo N ring counter. In the upper part of Figure 1, i.e., the modulo N ring counter, the $\sum_N$ denotes the integer addition modulo N, the symbol i denotes the content of the register of the counter which is updated with each clock. Thus, the register of the ring counter outputs cycles through the elements $0, 1, ..., N$ of the residue ring Z_N. That is, if the register of the counter has value i at time t, then its value at time $t+1$ is $(i+1) \bmod N$, here and hereafter the $x \bmod N$ is defined to be the least positive integer congruent to x modulo N. The semi-infinite sequence z^∞, i.e., the output sequence of the

---

[*] Supported partly by the Academy of Finland under Project 11281.

Modulo N ring counter

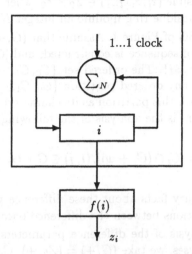

Fig. 1. The natural sequence generator

NSG, is defined by $z_i = f(i + i_0 \bmod N)$, where i_0 is the initial value of the register of the ring counter which is the key of the generator. The cryptographic function $f(x)$ and the modulus N are assumed not to be secret parameters of the generator.

For the binary natural sequence generator the following cryptographic analyses are equivalent:

1. differential analysis of the cryptographic function $f(x)$;
2. nonlinearity analysis of the cryptographic function $f(x)$;
3. autocorrelation analysis of the cryptographic function $f(x)$;
4. autocorrelation analysis of the output sequence;
5. two-bit pattern distribution analysis of the output sequence;
6. stability analysis of the mutual information $I(i; z_i z_{i+t-1})$, here and hereafter z^∞ denotes the output sequence of the NSG.
7. transdensity analysis of the additive stream cipher system with this NSG as the keystream generator by which we mean the analysis of the probability of agreement between two encryption resp. decryption transformations specified by two encryption resp. decryption keys [11].

By equivalence we mean one analysis results in another analysis.

We now prove the equivalences between the above seven analyses and show that an ideal difference property of the cryptographic function $f(x)$, by which we mean that the difference parameters defined below are approximately the same, ensures automatically an ideal nonlinearity of the cryptographic function $f(x)$, an ideal autocorrelation property of $f(x)$, an ideal autocorrelation property of the output sequence z^∞, an ideal two-bit pattern distribution property of the output sequence z^∞, and an ideal balance between the mutual information $I(i; z_i z_{i+t-1})$ for all possible $(z_i, z_{i+t-1}) \in Z_2 \times Z_2$, where t is arbitrary. In what follows Z_N denotes the residue ring modulo an integer N.

Consider now the NSG of Figure 1. Assume that $(G, +)$ is the abelian group over which the keystream sequence is constructed, and $|G| = n$. For each $i \in G$ let $C_i = \{x \in Z_N : f(x) = i\}$. The ordered set $\{C_0, C_1, \cdots, C_{n-1}\}$ is called the *characteristic class*. For any ordered partition $\{C_0, C_1, \cdots, C_{n-1}\}$ of Z_N, there exists a function $f(x)$ with this partition as its characteristic class. The differential analysis of the system is the analysis of the following *difference parameters*:

$$d_f(i, j; w) = |C_i \cap (C_j + w)|, (i, j) \in G \times G, \ w \in Z_N.$$

There are some elementary facts about these difference parameters [11], which represent some conservations between the difference parameters.

To see why the analysis of the difference parameters can be regarded as a kind of differential analyses, we take $(G, +) = (Z_2, +)$. Consider the input pairs (x, y) such that $x - y = a$, and consider the difference of the corresponding output pairs. Then we have the following expressions

$$\frac{|\{(x,y):f(x)-f(y)=1, \ x-y=a\}|}{|\{(x,y):x-y=a\}|} = \frac{d_f(0,1;a)}{N} + \frac{d_f(1,0;a)}{N}$$
$$\frac{|\{(x,y):f(x)-f(y)=0, \ x-y=a\}|}{|\{(x,y):x-y=a\}|} = \frac{d_f(0,0;a)}{N} + \frac{d_f(1,1;a)}{N},$$

These two expressions show that the difference parameters can be regarded as partial differentials or directional differentials of the function $f(x)$.

In what follows we prove the equivalences between the above seven analyses for the binary NSG (natural sequence generator).

Equivalence between differential and nonlinearity analyses: The nonlinearity analysis of the cryptographic function $f(x)$ refers to the analysis of the probability $p(f(x + a) - f(x) = b)$. It can be easily seen

$$\begin{aligned}
Np(f(x) - f(y) = 1) &= d_f(0, 1; a) + d_f(1, 0; a), \\
Np(f(x) - f(y) = 0) &= d_f(0, 0; a) + d_f(1, 1; a)
\end{aligned} \tag{1}$$

and

$$\begin{aligned}
2d_f(0, 0; -a) &= |C_0| - |C_1| + Np(f(x + a) - f(x) = 0), \\
2d_f(1, 1; -a) &= |C_1| - |C_0| + Np(f(x + a) - f(x) = 0), \\
2d_f(1, 0; -a) &= 2d_f(0, 1; -a) = N - Np(f(x + a) - f(x) = 0).
\end{aligned} \tag{2}$$

Then formulae (1) and (2) show the equivalence.

Equivalence between differential and autocorrelation analyses: The autocorrelation analysis of $f(x)$ refers to the analysis of the autocorrelation function

$$C_f(a) = \frac{1}{N} \sum_{x \in Z_N} (-1)^{f(x+a)-f(x)}.$$

It is easily verified

$$NC_f(a) = N - 4d_f(1, 0; a) \tag{3}$$

and

$$\begin{aligned}
4d_f(0, 0; a) &= 4|C_0| - N + NC_f(a), \\
4d_f(1, 1; a) &= 4|C_1| - N + NC_f(a), \\
4d_f(1, 0; a) &= 4d_f(0, 1; a) = N - NC_f(a).
\end{aligned} \tag{4}$$

Combining formulae (3) and (4) proves the equivalence between differential analysis and autocorrelation analysis of $f(x)$.

The autocorrelation analysis of the output binary sequence z^∞ refers to the analysis of the autocorrelation function

$$C_z(a) = \frac{1}{N} \sum_{i \in Z_N} (-1)^{z_{i+a}-z_i}.$$

Apparently by the definition of the NSG we have

$$C_z(a) = C_f(a), \quad \text{for each } a.$$

Thus, the above formulae (3) and (4) are also true we replace $C_f(a)$ with $C_z(a)$. This fact shows the equivalence between the differential analysis and the autocorrelation analysis of the output sequence z^∞.

Equivalence between differential and two-bit pattern distribution analyses: The two-bit pattern distribution analysis of z^∞ is concerned with how the two-bit patterns are distributed in a circle of length N in the sequence. For each fixed t with $0 < t \leq N - 1$ the vector (z_i, z_{i+t}) takes on elements of $Z_2 \times Z_2$ when i runs from 0 to $N-1$. Let $n[(z_i, z_{i+t}) = (a, b)]$ denote the number of times with which the vector (z_i, z_{i+t}) takes on $(a, b) \in Z_2 \times Z_2$ when i runs from 0 to $N - 1$. Then we have obviously

$$n[(z_i, z_{i+t}) = (a, b)] = d_f(a, b; -t). \tag{5}$$

Thus, for the binary NSG each difference parameter represents in fact the number times a two-bit pattern appears in a circle of length N of the binary output sequence z^∞.

Equivalence between differential and mutual information analyses:
Given two bits z_i and z_{i+t} of the output sequence of the binary NSG. It is cryptographically interesting to know how much information these two bits gives to the value of the register of the counter in the binary NSG at the time the output bit z_i was produced. It is easy to verify

$$I(i; z_i z_{i+t}) = \log_2 N - \log_2 d_f(z_i, z_{i+t}; -t) \cdot \text{ bits} \tag{6}$$

and

$$d_f(z_i, z_{i+t}; -t) = N 2^{-I(i; z_i z_{i+t})}, \tag{7}$$

where the mutual information $I(i; z_i z_{i+t})$ is measured in bits. Formulae (6) and (7) show clearly the equivalence. In addition they show that the difference parameters are in fact a measure of uncertainty.

Equivalence between differential and transdensity analysis: In a cipher system it is possible for two keys to determine the same encryption (resp. decryption) transformation. Even if the two transformation are distinct, it is cryptographically interesting to know the the probability of agreement between the two transformations. The control of this probability of agreement can prevent a cipher from any key approximation attack, that is, to use one key to decrypt the message encrypted by another key. Let E_k (resp. D_k) denote the encryption (resp. decryption) transformation specified by the key k. The analysis of the density (briefly, transdensity analysis) of a cipher system refers to the analysis of the probability of agreement $p(E_k(m) = E_{k'}(m))$, where m can be confined on plaintext blocks or without restriction [11].

For the additive binary stream cipher with the binary NSG as its keystream generator this probability can be expressed easily as

$$p(E_k = E_{k'}) = C_z(k - k' \bmod N) = C_f(k - k' \bmod N), \tag{8}$$

because of the additive structure of the additive stream cipher and the fact that the keystream sequences specified by all keys are shift versions of each other. Thus, the equivalence follows easily from formula (8).

So far we have proved the equivalences between differential and other six analyses. Thus, the equivalences among the seven analyses follows.

In addition, there is no tradeoff between all the above seven aspects and the linear complexity and its stability aspects for this generator (we will see this fact in later sections). This means that it is possible to design the NSG so that it has not only an ideal property for all the seven aspects in the usual senses, but also a large linear complexity and ideal linear complexity stability for the output sequence. It is because of these facts and that every periodic sequence can be produced by the natural sequence generator that the generator was called a natural one [11].

Formulae 1–8 clearly show that to ensure an ideal property for all the seven aspects, it suffices to control the difference property of the cryptographic function

$f(x)$. Thus, in what follows we will concentrate on the control of the difference property of $f(x)$, of the linear complexity, and of the sphere complexity of the output sequence of each specific NSG.

2   Cyclotomy and its cryptographic importance

The motivation of the investigation of cyclotomic numbers is related to the outstanding Waring problem, difference sets, and the solution of equations over finite fields [7, 9, 17]. Cyclotomic numbers invented by Gauss turn out to be quite valuable in the design and analysis of some keystream generators.

Let $N = df + 1$, be a odd prime and let θ be a fixed primitive element of Z_N. Denoting the multiplicative subgroup (θ^d) as D_0, we see that the coset decomposition of Z_N^* with respect to the subgroup D_0 is then

$$Z_N^* = \cup_{i=0}^{d-1} D_i,$$

where $D_i = \theta^i D_0$ for $0 \le i \le d-1$. The coset D_l is called the *index class l* [3] or *cyclotomic class l* [19]. Let $(l, m)_d$ denote the number of solutions (x, y) of the equation

$$1 = x - y, \quad (x, y) \in D_l \times D_m,$$

which were called *cyclotomic numbers* [2, 3, 8, 12], or equivalently,

$$(l, m)_d = |D_l \cap (D_m + 1)|.$$

Apparently, there are at most d^2 distinct cyclotomic numbers of order d and these numbers depend not only on N, d, l, m, but also on which of the $\phi(N-1)$ primitive elements of Z_N is chosen.

There are some elementary cyclotomic facts which are very important to our cryptographic applications, because they indicate several kinds of conservations between the cyclotomic numbers. They are the theoretical bases of the need of keeping the stability of local nonlinearities of some cryptographic functions.

We now see the meaning of the cyclotomic numbers from another viewpoint. From the definition we know the set $\{(l, m)_d : l = 0, 1, \cdots, d-1\}$ represents how the set $D_m + 1$ is distributed among the cyclotomic classes. Note

$$|D_l \cap (D_m + \theta^k)| = |D_{l+N-1-k} \cap (D_{m+N-1-k} + 1)|$$

for each k, we see that the d sets of numbers $\{(l, m)_d : l = 0, 1, \cdots, d-1\}$ for $m = 0, 1, \cdots, d-1$, represents also the distribution of the elements of any set $D_m + w$ with $w \neq 0$.

As seen above, cyclotomic numbers represent in fact the difference property of the partitions $\{D_0, D_1, \cdots, D_{d-1}\}$. So they should have connections with difference sets. Actually, the investigation of residue difference sets is the main motivation of the calculation of cyclotomic numbers [3, 19]. Now we see the cryptographic importance of cyclotomy.

Let the symbols as before. What we want to do now is to construct cryptographic functions from Z_N to an abelian group $(G, +)$ of d elements, where $G = \{g_0, g_1, \cdots, g_{d-1}\}$. Let

$$C_0 = D_0 \cup \{0\}, \quad C_i = D_i, \ i = 1, \cdots, d-1.$$

Without concerning the implementation problem, we define a function from Z_N to $(G, +)$ as: $f(x) = g_i$ iff $x \in C_i$.

If $i \cdot j \neq 0$, then we have

$$d_f(g_i, g_j; \theta^k) = (i + N - 1 - k, j + N - 1 - k)_d.$$

On the other hand, we have

$$d_f(g_0, g_0; \theta^k) = |(D_{N-1-k} \cup \{0\}) \cap (D_{N-1-k} \cup \{0\} + 1)|.$$

It follows that

$$0 \leq d_f(g_0, g_0; \theta^k) - (N - 1 - k, N - 1 - k)_d \leq 2.$$

Similarly, we have

$$0 \leq d_f(g_0, g_1; \theta^k) - (N - 1 - k, N - k)_d \leq 1.$$

and

$$0 \leq d_f(g_1, g_0; \theta^k) - (N - k, N - 1 - k)_d \leq 1.$$

Thus, we arrive at the conclusion that the difference parameters are almost the same as the cyclotomic numbers.

3   A basic theorem and main bridge

Before describing some binary cyclotomic generators, we introduce the sphere complexities and show why it is necessary to control the sphere complexity for those cyclotomic sequences described later.

Let x and y be finite sequences of length n over $GF(q)$, $WH(x)$ denote the Hamming weight, and $d_H(x, y) = WH(x - y)$, the Hamming distance between x and y. Let $O(x, y) = \{y : \ 0 < d_H(x, y) \leq u\}$ be the sphere without center x. The *sphere complexity* [10] is defined by

$$SC_u(x) = \min_{y \in O(x,u)} L(y).$$

here and hereafter $L(x)$ denotes the linear complexity or linear span of x.

Similarly, let s^∞ be a sequence of period N (not necessarily least period) over $GF(q)$. The sphere complexity [10] of periodic sequences is defined by

$$SC_u(s^\infty) = \min_{0 < v \leq u} \left[\min_{WH(t^N) = v, per(t^\infty) = N} L(s^\infty + t^\infty) \right],$$

where $per(t^\infty) = N$ denotes that t^∞ has a period N, and $t^N = t_0 t_1 ... t_{N-1}$.

That the control of the sphere complexity of keystream sequences for additive synchronous stream ciphers is cryptographically necessary follows from the fact that there is a polynomial-time algorithm which determines a LFSR with approximately the same output as the original keystream sequence, provided that the linear complexity of the keystream sequence has a bad stability. This algorithm can be roughly described as follows.

If the sphere complexity $SC_k(s^\infty) = l$ of the binary semi-infinite sequence s^∞ is small for some very small integer k, then theoretically the sequence s^∞ can be written as

$$s^\infty = t^\infty + w^\infty,$$

where s^∞, t^∞ and w^∞ all have a period N with respect to which the sphere complexity is concerned, and $L(t^\infty) = l$, and the Hamming weight $WH(w^N) \le k$. The task of this polynomial-time algorithm is to construct a LFSR of length l which produces the sequence t^∞ or another LFSR which outputs a sequence with the probability of agreement with the original sequence s^∞ no less than $1 - k/N$.

Suppose that a cryptanalyst gets a piece of the sequence s^∞, say S. Then the piece must be written as $S = T + W$, where T and W are the corresponding pieces of the periodic sequence t^∞ and w^∞ respectively. Since the k is very small, with a very high probability, which depends on the length of S and the pattern distribution of s^∞, it holds $S = T$. In this case if the length of S is large than $2l$, then applying the Berlekamp-Massey algorithm [14] to S will give an LFSR which produces the sequence t^∞ with the probability of agreement with s^∞ being no·less than $1 - k/N$.

If $S \ne T$, the Hamming weight of $S - T$ must be very small since k is very small. Then by changing a few bits in S the cryptanalyst gets T. However, he/she does not know the actual sequence $S - T$. But he/she can first get a number of sequences S_i by changing only one bit in the ith position of S for all i, in this way he/she gets m modified versions of S, where m is the length of S. Then apply the Berlekamp-Massey algorithm to each modified version to get a LFSR. After that use these LFSRs to decipher a long piece of ciphertext. If one LFSR has a probability of correct decipherment no less than $1 - k/N$, then the cryptanalyst accept this LFSR for approximating the original keystream generator. Otherwise changing two bits each time in S gives $m(m - 1)/2$ modified versions of S, then apply the Berlekamp-Massey algorithm to these modified version to see whether an acceptable LFSR is obtained. If not, try to modify three bits to get $m(m - 1)(m - 2)/6$ versions, and apply the Berlekamp-Massey algorithm to the modified versions again. Since k is very small, the cryptanalyst must get an acceptable LFSR after repeating the procedure a number of times. Since the complexity of the Berlekamp-Massey algorithm is of order $O(m^2)$, where m is the length of the input sequence, the complexity of this approximation algorithm must be polynomial. The smaller the k, the less the complexity of this approximation algorithm. This algorithm clearly shows the importance of the sphere complexity. It is quite clear that if $SC_6(s^\infty) = l$ is small, with the above algorithm a cryptanalyst must succeed in get a LFSR with the probability of agreement with the original generator larger or equal to $1 - 6/N$.

We describe the above algorithm here in order to show the cryptographic necessity of controlling the sphere complexity for our binary cyclotomic sequences for additive stream ciphering. The necessity of controlling the linear complexity of keystreams for additive stream ciphering follows from the efficient Berlekamp-Massey algorithm. After having shown the need for controlling the linear and sphere complexity for cyclotomic sequences, we now prove some theorems which will be needed when we control these complexities for those sequences.

Basic Theorem 1 *Let $N = p_1^{e_1} \cdots p_t^{e_t}$, where $p_1, \cdots, p_t$ are t pairwise distinct primes, q a positive integer such that $\gcd(q, N) = 1$. Then for each nonconstant sequence s^∞ of period N over $GF(q)$, we have*

1. *$L(s^\infty) \geq \max\{ord_{p_1}(q), \cdots, ord_{p_t}(q)\}$;*
2. *$SC_k(s^\infty) \geq \max\{ord_{p_1}(q), \cdots, ord_{p_t}(q)\}$,*
   if $k < \min\{\text{WH}(s^N), N - \text{WH}(s^N)\}$,

here and hereafter $L(s^\infty)$ and $SC_k(s^\infty)$ denote the linear and sphere complexity of the sequence respectively, $\text{WH}(x)$ the Hamming weight of x, and $s^N = s_0 s_1 \cdots s_{N-1}$.

Proof: Let K be a field of characteristic p, the n a positive integer not divisible by p, and ξ a primitive nth root of unity over K, the nth cyclotomic polynomial is defined by

$$Q_n(x) = \prod_{s=1, \gcd(s,n)=1}^{n} (x - \xi^s).$$

To prove the theorem, we need the following properties of the cyclotomic polynomial (see Lidl and Niederreiter [13] for proof).

1. $Q_n(x)$ is independent of the choice of ξ.
2. $\deg(Q_n(x)) = \phi(n)$.
3. The coefficients of $Q_n(x)$ belong to the prime subfield of K.
4. $x^n - 1 = \prod_{d|n} Q_d(x)$.
5. If $K = GF(q)$ with $\gcd(q, n) = 1$, then Q_n factors into $\phi(n)/d$ distinct monic irreducible polynomials in $K[x]$ of the same degree d, where d is the least positive integer such that $q^d = 1 \bmod n$, i.e., d is the order (or exponent) of q modulo n, denoted as $ord(q)$ modulo n or $ord_n(q)$.

It is easily seen that $ord_{p^k}(a) \geq ord_p(a)$ for any prime p and any positive integer a with $\gcd(a, p) = 1$. By assumptions and the above basic property 5 the polynomial $x^n - 1$ is equal to the product of $\phi(n)/d$ distinct monic irreducible polynomials over $GF(q)[x]$ of the same degree d, where d is the least positive integer such that $q^d = 1 \bmod n$, i.e., d is the order (or exponent) of q modulo

n. Since the minimum polynomial of each sequence of period N over $GF(q)$ divides $x^N - 1$ and $x^N - 1 = \prod_{n|N} Q_n(x)$, we consider the orders $ord_n(q)$ for each possible divisor n of N.

If n divides N, there are integers $h_{i_1}, ..., h_{i_s}$, where $1 \le h_{i_j} \le e_{i_j}$ for $1 \le j \le s$ and $1 \le s \le t$, such that $n = p_{i_1}^{e_{i_1}} \cdots p_{i_s}^{e_{i_s}}$. By the Chinese Remainder Theorem and the above conclusions

$$ord_n(q) = \mathrm{lcm}\{ord_{p_1^{e_1}}(q), \cdots, ord_{p_t^{e_t}}(q)\}$$
$$\ge \max\{ord_{p_1^{e_1}}(q), \cdots, ord_{p_t^{e_t}}(q)\}$$
$$\ge \min\{ord_{p_1}(q), \cdots, ord_{p_t}(q)\}.$$

Thus, the conclusions of this theorem follow. QED

One can see that the above lower bound is optimal. If $t = 1$ and $e_1 = 1$, then we have the general lower bound for sequences of a prime period. If $t = 1$, then it gives a lower bound for the linear complexity and sphere complexity of sequences with period being a prime power. Most of the theorems and corollaries in the paper are special cases of the above basic theorem, that is why we call it a basic theorem. We say that it is a bridge between number theory and stream ciphers because it makes a clear connections between the linear and sphere complexity of sequences and quite a number of number-theoretic problems such as primes of special forms (e.g., Sophie German primes, Stern primes, twin primes) and their distributions, primality testing, primitive roots and their distributions, and primitivity testing. Some of these connections will be made clear in Sections 4–7.

This basic theorem shows that it is usually quite easy to control the global linear and sphere complexities. However, it seems fairly difficult to control the local linear and sphere complexities. It is worthy to note that here we use the speciality of period to control the linear and sphere complexities, while some cryptographic functions are traditionally used to control the global linear complexity in the literature. Thus, we stress the importance of period.

4   Cyclotomic generator of order $2k$

Binary sequences with prime period are cryptographically attractive due to the following theorems about the linear and sphere complexities, which follow easily from Basic Theorem 1.

Theorem 1. *If N is prime, then for any nonconstant sequence s^∞ of period N over $GF(2)$ and over $GF(2^s \bmod N)$ with $\gcd(s, N-1) = 1$ and with $2^s \bmod N$ being a power of a prime,*

1. $L(s^\infty) \ge ord_N(2)$;
2. $SC_k(s^\infty) = \begin{cases} ord_N(2), & \text{if } k < \min\{\mathrm{WH}(s^N), N - \mathrm{WH}(s^N)\}; \\ 0, & \text{otherwise.} \end{cases}$

Proof: Setting $t = 1$ and $e_1 = 1$ in Basic Theorem 1 proves the theorem. QED

Theorem 2. *If $N = 4t + 1$ is prime and t is a odd prime, then for any nonconstant sequence s^∞ of period N over $GF(2)$ and over $GF(2^s \bmod N)$ with $\gcd(s, N - 1) = 1$ and with $2^s \bmod N$ being a power of a prime,*

1. *$L(s^\infty) = N$ or $N - 1$;*
2. *$SC_k(s^\infty) = \begin{cases} N \text{ or } N - 1, & \text{if } k < \min\{\mathrm{WH}(s^N), N - \mathrm{WH}(s^N)\}; \\ 0, & \text{otherwise.} \end{cases}$*

Proof: Recall that a is a primitive root modulo an integer N if and only if $\mathrm{ord}_N(a) = \phi(N)$, where $\phi(x)$ is the Euler function. Since both $N = 4t + 1$ and t are primes, it is seen that 2 is a primitive root of N. Then the conclusion of this theorem follows from Theorem 1. QED

Theorem 3. *Let $N = 4t - 1$ be a prime with t odd. If $(N-1)/2$ is prime (i.e., it is a Sophie Germain prime), then for any nonconstant sequence s^∞ of period N over $GF(2)$ and over $GF(2^s \bmod N)$ with $\gcd(s, N - 1) = 1$ and with $2^s \bmod N$ being a power of a prime,*

1. *$L(s^\infty) = N$ or $N - 1$;*
2. *$SC_k(s^\infty) = \begin{cases} N \text{ or } N - 1, & \text{if } k < \min\{\mathrm{WH}(s^N), N - \mathrm{WH}(s^N)\}; \\ 0, & \text{otherwise.} \end{cases}$*

Proof: By the special form of the prime N it is easy to see that 2 is a primitive root modulo N. Then the conclusion of this theorem follows from Theorem 1. QED

Theorem 4. *Let $N = 4t + 1$ be a prime with t odd and $t = t_1 t_2$, where t_1 and t_2 are primes. If*

$$2^{2t_1} \neq -1 \bmod N, \quad 2^{2t_2} \neq -1 \bmod N,$$

then for any nonconstant sequence s^∞ of period N over $GF(2^s \bmod N)$ (especially over $GF(2)$) with $\gcd(s, N - 1) = 1$ and with $2^s \bmod N$ being a power of a prime,

1. *$L(s^\infty) = N$ or $N - 1$;*
2. *$SC_k(s^\infty) = \begin{cases} N \text{ or } N - 1, & \text{if } k < \min\{\mathrm{WH}(s^N), N - \mathrm{WH}(s^N)\}; \\ 0, & \text{otherwise.} \end{cases}$*

Proof: Note that $\mathrm{ord}_N(2)$ divides $\phi(N) = N - 1 = 4t_1 t_2$ and that t_1 and t_2 are primes. It then follows by the two inequalities in the assumptions of the theorem that the order of 2 must be equal to $N - 1$. Combining this fact and Theorem 1 proves the theorem. QED

Theorem 5. *Let $N = 4t - 1$ be a prime with t odd and $2t - 1 = t_1 t_2$, where t_1 and t_2 are primes. If*

$$2^{t_1} \neq -1 \bmod N, \quad 2^{t_2} \neq -1 \bmod N,$$

then for any nonzero sequence s^∞ of period N over $GF(2)$ and over $GF(2^s \bmod N)$ with $\gcd(s, N - 1) = 1$ and with $2^s \bmod N$ being a power of a prime,

1. $L(s^\infty) = N$ or $N - 1$;
2. $SC_k(s^\infty) = \begin{cases} N \text{ or } N - 1, & \text{if } k < \min\{\text{WH}(s^N), N - \text{WH}(s^N)\}; \\ 0, & \text{otherwise.} \end{cases}$

Proof: With the same arguments as in Theorem 4 we see that the order of 2 modulo N is $N - 1$. Combing this fact and Theorem 1 yields the conclusion of this Theorem. QED

After having proved the above theorems, which are needed to control the linear complexity and sphere complexity of the cyclotomic sequences of order $2k$, we now describe the binary cyclotomic generator of order $2k$. Let prime $N = 2kf + 1$, and $D_0, D_1, \cdots, D_{2k-1}$ be the cyclotomic classes of order $2k$ defined as before. If we choose the mapping

$$H(x) = (i^{(N-1)/2k} \bmod N) \bmod 2$$

as the cryptographic function for the NSG of Figure 1, then we have the *binary cyclotomic generator of order $2k$*.

It is not difficult to verify that each difference parameter of the above cryptographic function $H(x)$ for the binary cyclotomic generator of order $2k$ can be expressed as the sum of k^2 cyclotomic numbers of order $2k$. Thus, if the cyclotomic numbers of order $2k$ have an ideal stability, the formulae in Section 1 show that we have an ideal property for the seven aspects described in Section 1 if k is small enough. Even if the cyclotomic numbers of order $2k$ are stable to an ideal extent, the largeness of the k may lead to a relatively bad difference property of the cryptographic function. Thus, only those generators derived from small k are cryptographically attractive.

Theorems 1–5 clearly show how to control the linear complexity and its stability for the output sequence of the binary cyclotomic generator of order $2k$. By Theorem 1, to control the linear complexity and its stability of the output sequences of the binary cyclotomic generators, it suffices to choose the prime N such that $\text{ord}_N(2)$ is large enough. The best choices for the primes are Sophie Germain primes, i.e., primes p such that $2p+1$ is also a prime, and Stern primes, i.e., primes $p = 4t + 1$ with t also prime.

If $k = 1$, then it is called the *binary cyclotomic generator* of order 2. The binary cyclotomic generators of order 2 can be further classified into DSC (difference set characterized) and ADSC (almost difference set characterized) generators which correspond the cases $N = -1 \bmod 4$ and $N = 1 \bmod 4$, respectively [11]. In the case $k = 1$ the output sequence of the DSC and ADSC generators are the 0-1 version of the Legendre sequences with a slight modification of the values for $\left(\frac{pi}{p}\right)$ for $i = 0, 1, \cdots$, and a proper choosing of the prime p.

A DSC generator with a Sophie Germain prime has the following cryptographic attributes: its output sequences have maximum linear complexity by Theorem 3; best autocorrelation property by the formula for cyclotomic numbers of order 2 or the difference set property of the set $f^{-1}(1)$ and formula (3), where $f(x) = (x^{(N-1)/2} \bmod N) \bmod 2$; best linear complexity stability by Theorem 3; the cryptographic function $f(x)$ has best nonlinearity with respect to

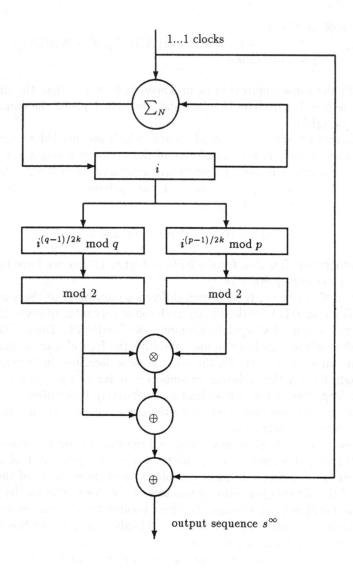

Fig. 2. The two-prime generator of order $2k$ including the twin-prime generator

the additions of Z_N and Z_2 by the difference set property of $f^{-1}(1)$ and formula (1). The ADSC generator with a Stern prime has almost the same cryptographic attributes.

5   Two-prime cyclotomic generator of order 2

Let p and q be two distinct odd primes with $\gcd(p-1, q-1) = 2$, and

$$R = \{0\}, \quad P = \{p, 2p, \cdots, (q-1)p\}, \quad Q = \{q, 2q, \cdots, (p-1)q\}.$$

Furthermore, let g be a fixed common primitive root of both primes p and q, $d = \gcd(p-1, q-1)$ and $de = (p-1)(q-1)$. Then it is proved in [21] there exits an integer x such that

$$Z_{pq}^* = \{g^s x^i : s = 0, 1, \cdots, e-1; \ i = 0, 1, \cdots, d-1\}.$$

The set Z_{pq}^* is also called the reduced residue system modulo $N = pq$. In Whiteman's generalized cyclotomy the *index class* or *cyclotomic class* D_i consists of e numbers and is defined by

$$D_i = \{g^s x^i : s = 0, 1, \cdots, e-1\}$$

and the generalized cyclotomic number $(i, j)_d$ by

$$(i, j)_d = |(D_i + 1) \cap D_j|.$$

There are d cyclotomic classes $D_0, \cdots, D_{d-1}$, which form a partition of Z_{pq}^*.

We now analyze the relation between the difference property of the partition of Z_{pq}^* and the generalized cyclotomic numbers. It is obvious that $x \in Z_{pq}^*$. Assume that the order of x modulo N is m. Then $m \geq d$. Let $w \in Z_{pq}^*$. Then there must exist two integers s and t with $0 \leq s \leq e-1$, $0 \leq t \leq d-1$ such that $w = g^s x^t$. Because $x^d = g^u$ for some fixed u such that $0 \leq u \leq d-1$, the difference parameter can be expressed as

$$d(i, j; w) = |(D_i + g^s x^t) \cap (D_j)|, \ 0 \leq i, \ j \leq d-1; \ w \in Z_N^*,$$
$$= |(D_{(m-t+i) \bmod d} + 1) \cap D_{(m-t+j) \bmod d}|$$
$$= ((m-t+i) \bmod d, (m-t+j) \bmod d)_d.$$

This means that for each $(i, j; w)$ with $0 \leq i, j \leq d-1$, $w \in Z_{pq}^*$, the difference parameter $d(i, j; w)$ is in fact one cyclotomic number. We will discuss the case for $w \notin Z_{pq}^*$ later.

As seen above, the index classes $D_0, \cdots, D_{d-1}$ is a partition of Z_N^*. Since

$$R = \{0\}, \quad P = \{p, 2p, \cdots, (q-1)p\}, \quad Q = \{q, 2q, \cdots, (p-1)q\},$$

the sets $D_0, \cdots, D_{d-1}$; R; P; Q form a partition of Z_N. To extend the partition of Z_N^* to Z_N, we have to study the difference property between the above sets. The following conclusions have been proven or are implied in [21]:

1. For any $r \in P \cup Q$, it holds

$$d(0, 1; r) = |(D_0 + r) \cap D_1| = (p-1)(q-1)/d^2. \tag{9}$$

2. For any $r \in P \cup Q$ and any $1 \leq k \leq d-1$, it holds

$$d(0, k; r) = |(D_0 + r) \cap D_k| = (p-1)(q-1)/d^2.$$

3. Let symbols as before, then

$$d(0, 0; r) = |(D_0 + r) \cap D_0|$$
$$= \begin{cases} (p-1)(q-1-d)/d^2, & r \in P, r \notin Q; \\ (q-1)(p-1-d)/d^2, & r \in Q, r \notin P. \end{cases}$$

Since $x \in Z_N^*$ and $x = g^u$ for some u with $0 \le u \le d-1$, for each r we have

$$
\begin{aligned}
d(i,j;r) &= |(D_i + r) \cap D_j| \\
&= |x^{d-i}(D_i + r) \cap x^{d-i}D_j| \\
&= d(0, (j+d-i) \bmod d; x^{d-i}r \bmod N).
\end{aligned}
$$

If $r \in P$ (or Q), then $x^{d-i} \in P$ (or Q).

For the partition $D_0, \cdots, D_{d-1}$ of Z_N^* and $w \neq 0$, combining the above results we obtain

$$
d(i,j;w) = \begin{cases}
(p-1)(q-1)/d^2, & i \neq j, \; w \in P \cup Q; \\
(p-1)(q-1-d)/d^2, & i = j, \; w \in P, \; w \notin Q; \\
(q-1)(p-1-d)/d^2, & i = j, \; w \in Q, \; w \notin P; \\
(i',j')_d \text{ for some } (i',j'), & \text{otherwise.}
\end{cases}
$$

In order to put the elements of R, P, Q to some of the D_i's to get a partition of Z_N with good difference property, we need the following result proved by Whiteman [21]:

$$
|D_0 \cap (Q+r)| = (p-1)/d; \quad \text{If } r \notin Q \cap R.
$$

To design the two-prime cyclotomic generator of order 2, we need functions from Z_{pq} to Z_2 with good nonlinearity with respect to the additions of the two rings. Due to the inspiration of Whiteman's result, we now consider the characteristic function of the partition $\{R \cup Q \cup D_0, \; P \cup D_1\} = \{C_0, C_1\}$ of Z_{pq}. In what follows in this section we assume that $d = \gcd(p-1, q-1) = 2$. To analyze the function, we need the generalized cyclotomic numbers of order 2 obtained by Whiteman [21].

Let symbols as before. If ff' is even, we have $(0,0) = (1,0) = (1,1)$ and two different cyclotomic numbers

$$
(0,0) = \frac{(p-2)(q-2)+1}{4}, \quad (0,1) = \frac{(p-2)(q-2)-3}{4}. \tag{10}
$$

If ff' is odd, we have $(0,1) = (1,0) = (1,1)$ and

$$
(0,0) = \frac{(p-2)(q-2)+3}{4}, \quad (0,1) = \frac{(p-2)(q-2)-1}{4}. \tag{11}
$$

We now analyze the difference property of the partition $\{C_0, C_1\}$ of Z_{pq}. Note that

$$
d_C(0,0;r) = |[(R+r) \cup (Q+r) \cup (D_0+r)] \cap [R \cup Q \cup D_0]|.
$$

Setting

$$
a(0,0;r) = |(Q+r) \cap Q| + |(Q+r) \cap D_0| + |(D_0+r) \cap Q| + |(D_0+r) \cap D_0|,
$$

we can prove

$$
0 \le d_C(0,0;r) - a(0,0,r) \le 2.
$$

So our task now is to estimate the $a(0,0;r)$ with $r \neq 0$. One simple fact is

$$|(Q+r) \cap Q| = \begin{cases} p-2, & r \in Q; \\ 0, & r \in P \cup Z_{pq}^*; \end{cases}$$

Note that if $r \in P$, then it is possible to have $Q + r \subset D_0$. Thus, for each r we have the following two apparent facts:

$$0 \leq |(Q+r) \cap D_0| \leq p-1;$$
$$0 \leq |Q \cap (D_0 + r)| \leq p-1.$$

It follows that

$$|(D_0 + r) \cup D_0| \leq a(0,0;r) \leq 3p - 4 + |(D_0 + r) \cup D_0|.$$

Setting

$$B = \max\{\frac{(p-2)(q-2)+3}{4}, \frac{(p-1)(q-3)}{4}, \frac{(p-3)(q-1)}{4}\}$$

and

$$C = \min\{\frac{(p-2)(q-2)-3}{4}, \frac{(p-1)(q-3)}{4}, \frac{(p-3)(q-1)}{4}\},$$

we get

$$C \leq a(0,0;r) \leq 3p - 4 + B,$$

and therefore

$$C \leq d_C(0,0;r) \leq 3p - 2 + B.$$

We can similarly prove that for each $r \neq 0$, it holds

$$C \leq d_C(1,1;r) \leq 3q - 4 + B.$$

In what follows we analyze $d_C(1,0;r)$ and $d_C(0,1;r)$. By definition we have

$$\begin{aligned} d_C(1,0;r) &= |(C_1+r) \cap C_0| = |[(P+r) \cup (D_1+r)] \cap (R \cup Q \cup D_0)| \\ &= |(P+r) \cap R| + |(P+r) \cap Q| + |(P+r) \cap D_0| \\ &\quad + |(D_1+r) \cap R| + |(D_1+r) \cap Q| + |(D_1+r) \cap D_0|. \end{aligned}$$

If $r \in P$, then by formula (9) we have

$$|(D_1 + r) \cap D_0| = (p-1)(q-1)/4.$$

In addition we have apparently

$$|(P+r) \cap Q| = |(P+r) \cap D_0| = |(D_1+r) \cap R| = 0$$
$$|(P+r) \cap R| = 1, \ 0 \leq |(D_1+r) \cap Q| \leq p-1.$$

Hence, we obtain in the case $r \in P$

$$1 + \frac{(p-1)(q-1)}{4} \leq d_C(1,0;r) \leq \frac{(p-1)(q-1)}{4} + p.$$

If $r \in Q$, we can similarly prove

$$\frac{(p-1)(q-1)}{4} \leq d_C(1,0;r) \leq \frac{(p-1)(q-1)}{4} + q - 1.$$

If $r \in Z_{pq}^* \setminus P \cup Q \cup R$, then by the formulae (10) and (11) we get

$$\frac{(p-2)(q-2)-3}{4} \leq |(D_1 + r) \cap D_0| \frac{(p-2)(q-2)+3}{4}.$$

In addition we have apparently

$$|(P+r) \cap R| = 0$$
$$0 \leq |(P+r) \cap Q| \leq \min\{p-1, q-1\}$$
$$0 \leq |(P+r) \cap D_0| \leq q - 1$$
$$0 \leq |(D_1 + r) \cap Q| \leq p - 1$$
$$0 \leq |(D_1 + r) \cap R| \leq 1.$$

It follows in this case that

$$\frac{(p-2)(q-2)-3}{4} \leq d_C(1,0;r)$$
$$\leq \frac{(p-1)(q-1)}{4} + \min\{p-1, q-1\} + \tfrac{3}{4}(p+q) + \tfrac{1}{2}.$$

Combining the results for the three cases, we obtain

$$\frac{(p-2)(q-2)-3}{4} \leq d_C(1,0;r) \leq \frac{(p-1)(q-1)}{4} + E,$$

where

$$E = \max\{p, q-1, \min\{p-1, q-1\} + \frac{3}{4}(p+q) + \frac{1}{2}\}.$$

Similarly, one can prove

$$\frac{(p-2)(q-2)-3}{4} \leq d_C(0,1;r) \leq \frac{(p-1)(q-1)}{4} + E,$$

Summarizing the above analysis, we obtain the difference property of the above cryptographic function $F_C(x)$, as described by the following theorem.

Theorem 6. *Let*

$$a = \frac{(p-2)(q-2)-3}{4}, \qquad d = \frac{(p-1)(q-1)}{4},$$

$$b = \max\{\frac{(p-2)(q-2)+3}{4}, \frac{(p-1)(q-3)}{4}, \frac{(p-3)(q-1)}{4}\}$$

and

$$c = \min\{\frac{(p-2)(q-2)-3}{4}, \frac{(p-1)(q-3)}{4}, \frac{(p-3)(q-1)}{4}\}.$$

then we have

$$c \leq d_C(0,0;r) \leq 3p-2+b,$$
$$c \leq d_C(1,1;r) \leq 3q-4+b,$$
$$a \leq d_C(1,0;r) \leq d+E,$$
$$a \leq d_C(0,1;r) \leq d+E,$$

for each $r \neq 0 \bmod N$, *where*

$$E = \max\{p, q-1, \min\{p-1, q-1\} + \frac{3}{4}(p+q) + \frac{1}{2}\}.$$

This theorem tells us that if $|p-q|$ is small enough, the cryptographic function $F_C(x)$ has an ideal difference property. Thus, the other six aspects are automatically ensured due to our formulae in Section 1. In this case, the facts that $|C_0| = (p-1)(q-1)/2 + q$ and $|C_1| = (p-1)(q-1)/2 + p - 1$, show that the function has also good balance. It is called a cyclotomic generator because the difference property and the nonlinearity of the above cryptographic function depend on the generalized cyclotomy developed by Whitman.

It is not difficult to see that the characteristic function of the partition $\{C_0, C_1\}$ can be expressed by

$$F_C(j) = \begin{cases} 1, & j \in R \cup Q; \\ 0, & j \in P; \\ (1 + (\frac{i}{p})(\frac{i}{q}))/2, & \text{otherwise.} \end{cases}$$

With this cryptographic function the binary two-prime cyclotomic generator of order 2 is depicted in Figure 2, where p and q are distinct odd primes and $k = 1$, the $\otimes$ and $\oplus$ denote bit multiplication and bit-XOR operations, and the other parts have the same meanings as those of Figure 1.

On the other hand, the two primes should be chosen properly for the purpose of controlling the linear complexity and the linear complexity stability of the output sequence. In fact we have generally the following result, which is a special case of Basic Theorem 1.

Theorem 7. *Let* $N = rs$ *be a product of two distinct primes, u an integer with* $\gcd(u, N) = 1$. *Then for any nonconstant sequence* s^∞ *of period N over $GF(u)$, it holds*

1. $L(s^\infty) \geq \min\{ord_r(u), ord_s(u)\}$;
2. $SC_k(s^\infty) \geq \min\{ord_r(u), ord_s(u)\}$, if $k < \min\{\text{WH}(s^N), N - \text{WH}(s^N)\}$,

where $SC_k(s^\infty)$ denotes the sphere complexity of the sequence, $\text{WH}(s^N)$ the Hamming weight of the finite sequence, $ord_r(u)$ the order of u modulo r.

Proof: Setting $t = 2$ and $e_1 = e_2 = 1$ in Basic Theorem 1 proves this theorem. QED

As consequences of the above theorem, we have the following corollaries:

Corollary 8. *Let* $r = 4t_1 + 1$, $s = 4t_2 + 1$, $r \neq s$. *If* r, s, t_1 *and* t_2 *are odd primes, then for any nonconstant binary sequence of period* $N = rs$,

1. $L(s^\infty) \geq \min\{r - 1, s - 1\}$
2. $SC_k(s^\infty) \geq \min\{r - 1, s - 1\}$, if $k < \min\{\text{WH}(s^N), N - \text{WH}(s^N)\}$.

Proof: Note that the proof of Theorem 2 shows that $ord_r(2) = r-1$ and $ord_s(2) = s - 1$, that is, 2 is a primitive root of both r and s. Combing these two facts and Theorem 7 proves this corollary. QED

Corollary 9. *Let* $r = 4r_1 - 1$, $s = 4s_1 - 1$, $(r - 1)/2$ *and* $(s - 1)/2$ *are all odd primes. If* $r > 5$ *and* $s > 5$, *then for each nonconstant binary sequence* s^∞ *of period* $N = rs$, *we have*

1. $L(s^\infty) \geq \min\{r - 1, s - 1\}$;
2. $SC_k(s^\infty) \geq \min\{r - 1, s - 1\}$; if $k < \min\{\text{WH}(s^N), N - \text{WH}(s^N)\}$.

Proof: The Proof of Theorem 3 shows that 2 is a common primitive root of r and s. Combining this and Theorem 7 yields the conclusion of this corollary. QED

Corollary 10. *Let* $r = 4r_1 + 1$, $s = 4s_1 - 1$. *If* $r, r_1, s, (s - 1)/2$ *all odd primes, then for each nonconstant binary sequence* s^∞ *of period* $N = rs$, *we have*

1. $L(s^\infty) \geq \min\{r - 1, s - 1\}$;
2. $SC_k(s^\infty) \geq \min\{r - 1, s - 1\}$; if $k < \min\{\text{WH}(s^N), N - \text{WH}(s^N)\}$.

Proof: The proofs of Theorems 2 and 3 show that 2 is common primitive root of both r and s. Combing this and Theorem 7 proves this corollary. QED

The above theorem and its three corollaries clearly show how to control the linear complexity and its stability for the output sequence of the binary two-prime generator of order 2.

Summarizing the above results, we see that the parameters should be chosen such that

1. p and q both are large enough with $\gcd(p - 1, q - 1) = 2$;
2. $|p - q|$ is small enough, compared with pq;
3. $ord_p(2)$ and $ord_q(2)$ both are large enough.

It should be made clear that the special properties of the primes p and q determine partly the quadratic partition of the primes, and thus contributes to the stability of the cyclotomic numbers, and consequently to the difference property and other six aspects described in Section 1.

Generally speaking, the two-prime generator is more flexible, due to the fact that we have much freedom to select the primes.

The best choices for the p and q are the twin primes. They ensures the best difference property and nonlinearity of the cryptographic function according to generalized cyclotomic numbers of order 2 (see [21]). In this case the output sequence of generator is the characteristic sequence of the twin-prime difference set with parameters $(N, h, \lambda) = (p(p+2), (N-1)/2, (N-3)/4)$ (see [21]). Note

$$n = h - \lambda = \begin{cases} (2t+1)^2, & \text{if } p = 4t+1; \\ 4t^2, & \text{if } p = 4t-1. \end{cases}$$

It follows from [11] that the linear complexity of the output sequence is N or $N-1$, provided that $p = 4t+1$. If $p = 4t+1$, then $N+2 = 4(t+1)-1$. It follows that p and $p+2$ has no common primitive root 2 if $p = 4t+1$. Nevertheless, the output sequence has best linear complexity.

If $p = 4t - 1$, it is possible for p and $p+2$ to have common primitive root 2. Assume that they have common primitive root 2, then by Theorem 7 we have

1. $L(z^\infty) \geq p - 1$;
2. $SC_k(z^\infty) \geq p - 1$, if $k < \min\{\text{WH}(z^N), N - \text{WH}(z^\infty)\}$.

Because the cryptographic function is the characteristic function of a twin-prime difference set, all the seven aspects described in Section 1 are optimal.

The output sequence of the two-prime cyclotomic generator of order 2 is an extension of 0-1 version of the Jacobi sequence in the sense that the values at the special sets Q, P and R are modified. In addition, the condition that $\gcd(p-1, q-1) = 2$ is essential to the generator. It is because of this condition that the generalized cyclotomic numbers of order 2 ensure an ideal difference property of the cryptographic function. Without this condition it cannot be called a cyclotomic generator.

6   Two-prime generator of order 4

To design cryptographic functions from Z_{pq} to Z_4, we follow the same approach as in the foregoing sections. Let $p = 4f + 1$ and $q = 4f' + 1$ with $\gcd(f, f') = 1$. Then $d = \gcd(p-1, q-1) = 4$ and $e = 4ff'$. Define the function

$$F(j) = \begin{cases} 1, & j \in \{0, q, 2q, \cdots, (p-1)q\}; \\ 0, & j \in \{p, 2p, \cdots, (q-1)p\}; \\ ((j^{(q-1)/4} \bmod q) \bmod 2) \oplus ((j^{(p-1)/4} \bmod p) \bmod 2) \oplus 1, & j \in Z_{pq}^*. \end{cases}$$

With this $F(x)$ we describe a generator based on the generalized cyclotomy of order 4, as depicted in Figure 2 with $k = 2$.

If we define the function $F^*(x)$ from Z_{pq}^* to Z_2 by

$$F^*(j) = ((j^{(q-1)/4} \bmod q) \bmod 2) \oplus ((j^{(p-1)/4} \bmod p) \bmod 2) \oplus 1, \ j \in Z_{pq}^*.$$

Then it is easy to know that $F^*(x)$ has characteristic set $C_1 = D_i \cup D_j \cup Q$, where $Q = \{0, q, 2q, \cdots, (p-1)q\}$, D_i and D_j are two of the four generalized cyclotomic classes developed by Whitman [21]. Thus, an ideal stability of the generalized cyclotomic numbers of order 4 ensures an ideal difference property and nonlinearity of the above function $F(x)$.

Fortunately, the generalized cyclotomic numbers of order four have an ideal stability [21]. Thus, it follows from the formulae in Section 1 and the relation between cyclotomic numbers and the difference parameters described in Section 2 that we have ensured an ideal property for all the seven aspects described in Section 1.

The control of the linear complexity and its stability for the output sequence of the binary two-prime cyclotomic generator of order 4 is the same as that of order 2. By Theorems 6 and 7 the parameters should be chosen such that

1. p and q both are large enough with $\gcd(p-1, q-1) = 4$;
2. $|p - q|$ is small enough, compared with pq;
3. $\text{ord}_p(2)$ and $\text{ord}_q(2)$ both are large enough.

7   Prime-square generator

Sequences with period of the square of a odd prime are cryptographically attractive due to the following theorems about their linear complexity and linear complexity stability which follow easily from Basic Theorem 1.

Theorem 11. *Let r be a odd prime, $N = r^e$ and q an integer with $\gcd(q, N) = 1$. Then for any nonconstant sequence of period N over $GF(q)$,*

1. $L(s^\infty) \geq \text{ord}_r(q)$;
2. $SC_k(s^\infty) \geq \text{ord}_r(q)$, *if $k < \min\{\text{WH}(s^N), N - \text{WH}(s^N)\}$.*

Proof: Setting $t = 1$ in Basic Theorem 1 proves this theorem. QED

Theorem 12. *Let r be a odd prime, $N = r^2$ and q a primitive root modulo r and r^2 does not divides $q^{r-1} - 1$, then for any nonzero sequence of period N over $GF(q)$,*

1. $L(s^\infty)$ *must be equal to one of $\{\sqrt{N}, \sqrt{N}-1, N-\sqrt{N}, N-\sqrt{N}+1, N-1, N\}$;*
2. $SC_k(s^\infty) \geq \sqrt{N} - 1$, *if $k < \min\{\text{WH}(s^N), N - \text{WH}(s^N)\}$.*

Proof: Since q is a primitive root of r and r^2 does not divides $q^{r-1} - 1$ by assumptions, it is known that q must be a primitive root of r^2 (for proof, see [1]). Thus, by the basic properties of cyclotomic polynomial presented in the proof of Basic Theorem 1 we know that the cyclotomic polynomials $Q_r(x)$ and $Q_{r^2}(x)$

are irreducible over $GF(q)$. Again from the properties of cyclotomic polynomials it follows

$$x^N - 1 = (x-1)Q_r(x)Q_{r^2}(x).$$

Note that $\deg(Q_r(x)) = r - 1$ and $\deg(Q_{r^2}(x)) = r(r-1)$ since q is a common primitive root of r and r^2. Combining these fact and the fact that the minimum polynomial of each sequence of period N over $GF(q)$ divides $x^N - 1$ proves this theorem. QED

Thus, the best primes p for binary sequences of period p^2 are the non-Wieferich primes with base q, i.e., those described by the above theorem, which is related to the *Fermat quotient* and some other number-theoretic problems. We prove these two theorems here because we need them to control the linear and sphere complexities of the output sequences of the prime-square generator.

If we choose one of the following functions

$$F_C(x) = \begin{cases} 1, & x \in R; \\ (x^{p(p-1)/2} \bmod p^2) \bmod 2, & \text{otherwise} \end{cases}$$

and

$$F_C(x) = \begin{cases} 0, & x \in R; \\ (x^{p(p-1)/2} \bmod p^2) \bmod 2, & \text{otherwise} \end{cases}$$

for the cryptographic function of Figure 1, where $N = p^2$, then we have the binary prime-square cyclotomic generator.

To describe the difference property of the above two cryptographic functions, we give a brief introduction to the cyclotomic numbers modulo p^2. Let p be a odd prime. By the Chinese Remainder Theorem there is a common primitive root α modulo both p and p^2. Setting $D_0 = (\alpha^2)$, a multiplicative subgroup of $Z_{p^2}^*$, and $D_1 = \alpha D_0$. Then the cyclotomic numbers of order 2 modulo p^2 are defined by

$$(l, m)_2 = |D_l \cap (D_m + 1)|, \ 0 \le l, m \le 1.$$

We need the following theorem, which was conjectured by the author and proved by D. Pei [16], to ensure an ideal property for all the seven aspects described in Section 1.

Theorem 13. *Let symbols as before. If $p = 3 \bmod 4$, we have*

$$(0,1) = (0,0) = (1,1) = \frac{p(p-3)}{4}, \ (1,0) = \frac{p(p-3)}{4} + p.$$

If $p = 1 \bmod 4$, we have

$$(0,1) = (1,0) = (1,1) = \frac{p(p-1)}{4}, \ (0,0) = \frac{p(p-1)}{4} - p.$$

Let $\{C_0, C_1\}$ be the characteristic class of the above function $F_C(x)$, then similar to Section 2 it is easily verified that the difference parameters of the above function are approximately the same as the four cyclotomic numbers defined for the modulus $N = p^2$. Thus, an ideal difference property and therefore an ideal property for the other six aspects are ensured by the formulae described in Section 1.

8   Behind the cyclotomic generators

There are several cryptographic ideas behind the construction of these cyclotomic generators. The first one is the order of choosing the design parameters for the generator. Contrary to the traditional approach, we first control the period of the output sequence. This will automatically ensure the linear complexity and its stability aspects only with the condition that the sequence is not a constant sequence. Then we choose the cryptographic function for other purposes. This approach is intended to avoid unnecessary tradeoffs.

The second cryptographic idea behind the design and analysis of cyclotomic generators is the idea of introducing good partners, in order to get a stable system. In particular, we search for pairs of period and finite field so that it is easy to control the linear complexity and its stability for those sequences over those fields with corresponding partner periods. This has been shown clearly by the theorems and corollaries concerning linear and sphere complexities. We say that such pairs work in harmony with respect to the aspects of linear complexity and its stability. For example, some Mersenne and Fermat primes are not good partners of the field $GF(2)$, since it is difficult to control the linear complexity and its stability for binary sequences with period of some Fermat and Mersenne primes. This is because that the order of 2 modulo these primes is quite small. From Basic Theorem 1 and its corollaries it is rational to use $\mathrm{ord}_N(q)$ as a measure on the partnership between an positive integer N and an integer q with respect to the linear and sphere complexity aspects when designing sequences of period N over $GF(q)$, where $\gcd(N, q) = 1$. We say that q and N are the *best partners* when q is a primitive root modulo N.

Another kind of partnership is to find an integer r which is a power of prime such that $\mathrm{lcm}\{\mathrm{ord}_{p_1}(r), \cdots, \mathrm{ord}_{p_h}(r)\}$ is large enough when designing sequences of period $N = p_1 \cdots p_h$ over $GF(r)$, where $p_1, \cdots, p_h$ are distinct primes (see Basic Theorem 1). We say that r is a *best common partner* of $p_1, \cdots, p_h$ if r is a common primitive root of these primes.

The third cryptographic idea is to use some techniques of ensuring "good + bad = good". With a simple argument each cryptographic function employed in the generators described in the foregoing sections can be expressed as

$$F(x) = H(G(x)),$$

where $G(x)$ is a mapping from Z_N to U which is a subgroup of the group $(Z_N^*, \cdot)$ with order d, and $H(x)$ a mapping from U to Z_d. The nonlinearity of $G(x)$ with respect to $(Z_N, +)$ and $(U, \cdot)$ is determined mainly by the (generalized) cyclotomic numbers of order d, which usually have an ideal stability; while the function $H(x)$ is almost linear (or with good linearity) with respect to $(U, \cdot)$ and $(Z_d, +)$. Thus, it is clear that one cryptographic idea behind the cyclotomic generators is

"GOOD + BAD = GOOD".

The fourth cryptographic idea is to use cryptographic functions $f(x)$ from an abelian group $(G_1, +)$ to another abelian group $(G_2, +)$ such that $|G_2|$ does not divide $|G_1|$. We say that such a function is *linearly non-approximatable*, since there is no linear mapping other than the zero constant function from $(G_1, +)$ to $(G_2, +)$. This technique makes any linear approximation attack with respect to the two operations out of sense. It should be pointed out that the nonlinearity definition based on the minimum correlation (measured in probability of agreement, or distance [15] between a function and all affine functions) is not rational in many cases. This definition makes no sense for the above cryptographic functions.

The fifth cryptographic idea is to make use of the relativity about nonlinearity and linearity. It is a common fact that the nonlinearity and linearity are relative to the operations considered, and that both linear components and nonlinear components should be employed in many cipher systems. To find out some cryptographic functions with good nonlinearity with respect to some operations, one may try to find some linear cryptographic function with respect to some other operations and use them in the context of the former operations. This is to say that bad things in one sense may be good ones in another sense, and one way to get goodness is to use badness in a proper way and proper context. To illustrate this philosophy, we first take the corresponding function $G(x) = x^{(p-1)/d} \bmod p$ used to construct the cyclotomic generator of order $2k$. Then $G(x)$ is linear with respect to $(Z_p^*, \cdot)$ and $(U, \cdot)$, where U is the multiplicative subgroup of Z_p^* with order d. But $G(x)$ has ideal nonlinearity with respect to $(Z_p, +)$ and $(U, +)$ if we define $G(0)$ to be any fixed element of U. And we use $G(x)$ in the context of the later pair of operations exactly. The same idea has been used for other generators.

To illustrate the relativity of linearity and nonlinearity, we prove the following theorem.

Theorem 14. *For every nonzero linear function $L(x)$ from $F = GF(q^m)$ to $K = GF(q)$ with respect to the additions of the two fields, its nonlinearity with respect to $(F^*, \times)$ and $(K, +)$ is optimal, as described by*

$$p(L(x) - L(x/\alpha) = b) = p(L(x(1 - \alpha^{-1})) = b) = \begin{cases} \frac{q^{m-1}-1}{q^m-1}, & \text{if } b = 0, \\ \frac{q^{m-1}}{q^m-1}, & \text{if } b \neq 0, \end{cases}$$

which holds for $\alpha \in F$ with $\alpha \neq 1$, and each $b \in K$.

Proof: For any nontrivial linear mapping $L(x)$ from F to K the kernel $L^{-1}(0)$ is an Abelian subgroup of $(F, +)$ and thus $L(x)$ takes on each element of K equally likely, that is, q^{m-1} times. Confining the linear mapping on $F^* = F \setminus \{0\}$ and using the above fact proves the theorem. QED

This theorem clearly shows the cryptographic importance of linear functions from $F = GF(q^m)$ to $K = GF(q)$ with respect to the additions of the two fields, especially the trace functions. Actually all the cryptographic functions for the cyclotomic generators are composition functions of linear functions with respect to different pairs of operations. It may be possible to employ this idea to design some block ciphers.

9   Some related number-theoretic problems

In the paper at least two bridges between some number-theoretic problems and the design and analysis of the natural sequence generator have been established. These bridges may play an important role in the interactions between number theory and stream ciphers, and particularly in the design and analysis of the cyclotomic generators described in this paper.

The first one is the bridge supported by Basic Theorem 1 and the theorems and corollaries concerning linear and sphere complexities. The reason for calling it a bridge has already made clear in Section 3. Obviously, there are more results which can be derived from Basic Theorem 1. If we go across this bridge from the stream cipher side, we shall encounter at least the following two basic number-theoretic problems when we are designing sequences of period N over $GF(q)$ with ideal linear and sphere complexities (see Basic Theorem 1).

Basic Problem 1 *Find large positive integers N's and positive integers q's which are powers of primes such that*

1. $\gcd(N, q) = 1$;
2. For any factor n of N, $ord_n(q) = \phi(n)$.

Basic Problem 2 *Find large positive integers N's and positive integers q's which are powers of primes such that*

1. $\gcd(N, q) = 1$;
2. N has a few factors;
3. For any factor n of N, $ord_n(q)$, a factor of $\phi(n)$, is as large as possible.

Attacking these two basic problems and many of its subproblems and variants will involve many, if not most, number-theoretic problems. Among them are the searching for special primes, such as Fermat primes, Mersenne primes, Stern primes, prime repunits, and twin primes, the distribution of special primes, primitivity testing, the distribution of primitive roots. These facts have already been shown clearly by all the theorems concerning the linear and sphere complexities proved in this paper. We need special primes for these generators, and have to know whether they exist. And if they exist, we have to know how to find large special primes.

We need not only twin primes for our twin-prime generator, but also special twin primes. The most interesting twin primes are those having a common primitive root 2 (the best common partner of the twins). However, some twins have, others don't. Problems as to which twins have the common primitive root 2 and how to find them are naturally important to the design of the twin-prime generators. By introducing sexes to twin we can solve one cryptographic problem for the twin-prime generator.

Let $(p, p+2)$ be a pair of twin primes and $p = \Xi(p) \bmod 4$, where $\Xi(p) = \pm 1$. Then we call $\Xi(p)$ the *sex characteristic* of the twins. If the twins $(p, p+2) = (4t-1, 4t+1)$ for some t, then we say that the twins have the same sex; otherwise, we say that they have different sexes.

In the above definitions, we say that twin primes $(p, p+2)$ have same sex, because in the expression of the form $4u \pm 1$, the u's for both p and $p+2$ are the same, and have therefore the same parity, if $p = 4t - 1$. If $p = 4t + 1$, then $p + 2 = 4(t+1) - 1$ and t and $t+1$ have different parities. That is why we call them twins with different sexes. This discussion has also proved the following property of twins.

Theorem (The Sex Principle of Twins) *If the smaller of the twins has sex characteristic* -1, *then the twins have the same sex; otherwise, they have different sexes.*

Theorem 15. *If p and $p + 2$ have the same sex, then it is possible for them to have common primitive root 2 (a common best partner); otherwise, they never have.*

Proof: With the help of the Law of Quadratic Reciprocity it is easy to see that a necessary condition for 2 to be a primitive root of a prime N is $N = 8k \pm 3$ for some k. Combining this fact and the definition of sexes proves the theorem. QED

Thus, the cryptographic importance of classifying twin primes into two classes according to their sexes for the design of the twin-prime generator clearly follows from the above theorem. Before searching for twins with the same sex, we have to know the distribution of the two classes. Solving this problem and searching for twin primes with the same sex are important design problems for the twin-prime generator.

The second bridge we set up when we design these cyclotomic generators is supported by the relations between the difference parameters of the cryptographic function $f(x)$, and the nonlinearity measure, the autocorrelation functions, mutual information, and two-bit pattern distributions, as described by the formulae in Section 1.

At one side of the second bridge are the autocorrelation property, the two-digit pattern distribution property of the output sequences, and the difference property and nonlinearity of the cryptographic functions of the NSG; while at the other side are cyclotomy-related problems and the Riemann Hypothesis for Curves over Finite Fields, which was proven to be true by Weil in 1948 [20].

We call it another bridge between number theory and stream ciphers because it makes a clear connection between the above set of cryptographic problems and the set of number-theoretic problems.

Related to cyclotomic numbers and their stability are the quadratic partition of primes and of some integers, the theory of quadratic forms, genus theory, class field theory, residue difference sets, group character, and character sums. In what follows we give a brief explanation to why these problems are related to the design of the cyclotomic generators. It is a pity that we cannot show here how some of these problems are related to the design and analysis of the cyclotomic generators since doing so must involve quite a number of number-theoretic concepts. However, pointing out some relations between design problems of cyclotomic generators and a number of number-theoretic problems might be helpful for those who are interested in these generators. Let us begin with the stability of cyclotomic numbers.

We now consider the binary cyclotomic generator of order 4 in Section 4. Let $N = 4f + 1$ be a chosen prime for the generator. Let $N = x^2 + 4y^2$, $x = 1 \bmod 4$, here y is two valued, depending on the choice of the primitive root [8]. There are five possible different cyclotomic numbers in the case f even; i.e., $(0,0)$, $(1,3)=(2,3)=(1,2)$, $(1,1)=(0,3)$, $(2,2)=(0,2)$, $(3,3)=(0,1)$ and

$$(0,0) = (p - 11 - 6x)/16,$$
$$(0,1) = (p - 3 + 2x + 8y)/16,$$
$$(0,2) = (p - 3 + 2x)/16,$$
$$(0,3) = (p - 3 + 2x - 8y)/16,$$
$$(1,2) = (p + 1 - 2x)/16.$$

For the case f odd, there are at most five distinct cyclotomic numbers, which are

$$(0,0) = (2,2) = (2,0) = (p - 7 + 2x)/16,$$
$$(0,1) = (1,3) = (3,2) = (p + 1 + 2x - 8y)/16,$$
$$(1,2) = (0,3) = (3,1) = (p + 1 + 2x + 8y)/16,$$
$$(0,2) = (p + 1 - 6x)/16,$$
$$\text{the rest} = (p - 3 - 2x)/16,$$

where $p = x^2 + 4y^2$ and $x = 1 \bmod 4$.

Although these formulas show a roughly ideal stability of cyclotomic numbers of order 4, the actual stability depends on the quadratic partition $p = x^2 + 4y^2$ and $x = 1 \bmod 4$. If there is a big difference between $|x|$ and $|y|$, then there is a considerable difference between the cyclotomic numbers. And it is not difficult to give examples to show there do exist primes such that there is a considerable difference between x^2 and y^2 in their quadratic partitions. Thus, choosing large special primes to ensure better stability of cyclotomic numbers is cryptographically necessary.

There are two approaches to do this. The first approach is to use traditional methods to get first large primes. Then use some special algorithm to get the partition and see whether it results in an ideal stability of cyclotomic numbers. This special algorithm should be related to classical problems about quadratic partitions such as the number of solutions of such a quadratic partition. Another approach is to consider directly for given n the set

$$\{x^2 + ny^2 : x, y \in Z\}.$$

If there are infinitely many primes in the set, we can search for the primes which give an ideal stability of cyclotomic numbers. With this approach the first problem we have to solve is whether there are infinite many primes in the above set. The Dirichlet theorem about primes in arithmetic progression does not help here. However, with some results from class field theory we can get a positive answer for each n [5].

If one has a quick look at each set of cyclotomic formulae known today, one will find that they have the same form. Behind this uniformity of all cyclotomic numbers is the Riemann Hypothesis for Curves over Finite Fields, which can be described as follows.

Riemann Hypothesis for Curves over Finite Fields: Suppose that $F(x, y)$ is a polynomial of total degree d, with coefficients in $GF(q)$ and with N zeros $(x, y) \in GF(q) \times GF(q)$. Suppose that $F(x, y)$ is absolutely irreducible, i.e., irreducible not only over $GF(q)$, but also over every algebraic extension thereof. Then

$$|N - q| \leq 2g\sqrt{q} + c_1(d),$$

where g is the genus of the curve $F(x, y) = 0$ and $c_1(d)$ is a constant depending on d.

This theorem, proven by Weil [20], not only indicates the uniformity of the form of cyclotomic formulae, but also can be used to set up bounds for pattern distributions in the output sequences of the cyclotomic generators. With the Weil Theorem one can see some cryptographic meanings of the genus of curves. For details about the Weil Theorem, the reader is referred to [18]. Since cyclotomic numbers are related to characters and character sums [8, 9, 12], the cyclotomic generators are naturally related to character and character sums. In fact many of the cryptographic functions $f(x)$ employed here are group characters.

One part of the design and analysis of these cyclotomic generators is to solve the following basic problem at the number-theory side of the second bridge.

Basic Problem 3 *Let N be a positive integer, and Z_N denote the residue ring modulo N. Find partitions $\{C_0, C_1\}$ of Z_N, i.e.,*

$$C_0 \cap C_1 = \Phi, \quad C_0 \cup C_1 = Z_N,$$

such that

$$|C_0| \approx N/2, \quad |C_1| \approx N/2, \qquad (12)$$
$$|C_i \cap (C_j + r)| \approx N/4$$

*for each nonzero r of Z_N, and each $i, j \in Z_2 = \{0, 1\}$. Here and hereafter $A \approx B$
means that $A = B \pm O(B^e)$ for some e with $0 \le e \le \frac{1}{2}$.*

For our cryptographic application we hope that $A \approx B$ means that A is as
near to B as possible. But for the flexibility of the solutions of the above basic
problem, we give such a definition of $A \approx B$.

The characteristic function of a partition $\{C_0, C_1\}$ of Z_N satisfying (12) ensures that it has an ideal difference property, and therefore an ideal property
for the other six aspects described in Section 1. But to ensure an ideal pattern
distribution property for various pattern lengths is to solve the following more
general basic problem.

Basic Problem 4 *Let $C = \{C_0, C_1\}$ be a partition of Z_N, $r_0, r_1, \cdots, r_{s-1}$ be s
pairwise distinct elements of Z_N, and*

$$D_{i_0 \cdots i_{s-1}}(r_0, \cdots, r_{s-1}) = \cap_{k=0}^{s-1}(C_{i_k} + r_k),$$
$$d_{i_0 \cdots i_{s-1}}(r_0, \cdots, r_{s-1}) = |\cap_{k=0}^{s-1}(C_{i_k} + r_k)|,$$

*where $i_0, \cdots, i_{s-1} \in Z_2$. Then $\{C_{i_0 \cdots i_{s-1}}(r_0, \cdots, r_{s-1}) : i_0, \cdots, i_{s-1} \in Z_2\}$ forms
a partition of Z_N.*
Find partitions $\{C_0, C_1\}$ of Z_N such that

$$d_{i_0 \cdots i_{t-1}}(r_0, \cdots, r_{t-1}) \approx \frac{N}{2^t} \tag{13}$$

*holds for each t with $1 \le t \le \lfloor \log_2 N \rfloor$, and for each set of pairwise distinct
elements $r_0, \cdots, r_{s-1}$ of Z_N.*

The conditions here include those of Basic Problem 3. They ensure that the
output sequence of the NSG employing the characteristic function of the parti-
tion has an ideal pattern distribution for each pattern with length t satisfying $1 \le t \le \lfloor \log_2 N \rfloor$, and also an ideal mutual information stability $I(i; z_{i_1} z_{i_2} \cdots z_{i_t})$. It
is noted that these *higher order difference parameters* $D_{i_0 \cdots i_{s-1}}(r_0, \cdots, r_{s-1})$ are
exactly measures on patterns distributions of binary sequences.

One class of solutions to Basic Problem 3 is those partitions $C = \{C_0, C_1\}$
of Z_N such that C_0 is a residue difference set of Z_N with $|C_1| \approx |C_0|$. However,
difference sets exist only for those $N = 3 \bmod 4$. For $N = 1 \bmod 4$ there may
exist some partitions $\{C_0, C_1\}$ of Z_N with almost the same difference property.
These are partitions $\{C_0, C_1\}$ of Z_N with C_1 being an almost difference set.

Let N be a odd number, and $D = \{d_1, \cdots, d_k\}$ a subset of an abelian group
$(G, +)$. If for each of half of the nonzero elements a's of Z_N, the equation

$$d_i - d_j = a$$

has exactly λ solutions (d_i, d_j) with d_i and d_j in D, and for each of the other
half exactly $\lambda + 1$ solutions, we call D an (N, k, λ) *almost difference set* (briefly,
a.d. set).

It is easy to see that N must be of the form $4t + 1$ if Z_N has an (N, k, λ) difference set. Almost difference sets are not easy to find. The cyclotomic numbers

of order 2 show that the quadratic residues modulo a prime $N = 1 \bmod 4$ form an $(N, (N-1)/2, (N-5)/4)$ a.d. set. For the biquadratic residues we have the following conclusion.

Theorem 16. *Let a prime $N = 4t + 1 = x^2 + 4y^2$ with $x = 1 \bmod 4$ and t being odd. Then the biquadratic residues modulo N form an $(N, t, (t-3)/4)$ a.d. set if and only if $x = 5$ or -3.*

Proof: By the formulae for cyclotomic numbers of order 4 presented in this section and the definition of a.d. sets we see that the biquadratic residues modulo N form an $(N, t, (t-3)/4)$ a.d. set if and only if

$$(0,0) - (1,1) = \frac{2x-7}{16} + \frac{3+2x}{16} = \frac{1}{4} = \pm 1,$$

which is equivalent to $x = 5$ or 3. This proves the theorem. QED

For other power residues it is not difficult to see whether they form a.d. sets by employing the cyclotomic numbers of various orders. In general, it is not easy to find a.d. residue sets. It is noted that difference sets and almost difference sets are employed in the cyclotomic generator of order 2 and in the twin-prime generator. So they are closely related to cyclotomic generators.

Another solution to Basic Problem 3 is to make use of power residues and cyclotomic numbers generally, as done in the foregoing sections. One advantage of the sets of power residues is that they form multiplicative groups, and this makes the implementation problem of characteristic functions of the corresponding partitions simple. There may exist other solutions to Basic Problem 3, which remains to be investigated.

The condition (12) can only give a very rough guarantee for other conditions in (13). The cyclotomic numbers and the Weil Theorem seem to indicate that partitions based on power residues give ideal solutions to Basic Problem 4. In fact we can set up bounds of order $N \pm O(\sqrt{N})$ for these higher order difference parameters $D_{i_0 \cdots i_{s-1}}(r_0, \cdots, r_{s-1})$ for $s \geq 3$ based on the Weil Theorem when N is a prime, but the bounds becomes more and more looser with the increase of s though the bounds remain in order $N \pm O(\sqrt{N})$.

We pointed out a number of number-theoretic problems here, since they are essential to the design and analysis of cyclotomic generators. For other type of generators there may be other related number-theoretic problems, such as the class numbers for imaginary quadratic fields which are related to properties of some number-theoretic generator [4].

10   Concluding remarks

Since we have controlled the difference property of the cryptographic functions and the linear and sphere complexities of the output sequences of the binary cyclotomic generators, the formulae in Section 1 and theorems and corollaries regarding the linear and sphere complexities show that these generators have the following properties:

1. the cryptographic function $f(x)$ has an ideal difference property;
2. the cryptographic function $f(x)$ has an ideal nonlinearity with respect to the additions of Z_N and Z_2;
3. the cryptographic function $f(x)$ has an ideal autocorrelation property;
4. the affine approximation of $f(x)$ with respect to $(Z_N, +)$ and $(Z_2, +)$ makes no sense, since there are only two trivial affine functions from Z_N to Z_2 for odd N;
5. the output sequence has an ideal autocorrelation property;
6. the output sequence has an ideal two-bit pattern distribution property;
7. the output sequence has ideal linear complexity and linear complexity stability;
8. the mutual information $I(i; z_i z_{i+t-1})$ has an ideal stability, here z^∞ denotes the output sequence of the NSG; and
9. the additive stream cipher system with this NSG as the keystream generator has an ideal density of encryption (resp. decryption) transformations, i.e., the probability of agreement between two encryption (resp. decryption) transformations specified by two keys is approximately $1/2$.

In fact we can calculate exact values of measures (such as autocorrelation values, the mutual information) for the above aspects based on the formulae in Section 1 if we have formulae for the difference parameters. For example measures for the above aspects for the cyclotomic generator of order 2 can be expressed exactly in terms of N, the modulus for the modulo N ring counter. If we have bounds for the difference parameters, then using the formulae in Section 1 gives bounds for measures on the above aspects.

In addition, the Weil Theorem and the formulae of cyclotomic numbers seem to indicate that the output sequences of these cyclotomic generators have an ideal distribution property for any pattern with length $1 \leq l \leq \lfloor \log_2 N \rfloor$.

In this paper we consider only some binary cyclotomic generators. It is possible to extend the results for binary cyclotomic generators to cyclotomic generators over other fields. For the linear complexity and sphere complexity of periodic sequences over other fields we have similar results.

The performances of these cyclotomic generators are basically of the same order, since they are all based on the exponentiation-modulo-N operation. It seems that these generators are not fast, but with a fast exponentiation algorithm it is possible to get a reasonable performance if the moduli are not too large. For the time being moduli having about 64 bits seems enough for the generators. Though the generators are not fast, they might have an ideal security. Thus, trading performance for security might be possible.

Finally we mention that some related work about some randomness aspects of the Legendre and Jacobi sequences was done by Damgård [6].

Acknowledgment: I would like to thank Arto Salomaa for several discussions on this topic with me, and the reviewers for helpful comments.

References

1. T. M. Apostol, *Introduction to Analytic Number Theory*, Springer-Verlag, 1976.
2. L. D. Baumert and H. Fredricksen, *The Cyclotomic Number of Order Eighteen with Applications to Difference Sets*, Math. Comp. 21, 1967, pp. 204-219.
3. L. D. Baumert, *Cyclic Difference Sets*, Lecture Notes in Mathematics, vol. 182, Springer-Verlag, 1971.
4. T. W. Cusick, *Properties of the X^2 mod N generator*, to appear in IEEE Trans. Inform. Theory, 1995.
5. D. A. Cox, *Primes of the Form $x^2 + ny^2$: Fermat, Class Field Theory, and Complex Multiplication*, John Wiley & Sons, 1989.
6. I. Damgård, *On the Randomness of Legendre and Jacobi Sequences*, Advances in Cryptology: Crypto'88, S. Goldwasser (Ed.), LNCS 403, Springer-Verlag, 1990, pp. 163-172.
7. J.-M. Deshouillers, *Waring's Problem and the Circle-Method*, in Number Theory and Applications, R. A. Mollin Eds., Kluwer Academic Publishers, 1989, pp. 37-44.
8. L. E. Dickson, *Cyclotomy, Higher Congruences, and Waring's Problem*, Amer. J. Math. 57, 1935, pp. 391-424, pp. 463-474.
9. L. E. Dickson, *Solution of Waring's Problem*, Amer. J. Math. 58, 1936, pp. 530-535.
10. C. Ding, G. Xiao, and W. Shan, *The Stability Theory of Stream Ciphers*, LNCS 561, Springer-Verlag, 1991.
11. C. Ding, *The Differential Cryptanalysis and Design of the Natural Stream Ciphers*, Fast Software Encryption: Proc. of the 1993 Cambridge Security Workshop, R. Anderson (Ed.), LNCS 809, Springer-Verlag, 1994, pp. 101-115.
12. E. Lehmer, *On the Number of Solutions of $u^k + D = w^2$ mod p*, Pacific J. Math. 5, 1955, pp. 103-118.
13. R. Lidl, H. Niederreiter, *Finite Fields*, in Encyclopedia of Mathematics and Its Applications, vol. 20, Addison-Wesley, 1983.
14. J. L. Massey, *Shift-Register Synthesis and BCH Decoding*, IEEE Trans. Inform. Theory, vol. IT-15, January, 1969, pp. 122-127.
15. W. Meier, O. Staffelbach, *Nonlinearity Criteria for Cryptographic Functions*, LNCS 434, Advances in Cryptology, Springer-Verlag, 1990, pp. 549-562.
16. D. Pei, *Personal communications*, Jan. 1994.
17. S. Pillai. *On Waring's Problem*, I. Ind. Math. Soc. 2, 1933, pp. 16-44.
18. W. M. Schmidt, *Equations over Finite Fields: An Elementary Approach*, Lecture Notes in Mathematics, vol. 536, Springer-Verlag, 1976.
19. T. Storer, *Cyclotomy and Difference Sets*, Marham, Chicago, 1967.
20. A. Weil, *Sur les Courbes Algébriques et les Variétés qui s'en Déduisent*, Actualités Sci. Ind. No. 1041.
21. A. L. Whiteman, *A Family of Difference Sets*, Illinois J. Math. 6, 1962, pp. 107-121.

Construction of Bent Functions and Balanced Boolean Functions with High Nonlinearity

Hans Dobbertin

German Information Security Agency
P.Ö. Box 20 10 63, D-53133 Bonn, Germany

dobbertin@skom.rhein.de

Abstract. A general explicit construction of bent functions is described, which unifies well known constructions due to Maiorana-McFarland and Dillon as two opposite extremal cases. Within this framework we also find new ways to generate bent functions. Then it is shown how the constructed bent functions can be modified in order to obtain highly nonlinear balanced Boolean functions. Although their nonlinearity is the best known so far, it remains open whether this bound can still be improved.

1 Introduction

Boolean functions form important components of various practical cryptographic algorithms. One basic criterion for their design is nonlinearity. The significance of this aspect has again been demonstrated by the recent development of linear cryptanalysis by Matsui [5] and others.

Loosely speaking, bent functions are Boolean functions achieving the highest possible nonlinearity uniformly. In view of the Parseval equation this definition implies that they exist only for an even number of variables.

Bent functions were introduced by Rothaus [8] in 1976. They turned out to be rather complicated combinatorical objects. While a concrete description of all bent functions is elusive, there are two well-known explicit constructions of special bent functions due to Maiorana-McFarland [6] and Dillon [4].

In the next section a general construction of normal bent functions is described (triple construction, see Definition 1 and Lemma 2), where we call a Boolean function with $2n$ variables normal if it is constant on a n-dimensional affine subspace. Within the framework of the triple construction we obtain the bent functions of Maiorana-McFarland and of Dillon as the two opposite extremal cases, and we also find new ways to construct bent functions (see Theorem 4).

Depending on the conditions of the concrete application, it has often to be considered as a defect from the cryptographic point of view that bent functions are necessarily non-balanced. In the third section it is shown how normal bent functions can be modified in order to get highly nonlinear balanced Boolean functions with $2n$ variables (see Proposition 8). We conclude that, in the balanced case, the maximal nonlinearity for $2n$ variables is connected by a certain inequality with the maximal nonlinearity for n variables (see Theorem 9). However, the challenging problem to determine the maximal nonlinearity of balanced Boolean functions precisely remains open.

2   Normal Bent Functions

The **Walsh transformation** of a Boolean function g defined on a finite vector space V over $GF(2)$ with a scalar product is denoted by g^W:

$$g^W(a) = \sum_{x \in V} (-1)^{g(x) + \langle a, x \rangle}.$$

The Walsh transformation is a very powerful tool for analyzing Boolean functions. Note that the set of values occuring as Walsh coefficients is independent of the choice of the scalar product. Recall that a **bent function** f on a $2n$-dimensional vector space V over $GF(2)$ is defined by the property

$$f^W(z) = \pm 2^n \text{ for all } z \in V.$$

We call a Boolean function f with $2n$ variables **normal**, if there is an affine subspace with dimension n, on which f is constant. In the following we shall be concerned with normal bent functions. Starting point of our investigation is the following fact (see Lemma 7, Section 3):

Let f be a normal bent function on a $2n$-dimensional vector space V over $GF(2)$ such that the restriction of f on an affine subspace U of dimension n is constant. Then the restriction of f on each proper co-set of U is balanced.

Thus if one wants to construct a normal bent function then this means essentially that one has to construct a suitable collection $(f_y)_{y \in W \setminus \{y_0\}}$ ($y_0 \in W$ fixed) of balanced Boolean functions on a n-dimensional vector space W over $GF(2)$. The Boolean function on W^2 corresponding to such a collection is defined as

$$f(x, y) = \begin{cases} f_y(x) \text{ for } y \neq y_0 \\ \\ \text{constant, otherwise.} \end{cases}$$

For the following the value, 0 or 1, of the constant is not important. Hence we assume that it is 0. If instead of f we consider its support $T = \text{supp} f$ then the above setting means

$$T = \bigcup_{y \in W \setminus \{y_0\}} S_y \times \{y\},$$

where $\text{supp} f_y = S_y \subseteq W$ und $\#S_y = 2^{n-1}$.

It is a natural idea to endow W with a field structure, to choose some fixed subset S of W with $\#S = 2^{n-1}$, and to define the S_y as a permutation of sets of the form

$$yS + c_y.$$

This leads to the following construction of Boolean functions:

Definition 1 (Triple Construction). *Let L be the field $GF(2^n)$. Choose*

$$\sigma : L \longrightarrow GF(2) \ balanced,$$
$$\phi : L \longrightarrow L \ bijective,$$
$$\psi : L \longrightarrow L \ arbitrary.$$

The Boolean function $f = f_{\sigma,\phi,\psi}$ on L^2 associated to the triple (σ, ϕ, ψ) is defined as follows:

$$f(x, \phi(y)) = \begin{cases} \sigma\left(\frac{x + \psi(y)}{y}\right) & \text{for } y \neq 0 \\ \\ 0 & \text{otherwise.} \end{cases}$$

The support of f is

$$\operatorname{supp} f = \bigcup_{y \in L^*} (yS + \psi(y)) \times \{\phi(y)\},$$

where $S = \operatorname{supp} \sigma$.

We call (σ, ϕ, ψ) a **bent triple** if the associated Boolean function $f_{\sigma,\phi,\psi}$ is a bent function.

In the sequel, with respect to the Walsh coefficients of a Boolean function g, we refer to the scalar product

$$\langle x, y \rangle = \operatorname{Tr}(xy)$$

if g is defined on the field L, and to the scalar product

$$\langle (x, u), (y, v) \rangle = \operatorname{Tr}(xy + uv)$$

if g is defined on L^2.

Lemma 2. *Suppose that σ, ϕ, ψ and $f = f_{\sigma,\phi,\psi}$ are given as specified in the triple construction (Definition 1).*

1. *For all $b \in L$ we have $f^W(0, b) = (-1)^{\operatorname{Tr}(b\phi(0))} 2^n$. For $a, b \in L$, $a \neq 0$ set*

$$\Gamma_{a,b}(x) = \operatorname{Tr}\left(a\psi(x/a) + b\phi(x/a)\right).$$

Then

$$f^W(a, b) = \sum_{x \in L} (-1)^{\Gamma_{a,b}(x)} \sigma^W(x).$$

2. *Let $T_\sigma \subseteq L$ denote the affine subspace generated by $\operatorname{supp} \sigma^W$. The following condition implies that (σ, ϕ, ψ) is a bent triple:*

$$\phi \text{ and } \psi \text{ are affine}^1 \text{ on } aT_\sigma \text{ for all } a \in L^*. \tag{1}$$

---

[1] ϕ is said to be *affine on a affine subspace* T if there is an affine mapping which coincides on T with ϕ, or equivalently if $\phi(u) + \phi(v) + \phi(w) = \phi(u + v + w)$ for all $u, v, w \in T$.

Proof. 1. We have

$$f^W(a,b) = \sum_{x,y \in L} (-1)^{f(x,y)+\langle (a,b),(x,y)\rangle} = \sum_{x,y}(-1)^{f(x,\phi(y))+\mathrm{Tr}(ax+b\phi(y))}$$

$$= \sum_{x}\sum_{y\neq 0}(-1)^{\sigma\left(\frac{x+\psi(y)}{y}\right)+\mathrm{Tr}(ax+b\phi(y))} + (-1)^{\mathrm{Tr}(b\phi(0))}\sum_{x}(-1)^{\mathrm{Tr}(ax)}$$

By the substitution $z = (x + \psi(y))/y$ we get

$$\sum_{x}\sum_{y\neq 0}(-1)^{\sigma\left(\frac{x+\psi(y)}{y}\right)+\mathrm{Tr}(ax+b\phi(y))} = \sum_{y\neq 0}\sum_{z}(-1)^{\sigma(z)+\mathrm{Tr}(ayz+a\psi(y)+b\phi(y))}$$

$$= \sum_{y\neq 0}(-1)^{\mathrm{Tr}(a\psi(y)+b\phi(y))}\sigma^W(ay)$$

Since σ is balanced it follows that $\sigma^W(0) = 0$. In the above sum over y we therefore can add the case $y = 0$, and for $a = 0$ we conclude

$$f^W(0,b) = (-1)^{\mathrm{Tr}(b\phi(0))}2^n.$$

In the following suppose $a \neq 0$. Then $\sum_{x}(-1)^{\mathrm{Tr}(ax)} = 0$, and it follows that

$$f^W(a,b) = \sum_{y}(-1)^{\mathrm{Tr}(a\psi(y)+b\phi(y))}\sigma^W(ay)$$

$$= \sum_{x}(-1)^{\mathrm{Tr}(a\psi(x/a)+b\phi(x/a))}\sigma^W(x).$$

2. Let g and Γ be Boolean functions on L, where Γ is affine. Then obviously

$$\sum_{x \in L}(-1)^{\Gamma(x)}g^W(x) = \pm 2^n.$$

Of course this equation already holds if Γ is affine on T_g, the affine subspace generated by $\mathrm{supp}\, g^W$.

Condition (1) assures that all mappings $\Gamma_{a,b}$ are affine on T_σ. From 1. it follows that $f^W(a,b) = \pm 2^n$ for all $a, b \in L$. $\qquad\square$

In order to state the main result of this section we need a preparing lemma.

Lemma 3. *Let U be a subspace of the vector space $V = G(2)^n$, and $y_0 \in V$. Then there is an onto linear mapping $\rho : V \longrightarrow U$ such that a one-to-one correspondence between all Boolean functions $\sigma : V \longrightarrow \mathrm{GF}(2)$ with*

$$\mathrm{supp}\, \sigma^W \subseteq y_0 + U$$

and all Boolean functions τ on U is given by setting

$$\sigma(x) = \tau\rho(x) + \langle x, y_0\rangle.$$

Moreover, all σ are balanced if and only if $y_0 \notin U$.

Proof. We have the formula

$$(\sigma\Lambda)^W(a) = \sigma^W((\Lambda^*)^{-1}(a)),$$

where $\Lambda : V \longrightarrow V$ is a bijective linear mapping, and Λ^* is the adjoint of Λ, i.e. $\langle \Lambda(x), y \rangle = \langle x, \Lambda^*(y) \rangle$. On the other hand $(\sigma + \ell_{y_0})^W(a) = \sigma^W(a + y_0)$, for the linear mapping $\ell_{y_0}(x) = \langle x, y_0 \rangle$. Thus the first assertion has to be shown only for the case that U is generated by unit vectors and $y_0 = 0$. But this case is easily verified.

The last statement is obvious, since σ is balanced iff $\sigma^W(0) = 0$. □

Maiorana-McFarland and Dillon gave two different constructions of bent functions. We say these bent functions are of *Maiorana-McFarland type (MM-type)* and *Dillon type (D-type)*, respectively. Their definitions can be found in the proof of the following theorem.

Theorem 4. *Let $L = \mathrm{GF}(2^n)$, and let (σ, ϕ, ψ) be given as described in the triple construction.*

1. *If σ is affine then (σ, ϕ, ψ) is a bent triple for arbitrary ϕ, ψ. In this case one obtains precisely the bent functions of MM-type.*
2. *Conversely if ϕ and ψ are affine then (σ, ϕ, ψ) is a bent triple for arbitrary σ. In this case $f_{\sigma,\phi,\psi}$ and $f_{\sigma,\mathrm{id},0}$ are affinely equivalent. The bent functions of D-type are precisely the functions of the form $f_{\sigma,\mathrm{id},0}$.*
3. *Besides the two opposite extremal cases 1. and 2. there are further bent triples:*
   Suppose $\phi(x) = x^d$, $\psi(x) = x^{d'}$ (or $\psi = 0$) for $d, d' < 2^n - 1$, and let a non-trivial subspace U of L and $y_0 \in L \setminus U$ be given such that the following conditions are satisfied:
   (a) ϕ is bijective, i.e. d is relatively prime to $2^n - 1$,
   (b) ϕ and ψ are not affine, i.e. d and d' are not powers of 2,
   (c) ϕ and ψ are affine on $y_0 + U$.
   Define $\sigma : L \longrightarrow \mathrm{GF}(2)$ as a non-affine balanced Boolean function such that the support of σ^W is a subset of $y_0 + U$. This means that σ is of the form

$$\sigma(x) = \tau\rho(x) + \mathrm{Tr}(xy_0),$$

where $\rho : L \longrightarrow U$ is an onto linear mapping chosen according to Lemma 3, and τ is an arbitrary non-affine Boolean functions on U.
Then (σ, ϕ, ψ) is a bent triple, which is neither of D- nor of MM-type. The explicit definition of the corresponding bent function is

$$f(x, y^d) = \begin{cases} \tau\rho\left(\frac{x}{y} + y^{d'-1}\right) + \mathrm{Tr}\left((\frac{x}{y} + y^{d'-1})y_0\right) & \text{if } y \neq 0, \\ 0 & \text{otherwise.} \end{cases}$$

Two concrete examples are given after the proof of this theorem.

Proof. 1. Obviously σ is affine if and only if $T_\sigma = \operatorname{supp}\sigma^W$ is a singleton. Thus the first claim follows immediately by Lemma 2.2.

We identify L with $\mathrm{GF}(2)^n$. A bent function $g : L^2 \longrightarrow \mathrm{GF}(2)$ is of **MM-type** if it is of the form

$$g(x, y) = \langle x, \pi(y)\rangle + h(y),$$

where π is a bijection on L, h is an arbitrary Boolean function on L and $\langle\cdot,\cdot\rangle$ is the canonical scalar product on $\mathrm{GF}(2)^n$. W.l.o.g. we can assume $\pi(0) = 0$ and $h(0) = 0$.

Let $\varrho : L \longrightarrow L$ be the bijective linear mapping defined by the equation

$$\langle x, y\rangle = \mathrm{Tr}(x\varrho(y))$$

for all $x, y \in L$. Now define ϕ such that the equation

$$\phi^{-1}(y) = \frac{1}{\varrho\pi(y)} \quad (y \in L^*)$$

holds, and choose ψ with the property

$$\mathrm{Tr}\left(\frac{\psi(y)}{y}\right) = h\phi(y) \quad (y \in L^*).$$

Then as desired we have

$$g = f_{\mathrm{Tr},\phi,\psi}.$$

Similarly we see that every $f_{\sigma,\phi,\psi}$ with linear σ is of MM-type.

2. Again the first assertion follows immediately from Lemma 2.2. If ϕ and ψ are affine then $f_{\sigma,\mathrm{id},0}\Lambda = f_{\sigma,\phi,\psi}$ for the bijective affine mapping

$$\Lambda : (x, y) \longmapsto \left(x + \psi\phi^{-1}(y), \phi^{-1}(y)\right).$$

It remains to verify that the bent functions of D-type are precisely the $f_{\sigma,\mathrm{id},0}$:

Let $E = \mathrm{GF}(2^{2n})$. We consider E as field extension of $L = \mathrm{GF}(2^n)$. Choose an $\alpha \in E$ with $E = L[\alpha]$. We set

$$\bar{x} = x^{2^n} \quad (x \in E),$$

for the Frobenius automorphism of $E : L$. The mapping

$$h : \begin{cases} x \longmapsto \bar{x}/x \\[4pt] E^* \longrightarrow E^* \end{cases}$$

is a multiplicative homomorphism with image

$$H = \{z \in E^* : z\bar{z} = 1\}.$$

The kernel of h is L^*. Therefore H is a representation system for the elements of the factor group E^*/L^*, i.e. the sets of the form xL^* $(x \in E^*)$. In view of $H \cong E^*/L^*$ we have

$$\#H = 2^n + 1.$$

Set $H_1 = H \setminus \{1\}$. The bent functions of **D-type** are precisely the characteristic functions of sets of the form

$$D(Z) = \bigcup_{z \in Z} zL^*,$$

where $Z \subseteq H_1$ has exactly 2^{n-1} elements. (The condition $\#Z = 2^{n-1}$ assures that $\chi_{D(Z)}$ is a bent function.)

A bijection between H_1 and L is defined by

$$\Upsilon : z \longmapsto \frac{\overline{\alpha}z + \alpha\overline{z}}{z + \overline{z}}.$$

If we identify L^2 and E by setting $(x, y) = x + y\alpha$ then for $S = \Upsilon[Z]$

$$D(Z) = \bigcup_{z \in Z} zL^* = \bigcup_{y \in L^*} yS \times \{y\} = \operatorname{supp} f_{\sigma, \mathrm{id}, 0}, \tag{2}$$

where σ is the characteristic function of S.

Remark. As we have seen there is an interesting analogy between the field extensions $E : L$ and $\mathbb{C} : \mathbb{R}$. For instance, (2) can be interpreted as the change from the representation of D in "polar coordinates" to the representation in "cartesian coordinates."

3. If the power functions ϕ and ψ are affine on T_σ, the affine subspace generated by the support of σ^W, then of course they are affine on aT_σ for all $a \in L^*$. Thus (σ, ϕ, ψ) is a bent triple by Lemma 2.2 if the conditions (a) – (c) are satisfied. $\quad\square$

We want to give two concrete examples how the exponents d, d' and the affine subspace $y_0 + U$ can be choosen such that the conditions (a) – (c) in Theorem 4.3 are satisfied.

Example 1. Assume that n is not prime and not a power of 2, i.e. $n = mr$ with $r > 1$ and odd $m > 1$. Let $d = 2^r + 1$, $U = K = \mathrm{GF}(2^r)$, and $y_0 \in L \setminus U$. Then for all $x \in y_0 + U$ we have

$$\begin{aligned}
x^d &= ((x + y_0) + y_0)^{2^r + 1} \\
&= (x + y_0)^2 + (x + y_0)y_0 + (x + y_0)y_0^{2^r} + y_0^{d} \\
&= x^2 + \left(y_0 + y_0^{2^r}\right)x,
\end{aligned}$$

i.e. ϕ is linear on $y_0 + U$. It remains to show that $2^r + 1$ and $2^n - 1$ are relatively prime. In fact

$$(2^r + 1)(2^{(m-1)r-1} - 2^{(m-2)r-1} + \dots - 2^{r-1} + 2^{n-1}) \equiv 1 \pmod{2^n - 1}.$$

Note that the surjective linear mapping ρ from L onto $U = K$ can here be taken as the orthogonal[2] projection $\rho = \mathrm{Tr}_{L:K}$, that is

$$\sigma(x) = \tau \mathrm{Tr}_{L:K}(x) + \mathrm{Tr}(xy_0),$$

---

[2] That is $\mathrm{Tr}((x + \rho(x))u) = 0$ for all $x \in L$, $u \in U$.

where τ is any non-affine mapping on K. To complete the definition of the bent triples set $\psi = 0$.

Example 2. Assume that $n = 2r > 2$ is even. Set

$$d = (2^r + 2)(1 + 2^2 + 2^4 + \dots + 2^{2s}) + 1.$$

Then

$$d \equiv 2^{2(s+1)} \pmod{2^r - 1},$$

and consequently the restriction of $\phi(x) = x^d$ onto $K = \mathrm{GF}(2^r)$ is linear. Similarly set

$$d' = (2^r + 2)(1 + 2^2 + 2^4 + \dots + 2^{2s'}) + 1.$$

Now let U be any hyperplane of K and $y_0 \in K \setminus U$. Then the conditions (b) and (c) of Theorem 4.3 are fulfilled. We do not know when in general ϕ is bijective. But at least for $s = 0$ this is always the case, since

$$(2^r + 3)(3 \cdot 2^{n-3} - 2^{r-3}) \equiv 1 \pmod{2^n - 1}.$$

To describe the surjection from L onto U for a concrete example, suppose that

$$U = \ker \mathrm{Tr}_K$$

and r is odd. Then again the orthogonal projection can be taken as ρ, i.e. $\rho(x) = \mathrm{Tr}_{L:K}(x) + \mathrm{Tr}(x)$, and consequently

$$\sigma(x) = \tau(\mathrm{Tr}_{L:K}(x) + \mathrm{Tr}(x)) + \mathrm{Tr}(xy_0).$$

It is easy to find further similar examples of bent triples, where ϕ and ψ are power functions.

According to Theorem 4.3 we have found a new construction of bent functions. But of course this does not mean that the resulting bent functions are really new in the sense that they cannot be derived from already known ones using simple modifications. For instance, the following constructions alter a given bent function f on $V = \mathrm{GF}(2)^{2n}$ into another bent function g on V:

1. "affine modification," i.e. $g = f\Lambda$, where $\Lambda : V \longrightarrow V$ is bijective and affine,
2. "affine addition," i.e. $g = f + \ell$, where $\ell : V \longrightarrow \mathrm{GF}(2)$ is affine,
3. "dualizing," i.e. $g = f^*$, where the *dual* bent function f^* associated to f is defined by the equation

$$f^W = 2^n (-1)^{f^*},$$

Also, the direct sum of bent functions is again a bent function. Another kind of modifications are those which require certain conditions to guarantee that the resulting function is bent again. A lot of them are known in the literature, the most simple is

4. "skipping," i.e. setting $g = f + \chi_U$, where f is bent function with $2n$ variables, and U is an affine subspace with dimension n such that f is constant on U.

Note that skipping can be applied precisely to normal bent functions.

Let $\mathcal{D}$, $\mathcal{M}$ and $\mathcal{N}$ denote the class of all bent functions of D-type, MM-type and the type constructed in Theorem 4.3, respectively. For any class $\mathcal{B}$ of bent functions let $\widetilde{\mathcal{B}}$ denote its completion under *equivalence* (i.e. modifications of type 1. and 2.), and let $\overline{\mathcal{B}}$ denote its completion under forming direct sums, the above modifications 1. – 3. and known "conditional" modifications such as for instance 4.

The class of all explicitly known bent functions so far can now be written as

$$\overline{\mathcal{D} \cup \mathcal{M}}.$$

Thus if we want to show that the bent functions of Theorem 4.3 are *new* in the strongest sense of the word then we would have to establish that $\mathcal{N}$ is not included in $\overline{\mathcal{D} \cup \mathcal{M}}$. However, this seems to be a very difficult problem. The weaker statement $\mathcal{N} \not\subseteq \widetilde{\mathcal{D} \cup \mathcal{M}}$ is a desirable first step. The probably easier half of this statement is shown next. To this end we have proven by computation of explicit examples:

Lemma 5. *The class $\mathcal{N}$ contains bent functions with non-degenerated second derivation.*

Proposition 6. *The class $\mathcal{N}$ is not contained in $\widetilde{\mathcal{M}}$.*

Proof. To seperate $\mathcal{D}$ from $\widetilde{\mathcal{M}}$, Dillon [4] showed that the bent functions of MM-type have a degenerated second derivation, while this in general is not true for bent functions of D-type. In view of Lemma 5 we can argue in the same way. □

Remark. Carlet [2] has shown that a generalized form of skipping applied to bent functions of MM-type leads out of $\widetilde{\mathcal{PS} \cup \mathcal{M}}$, where $\mathcal{PS}$ denotes the class of all bent functions, which are "partial spreads." ($\mathcal{PS}$ includes $\mathcal{D}$ as the subclass of its concretely known examples.)

3   Balanced Boolean Functions

The **spectral radius** of a Boolean function $f : \mathrm{GF}(2)^m \longrightarrow \mathrm{GF}(2)$ is

$$R_f = \max\{\,|f^W(a)|. : a \in \mathrm{GF}(2)^m\}.$$

R_f can be considered as a measure for the linearity of f. Thus if we are interested in Boolean functions with high nonlinearity then we have to look for f's with small R_f.

The **Parseval equation** states that the square sum over the Walsh coefficients of a Boolean function with m variables equals to 2^{2m}. Hence we have

$$R_f \geq 2^{m/2},$$

and in the even case $m = 2n$ this lower bound is achieved by bent functions, which can be characterized by the property that their Walsh spectrum consists only of the values $\pm 2^n$. (The odd case is known to be difficult; see [7].)

However, bent functions are not balanced. Thus if we ask for balanced functions with high nonlinearity then it is a natural idea to obtain them by "making a bent function balanced," where the spectral radius is increased as less as possible. In the following it will be shown that this idea works for all *normal* bent functions. In this way we can construct balanced functions with almost the same nonlinearity as bent functions.

First we prove the following lemma (cf. [2], Lemma 1):

Lemma 7. *Let* $W = \mathrm{GF}(2)^n$ *and* $V = W^2$. *Let* f *be a normal bent function on* V. *That is w.l.o.g.* $f(x, 0) = 0$ *for all* $x \in W$. *Then*[3] $f^W(0, b) = 2^n$ *for all* $b \in W$. *Moreover for each fixed* $y \in W \setminus \{0\}$ *the function*

$$f_y : \begin{cases} W \longrightarrow \mathrm{GF}(2) \\ x \longmapsto f(x, y) \end{cases}$$

is balanced.

Proof. Set for $y \in W$

$$F(y) = \begin{cases} f_y{}^W(0) = \sum_x (-1)^{f(x,y)} & \text{for } y \neq 0 \\ 0 & \text{otherwise.} \end{cases}$$

We compute

$$\sum_b f^W(0, b) = \sum_{b,x,y} (-1)^{f(x,y)+\langle b,y \rangle}$$

$$= \sum_{b,x} (-1)^{f(x,0)+\langle b,0 \rangle} + \sum_{y \neq 0} F(y) \left(\sum_b (-1)^{\langle b,y \rangle} \right)$$

$$= 2^{2n} + 0 = 2^{2n}.$$

In view of $f^W(0, b) = \pm 2^n$ we conclude $f^W(0, b) = 2^n$. Hence

$$2^n = f^W(0, b) = 2^n + \sum_{y \neq 0} F(y)(-1)^{\langle b,y \rangle},$$

and therefore $\widehat{F}(b) = \sum_y F(y)(-1)^{\langle b,y \rangle} = 0$ for all $b \in W$. Consequently $F = 2^{-n} \widehat{\widehat{F}} = 0$. $\square$

---

[3] We use the canonical scalar product to define the Walsh coefficients.

Remark. Lemma 7 shows that if f is a normal bent function then also its dual f^* is normal.

Proposition 8. *Let $W = GF(2)^n$ and $V = W^2$. Let f be a normal bent function on V. That is w.l.o.g. $f(x,0) = 0$ for all $x \in W$. Furthermore let a balanced function $\theta : W \longrightarrow GF(2)$ be given. Set for $x, y \in W$*

$$\Theta(x,y) = \begin{cases} f(x,y), & \text{if } y \neq 0 \\ \\ \theta(x), & \text{otherwise.} \end{cases}$$

Then Θ is balanced and we have

$$\Theta^W(a,b) = \begin{cases} f^W(a,b) + \theta^W(a), & \text{if } a \neq 0 \\ \\ 0, & \text{otherwise.} \end{cases}$$

In particular it follows that

$$R_\Theta = 2^n + R_\theta.$$

Proof. We have

$$\Theta^W(a,b) = \sum_{x,y}(-1)^{\Theta(x,y)+\langle a,x\rangle+\langle b,y\rangle}$$

$$= \sum_x(-1)^{\theta(x)+\langle a,x\rangle} + \sum_{x,y}(-1)^{f(x,y)+\langle a,x\rangle+\langle b,y\rangle} - \sum_x(-1)^{\langle a,x\rangle}$$

$$= \theta^W(a) + f^W(a,b) - \sum_x(-1)^{\langle a,x\rangle}.$$

This proves the first assertion, since $f^W(0,b) = 2^n$ by Lemma 7.

If we apply Lemma 7 to the dual bent function of f then in particular we see that for each fixed $a \in W \setminus \{0\}$ both values 2^n and -2^n are attained by $f^W(a,b)$. This implies $R_\Theta = 2^n + R_\theta$. □

In order to discuss the implications of Proposition 8 we introduce the notations

$$R(m) = \min\{R_f \mid f : GF(2)^m \longrightarrow GF(2)\},$$
$$RB(m) = \min\{R_f \mid f : GF(2)^m \longrightarrow GF(2) \text{ balanced}\}.$$

Theorem 9. $RB(2n) \leq 2^n + RB(n)$.

Proof. Use Proposition 8 with some θ such that $R_\theta = RB(n)$. □

For even $m = 2^s u$, u odd, one concludes by an inductive application of Theorem 9:

Corollary 10. $RB(m) \leq 2^{m/2} + 2^{m/4} + 2^{m/8} + \dots + 2^u + RB(u)$.

On the other hand one easily verifies that

$$RB(u) \leq 2^{\frac{u+1}{2}}.$$

Using this fact and a lower bound basically derived from the Parseval equation we obtain for $m \geq 4$

$$2^{m/2} + 4 \leq RB(m) \leq 2^{m/2} + 2^{m/4} + 2^{m/8} + \dots + 2^u + 2^{\frac{u+1}{2}}. \tag{3}$$

Independently the upper bound of (3) has been found by Seberry, Zhang and Zheng (see Theorem 1 of [9]). For instance it yields

$$132 \leq RB(14) \leq 144. \tag{4}$$

For $u = 1, 3, 5$ and 7 it is known that $R(u) = RB(u) = 2^{(u+1)/2}$. But in 1983 Patterson and Wiedemann [7] showed that

$$R(15) \leq 216 = \frac{27}{32} 2^{\frac{15+1}{2}}. \tag{5}$$

We can derive from this fact a similar result for balanced Boolean functions (cf. Theorem 2 of [9]). In fact note that for the spectral radius of the direct sum of Boolean functions f and g we have the formula $R_{f \oplus g} = R_f R_g$. Hence

$$RB(n + m) \leq RB(n) R(m) \tag{6}$$

for all n and m, since $f \oplus g$ is balanced if f is balanced. Thus by (4), (5) and (6)

$$RB(29) \leq RB(14) R(15) \leq 144 \cdot 216 = \frac{243}{256} 2^{\frac{29+1}{2}}.$$

More generally this implies

$$RB(u) < 2^{\frac{u+1}{2}} \quad \text{for all odd } u \geq 29,$$

since $RB(u + 2) \leq RB(u)R(2) = 2\,RB(u)$. Note that therefore Corollary 10 is stronger than the upper bound of (3). We close our investigation with two conjectures:

Conjecture A. *The recursive inequality given in Theorem 9 is sharp:*

$$RB(2n) = 2^n + RB(n).$$

Conjecture B. *For odd m we have the asymptotic formulas*

$$RB(m) \approx R(m) \approx 2^{m/2}.$$

4   Applications

Theorem 4 unifies and extends the known constructions of bent functions. Proposition 8 describes how we can get balanced Boolean functions with highest nonlinearity known so far. On the other hand, the generation of bent functions with $2n$ variables by Theorem 4 requires a balanced Boolean function σ with n variables. If n is even, one can choose σ highly nonlinear according to Proposition 8, and then use Theorem 4.2. That is, *applications of Theorem 4 and Proposition 8 can be linked together inductively.*

Adding also the forming of direct sums, affine modifications, affine addition, dualizing and (generalized) skipping as ingredients, all this can be considered as a cooking book for both, the generation of bent functions and of highly nonlinear balanced Boolean functions with an even number of variables.

Concluding Remarks

1. There are certainly more ways to construct "non-standard" bent triples (σ, ϕ, ψ) than those given in Theorem 4.3. The smaller the support of the Walsh transformation of σ becomes, the more freedom we get for ϕ and ψ. We have restricted ourselves to power functions, because this seems to be the simplest non-trivial case.

2. Another class of explicitly constructable bent functions has recently been found by Carlet [3]. This class can also be derived from a slightly modified version of the triple construction, as will be shown in a forthcoming paper.

3. Simon Blackburn [1] mentioned a simple counting argument showing that there are non-normal Boolean functions with $2n$ variables for $n \geq 6$. In fact, for each affine subspace S of dimension n there are precisely

$$b_n = 2 \cdot 2^{2^{2n} - 2^n}$$

Boolean functions, which are constant on S. On the other hand it is well-known that there are exactly

$$s_n = \prod_{i=0}^{n-1} \frac{2^{2n} - 2^i}{2^n - 2^i} \approx 2^{n^2}$$

subspaces of dimension n, i.e. the number of affine subspaces of dimension n is $a_n = 2^n s_n$. Thus an upper bound for the number of normal Boolean functions on $\mathrm{GF}(2)^{2n}$ is given by

$$u_n = a_n b_n = 2^{2^{2n} - 2^n + n + 1} s_n \approx 2^{2^{2n} - 2^n + n^2 + n + 1}.$$

However, as one easily verifies, u_n is smaller than $2^{2^{2n}}$, the number of all Boolean functions on $\mathrm{GF}(2)^{2n}$, for $n \geq 6$. It remains open whether there are non-normal bent functions.

References

1. Blackburn, S.: private communication, 1995.
2. Carlet, C.: Two new classes of bent functions. Advances in Cryptology, Eurocrypt '93, Lecture Notes in Computer Science 765, Springer–Verlag 1994, pp. 77–101.
3. Carlet, C.: Generalized partial spreads. IEEE Transactions on Information Theory (to appear).
4. Dillon, J. F.: Elementary Hadamard difference sets. Proceedings of the Sixth Southeastern Conference on Combinatorics, Graph Theory and Computing, Boca Raton, Florida, Congressus Numerantium No. XIV, Utilitas Math., Winnipeg, Manitoba, 1975, pp. 237 – 249.
5. Matsui, M.: Linear cryptanalysis method for DES cipher. Advances in Cryptology, Eurocrypt '93, Lecture Notes in Computer Science 765, Springer–Verlag 1994, pp. 386 – 397.
6. McFarland, R. L.: A family difference sets in non-cyclic groups. J. Combinatorical Theory, Ser. A, 15 (1973), pp. 1 – 10.
7. Patterson, N. J., Wiedemann, D. H.: The covering radius of the $(2^{15}, 16)$ Reed-Muller code is at least 16276. IEEE Transactions on Information Theory 29 (1983), pp. 354 – 356.
8. Rothaus, O. S.: On "bent" functions. J. Combinatorical Theory, Ser. A, 20 (1976), pp. 300 – 305.
9. Seberry, J., Zhang, X., Zheng, Y.: Nonlinearity and propagation characteristics of balanced Boolean functions. Information & Computation (to appear).

Additive and Linear Structures of Cryptographic Functions

Xuejia Lai

R³ Security Engineering AG, Aathal, Switzerland
lai@r3.ch

Abstract. In the design and analysis of cryptographic algorithms, exploiting the structures of such algorithms is an important aspect. In this paper, additive and linear structures of functions from $GF^n(q)$ to $GF^m(q)$ will be considered. A function f is said to have an additive structure if there is a non-zero vector a, such that $f(x + a) - f(x)$ remains invariant for all x. Such a vector a is called an additive translator of the function f. A function f is said to have a linear structure if f has an additive translator a and if $f(x + ca) - f(x) = c(f(a) - f(0))$ for all c in $GF(q)$. We call this a a linear translator of f. We show how to use such additive and linear structures to simplify the expression of the function f. It is shown that function f has r linearly independent linear translators if and only if there is a non-singular linear transformation such that the composition of this linear transformation with the original function gives a function that is the sum of a linear function of r variables and some function of the other $n - r$ variables. In particular, when q is a prime, then any additive translator is a linear translator, which implies that f becomes a sum of an r-variable linear function and an $n - r$-variable function if and only if f has r linearly independent additive translators. Moreover, for an invertible function f, there is a one-to-one relationship between the linear translators of f and the linear translators of its inverse function.

1. Introduction

Linear structures of block ciphers have been investigated for their cryptanalytic significance. According to Evertse [2], "a block cipher has a linear structure if there are subsets of P, K and C of plaintext bits, key bits and ciphertext bits of this block cipher, respectively, such that for *each* plaintext and *each* key, a simultaneous change of all plaintext bits in P and all key bits in K has the *same* effect on the exclusive-or sum of the bits in C of the corresponding ciphertext." One well-known example [3] of such linear structures is the DES function for which the complementation of all plaintext bits and key bits results in the complementation of all ciphertext bits. It has been pointed out in [1, 2, 3, 6] that block ciphers with linear structures are vulnerable to attacks much faster than exhaustive key search. In [4], Meier and Staffelbach considered the nonlinearity of Boolean functions in terms of their Hamming distance to functions having a linear structure and constructed Boolean functions that have maximum distance to functions with a linear structure.

In this paper we consider the additive and linear structures of functions from an n dimensional vector space F^n to an m dimensional vector space F^m, where $F = GF^n(q)$ is a finite field. In Section 1, such structures will be studied in terms of *additive* and *linear translators*, where it is shown that the set of all additive translators of a function forms an additive subgroup of the domain vector space and that the set of all linear translators of a function forms a linear subspace. The main result is in Section 3 where it is shown that the linear subspace of linear translators of an n-variable function has dimension r if and only if that there is a non-singular linear transformation on the variables of the function such that the composition of this linear transformation with the original function gives a function that is the sum of a linear function in r variables and some function of $n - r$ variables which has no linear structure. In particular, if F is a prime field, i.e., a field of p elements where p is a prime, an n-variable function has r linearly independent additive translators if and only if there is a nonsingular matrix A such that $f(xA)$ is the sum of a linear function of r components of x and some function of the other $n - r$ components of x. In Section 4 it is shown that the dimension of the subspace of linear translators remains the same for every function obtained by the composition of a non-singular linear transformation and the original function. It is then shown that for a nonlinear function the sum of its nonlinear degree and the dimension of the subspace of its linear translators is upper bounded by the number n of variables. Finally, we show in Section 5 that there is a one-to-one relationship between the linear structure of an invertible function and the linear structure of its inverse function. An additional remark is made in Section 6 which interprets the additive structures in terms of differentials used in differential cryptanalysis.

2. Additive and linear structures of functions over finite fields

Throughout this paper, elements of F^n will be denoted as x or $(x_1, x_2, \ldots, x_n)$, where $x_i \in F$ and where $x_1, x_2, \ldots, x_n$ are the coordinates of x under the canonical basis of F^n :

$$e_1 = (1, 0, \ldots, 0), \quad e_2 = (0, 1, 0, \ldots, 0), \cdots, \quad e_n = (0, \ldots, 0, 1), \qquad (1)$$

that is, $x = x_1 e_1 + x_2 e_2 + \cdots + x_n e_n$. A function $f : F^n \mapsto F^m$ will be denoted as $f(x)$ or as $f(x_1, x_2, \ldots, x_n)$ and will be called an n-variable function.

Definition.   A function $f : F^n \mapsto F^m$ is said to have an *additive structure* if there is a non-zero vector a in F^n, such that $f(x + a) - f(x)$ is invariant for all x in F^n. Such a vector a will be called an *additive translator* of the function f. A function f is said to have a *linear structure* if there is a non-zero vector a in F^n, such that

$$f(x + ca) - f(x) = c(f(a) - f(0)) \qquad (2)$$

for all c in F, such a vector a will be called a *linear translator* of the function f. By setting $c = 1$, we see that any linear translator is always an additive translator. (By way of convention, we consider the all-zero vector $\mathbf{0}$ to be a linear translator for any function.)

By choosing $x=0$, we see that $a = (a_1, \ldots, a_n)$ is an additive translator of function f if and only if

$$f(x_1+a_1, x_2+a_2, \ldots, x_n+a_n) = f(x_1, x_2, \ldots, x_n) + f(a_1, a_2, \ldots, a_n) - f(0, 0, \ldots, 0)$$

or in vector notation,

$$f(x + a) = f(x) + f(a) - f(0), \tag{3}$$

for all $x=(x_1, \ldots, x_n)$ in F^n. Similarly, we obtain from (2) that

$$f(ca) = cf(a) - (c - 1)f(0) \tag{4}$$

holds for all c in F if a is a linear translator of f.

Theorem 1.   *For a function $f : F^n \mapsto F^m$, the set of all additive translators of f forms an additive subgroup of F^n, and the set of all linear translators of f forms a linear subspace of F^n.*

Proof.   Let a, b be two additive translators of function f. Set $x = b$ in (3), we have

$$f(a + b) = f(a) + f(b) - f(0). \tag{5}$$

Thus, for all x in F^n,

$$\begin{aligned} f(x + a + b) &= f(x + a) + f(b) - f(0) \\ &= f(x) + f(a) + f(b) - f(0) - f(0) \\ &= f(x) + f(a + b) - f(0), \end{aligned}$$

so that $a + b$ is also an additive translator of f. Thus, the set of additive translators is an additive subgroup.

Since a linear translator is always an additive translator, it remains to show that if a is a linear translator of function f then for any c_0 in F, $b = c_0 a$ is also a linear translator. For any c in F, because a is a linear translator,

$$\begin{aligned} f(x + cb) - f(x) &= f(x + cc_0 a) - f(x) = c[c_0(f(a) - f(0))] \\ &= c[f(c_0 a) - f(0)] = c(f(b) - f(0)) \end{aligned}$$

where the third equality is obtained from (4). Thus, $b = c_0 a$ is indeed a linear translator of F, so that the set of linear translators is a linear subspace. $\quad\square$

Now we show that over a prime field additive translator and linear translator are the same concept. This explains the reason that 'additive' structure was often referred to as 'linear' structure in the earlier literature because the most often considered operation in cryptography is the bitwise-XOR, i.e., over $GF(2)$. In particular, the set of additive translators forms a linear subspace for functions over prime fields. Later we will show by example that this is not the case over a general finite field.

Theorem 2.   *If $F = GF(p)$, a finite field of p elements where p is a prime, then an additive translator is also a linear translator.*

Proof.   Let a be an additive translator of f. By choosing $b = a$, $b = 2a$, ..., $b = (p-1)a$, it follows from Theorem 1 that for each element c of the prime field F, ca is also an additive translator of f. By using (5) repeatedly, we obtain

$$
\begin{aligned}
f(ca) &= f((c-1)a + a) \\
&= f((c-1)a) + f(a) - f(0) \\
&= f((c-2)a) + 2f(a) - 2f(0) \\
&= f(a) + (c-1)f(a) - (c-1)f(0) \\
&= cf(a) - (c-1)f(0).
\end{aligned}
$$

In (2) setting $x=0$ and using the above equality, we have

$$
f(x + ca) - f(x) = f(ca) - f(0) = cf(a) - cf(0)
$$

which shows that a is a linear translator of f.   □

3. Using linear structures to simplify functions

Now we show how to use linear structures to simplify the expression of a function. Let L_f denote the set of all linear translators of function f.

Theorem 3.   *Let f be a function from F^n to F^m. Then there exists an $n \times n$ invertible matrix A over F such that*

$$
g(x_1, \ldots, x_n) = f((x_1, \ldots, x_n)A) = x_1 v_1 + \cdots + x_r v_r + g^*(x_{r+1}, \ldots, x_n) \quad (6)
$$

where $v_i \in F^m$, for $1 \le i \le r$ and g^* is some non-constant function of $n - r$ variables which has no linear structure if and only if the set of linear translators of f forms an r-dimensional subspace of F^n.

Proof.   First, suppose that L_f is an r-dimensional subspace. Let $a_1, \ldots, a_r$ be r linearly independent linear translators that form a basis for L_f. $a_1, \ldots, a_r$ can be extended to a basis of the vector space F^n : $a_1, \ldots, a_r, a_{r+1}, \ldots, a_n$. Let A be the linear transformation from the canonical basis (1) $e_1, \ldots, e_n$ to the basis $a_1, \ldots, a_n$. Then A is non-singular and $a_i = e_i A$. We have

$$
g(x_1, ..., x_n) = f((x_1, ..., x_n)A) = f((x_1 e_1 + \cdots + x_n e_n)A) = f(x_1 a_1 + \cdots + x_n a_n).
$$

Because L_f is a subspace of F^n, $x_1 a_1$ is also a linear translator so we may use (3) and (4) to obtain

$$
\begin{aligned}
f(x_1 a_1 + \cdots + x_n a_n) &= f(x_2 a_2 + \cdots + x_n a_n) + f(x_1 a_1) - f(0) \\
&= x_1 f(a_1) - (x_1 - 1)f(0) - f(0) + f(x_2 a_2 + \cdots + x_n a_n) \\
&= x_1(f(a_1) - f(0)) + f(x_2 a_2 + \cdots + x_n a_n).
\end{aligned}
$$

Proceeding similarly for the linear translators $a_2, \ldots, a_r$ gives

$$g(x_1, ..., x_n) = x_1 v_1 + \cdots + x_r v_r + g^*(x_{r+1}, \ldots, x_n),$$

where $v_i = f(a_i) - f(0)$, for $i = 1, ..., r$, and where

$$g^*(x_{r+1}, \ldots, x_n) = f(x_{r+1} a_{r+1} + \cdots + x_n a_n)$$

is a function of $n - r$ variables. Moreover, if $(b_{r+1}, \ldots, b_n) \neq (0, \ldots, 0)$ is a linear translator of function g^*, then $b = (0, \ldots, 0, b_{r+1}, \ldots, b_n)$ is a linear translator of the function g, and b is linearly independent of $e_1, \ldots, e_r$ so that bA is linearly independent of $a_1, \ldots, a_r$. But then bA will be a linear translator of f because

$$f(xA + cbA) = g(x + cb) = g(x) + c(g(b) - g(0)) = f(xA) + c(f(bA) - f(0))$$

for all xA in F^n. This contradicts the assumption that $\dim L_f = r$. Therefore, function g^* has no linear structure.

Conversely, suppose that A is a non-singular transformation such that

$$g(x_1, \ldots, x_n) = f((x_1, \ldots, x_n)A) = x_1 v_1 + \cdots + x_r v_r + g^*(x_{r+1}, \ldots, x_n)$$

where g^* has no linear structure. First we show that $e_1, \ldots, e_r$ are linear translators of g. For $1 \leq i \leq r$ and for all c in F^n,

$$
\begin{aligned}
g(x + ce_i) &= g(x_1, \ldots, x_{i-1}, x_i + c, x_{i+1}, \ldots, x_n) \\
&= x_1 v_1 + \cdots + (x_i + c)v_i + \cdots + x_r v_r + g^*(x_{r+1}, \ldots, x_n) \\
&= x_1 v_1 + \cdots + x_r v_r + g^*(x_{r+1}, \cdots, x_n) + cv_i \\
&= g(x) + cv_i,
\end{aligned}
\tag{7}
$$

By setting $x = 0$ in the above equation, we obtain that $g(ce_i) = g(0) + cv_i$. For $c = 1$, we have $v_i = g(e_i) - g(0)$. Thus,

$$g(x + ce_i) - g(x) = cv_i = c(g(e_i) - g(0)), \tag{8}$$

which implies that e_i is a linear translator of g.

Now we show that every linear translator of function g is a linear combination of $e_1, \ldots, e_r$. Let $b = (b_1, ..., b_n)$ be a linear translator of g, then for all c in F,

$$g(x + cb) = g(x) + c(g(b) - g(0)). \tag{9}$$

The left side of the above equation is

$$(x_1 + cb_1)v_1 + \cdots + (x_r + cb_r)v_r + g^*(x_{r+1} + cb_{r+1}, \ldots, x_n + cb_n)$$

and the right side is

$$x_1 v_1 + \cdots + x_r v_r + g^*(x_{r+1}, ..., x_n) + c[b_1 v_1 + \cdots + b_r v_r + g^*(b_{r+1}, ..., b_n) - g^*(0, ..., 0)].$$

Thus, (9) is equivalent to

$$g^*(x_{r+1} + cb_{r+1}, ..., x_n + cb_n) = g^*(x_{r+1}, ..., x_n) + c[g^*(b_{r+1}, ..., b_n) - g^*(0, ..., 0)],$$

that is, $b = (b_1, ..., b_n)$ is a linear translator of function g if and only if $(b_{r+1}, ..., b_n)$ is a linear translator of function g^*. From the assumption that g^* has no linear structure we obtain $b_{r+1} = \cdots = b_n = 0$. Therefore, every linear translator of function g is a linear combination of $e_1, \ldots, e_r$. This completes the proof of the theorem.   □

By applying Theorem 1 and 2, Theorem 3 implies:

Corollary 4.   *Let F be a finite field of p elements where p is a prime. Then a function $f : F^n \mapsto F^m$ has r linearly independent additive translators if and only if there exists an $n \times n$ invertible matrix A over F such that*

$$g(x_1, \ldots, x_n) = f((x_1, \ldots, x_n)A) = x_1 v_1 + \cdots + x_r v_r + g^*(x_{r+1}, \ldots, x_n) \quad (10)$$

where $v_i \in F^m$, for $1 \leq i \leq r$ and g^* is some non-constant function of n-r variables without additive structure.

Example 1.   Consider the case that $n = 3, m = 1$ and $F = GF(2)$. For the following function of 3 variables,

$$f(x_1, x_2, x_3) = x_1 x_2 + x_1 x_3 + x_2 x_3 + x_2 + x_3,$$

the vector (111) in F_2^3 is a linear translator of f:

$$f(x_1 + 1, x_2 + 1, x_3 + 1) = f(x_1, x_2, x_3) + 1.$$

Let $A = \begin{bmatrix} 1\,1\,1 \\ 1\,0\,1 \\ 1\,1\,0 \end{bmatrix}$, then $\det A \neq 0$, and

$$f((x_1, x_2, x_3)A) = f(x_1 + x_2 + x_3, x_1 + x_3, x_1 + x_2) = x_1 + x_2 x_3.$$

Remark. We have proved in Theorem 2 that if F is a prime field, then the set of additive translators of a function forms a linear subspace. It was shown in [5, Proposition 3] that, for a quadratic function f, the set of additive translators is a linear subspace of F^n for any finite field F. In general, however, this result is not true. The following example due to Nyberg shows that there exist functions for which the set of additive translators is not a linear subspace.

Example 2.   Consider the following function from $(GF(2^2))^3$ to $GF(2^2)$:

$$f[X, Y, Z] = f[(x_1, x_2), (y_1, y_2), (z_1, z_2)] = (x_1 x_2 y_1 + y_2, z_1 z_2)$$

where x_i, y_i, z_i are in $GF(2)$. f has only one nonzero additive translator

$$[(0, 0), (0, 1), (0, 0)].$$

Thus, the set of additive translators cannot be a subspace because any non-zero subspace contains at least four elements.   □

Remark. The result of Theorem 3 implies that it is not always possible to transform a function with r linearly independent *additive* translators into the form as the sum of a linear function of r variables and some function of $(n - r)$ variables even if the set of additive translators is a subspace, as shown by the following example. [This example further implies that Proposition 4 in [5] is not true.]

Example 3.   Consider again the function

$$f(x_1, x_2, x_3) = x_1 x_2 + x_1 x_3 + x_2 x_3 + x_2 + x_3,$$

which is of the same form as in Example 1 but is over the field $F = GF(2^2)$. It is easy to verify that the vectors $\alpha(111), \alpha \in F$ are the only additive translators of f and that they form a linear subspace. However, for $\alpha \in GF(2^2)$ such that $\alpha^2 \neq \alpha$,

$$f(\alpha(1, 1, 1)) = f(\alpha, \alpha, \alpha) = \alpha^2 \neq \alpha = \alpha f(1, 1, 1),$$

so that α is not a linear translator of f. Theorem 3 then implies that the function f cannot be transformed into the form

$$g(x_1, x_2, x_3) = x_1 a_1 + g^*(x_2, x_3).$$

4. Linear Structures and nonlinearity

Theorem 3 implies that for any function g obtained from a function f by a non-singular linear transformation on the variables of f, $\dim(L_g) = \dim(L_f)$. That is, $\dim(L_f)$ is invariant under linear transformations on the variables of f. Thus, $\dim(L_f)$ provides a useful measure of the "partial" linearity of function f. Another such invariance is the nonlinear degree (or nonlinear order) of f. For a function from F^n to F^m, its component functions (from F^n to F) can be written in the form of multivariable polynomials. We define the degree of monomial $x_1^{i_1} \cdots x_n^{i_n}$ to be $\sum_1^n i_j$ and the degree of a polynomial as the maximum of the degrees of the monomials occurring in the polynomial. The (total) degree of f is then defined as the maximum of the degrees of its component polynomials. The following Lemma shows that the nonlinear degree (or nonlinear order) of f is also invariant under nonsingular linear transformations on its variables.

Lemma 5. *Let*

$$g(x_1, \ldots, x_n) = f((x_1, \ldots, x_n)A)$$

where A is a nonsingular $n \times n$ matrix over F, then $\deg(f) = \deg(g)$.

Proof.   We show first that $\deg(g) \leq \deg(f)$, the equality then follows from the invertibility of A. Let $x_1^{i_1} \cdots x_n^{i_n}$ be a monomial of f with degree $\sum_1^n i_j$. Under the transformation $A = \{a_{ij}\}$, the monomial becomes

$$(a_{11}x_1 + a_{21}x_2 + \cdots + a_{n1}x_n)^{i_1} \cdots (a_{1n}x_1 + a_{2n}x_2 + \cdots + a_{nn}x_n)^{i_n}.$$

Using the fact that $\deg(f_1 + f_2) \leq \max(\deg(f_1), \deg(f_2))$ and $\deg(f_1 f_2) \leq \deg(f_1) + \deg(f_2)$, we obtain

$$
\begin{aligned}
\deg\ &[(a_{11}x_1 + \cdots + a_{n1}x_n)^{i_1} \cdots (a_{1n}x_1 + \cdots + a_{nn}x_n)^{i_n}] \\
&\leq \deg[(a_{11}x_1 + \cdots + a_{n1}x_n)^{i_1}] + \cdots + \deg[(a_{1n}x_1 + \cdots + a_{nn}x_n)^{i_n}] \\
&\leq i_1 \deg(a_{11}x_1 + \cdots + a_{n1}x_n) + \cdots + i_n \deg(a_{1n}x_1 + \cdots + a_{nn}x_n) \\
&= \sum_1^n i_j = \deg(x_1^{i_1} \cdots x_n^{i_n}),
\end{aligned}
$$

that is, the degree of each monomial will not increase under the transformation A. Thus, the degree of each component polynomial of f will not increase under the transformation A, which implies that $\deg(g) \leq \deg(f)$. On the other hand, $f(x_1, \ldots, x_n) = g((x_1, \ldots, x_n)A^{-1})$ because A is invertible, so the above argument implies that $\deg(f) \leq \deg(g)$. Therefore, $\deg(f) = \deg(g)$.   □

Now applying Theorem 3 to a nonlinear function f with $r = \dim(L_f)$, we know that f can be transformed into the form:

$$
g(x_1, \ldots, x_n) = x_1 v_1 + \cdots + x_r v_r + g^*(x_{r+1}, \ldots, x_n).
$$

Therefore,

$$
\deg(f) = \deg(g) = \deg(g^*) \leq n - r = n - \dim(L_f).
$$

Thus, we obtain the following result.

Theorem 6.   *For a nonlinear function $f : F^n \mapsto F^m$,*

$$
\deg(f) + \dim(L_f) \leq n.
$$

In particular, a nonlinear function of full degree has no linear structure. Note that for a linear function f, $\deg(f) = 1$ and $\dim(L_f) = n$.

Example 4.   Consider again the Boolean function in Example 1. Because that the function f has non-linear degree 2, the above result implies that

$$
1 \leq \dim(L_f) \leq 3 - \deg(f) = 1.
$$

Thus, $(1, 1, 1)$ is the only non-zero linear translator of the function f.

5. Linear structures of invertible functions

In cryptography, invertible functions are of special interest. For example, a block cipher with a given key is an invertible function. Another example is that the S-boxes in the DES consist of invertible functions from F_2^4 to F_2^4. The next result shows the relationship between the linear structures of an invertible function and that of the inverse function.

Theorem 7. Let $f : F^n \mapsto F^n$ be an invertible function, where F is an arbitrary finite field. Then a is a linear (additive) translator of f if and only if $d = f(a) - f(0)$ is a linear (additive) translator of f^{-1}. Moreover, the function f and its inverse f^{-1} have the same number of linearly independent linear translators.

Proof. For the first part of Theorem, we show first the case for linear translators. Denote $y = f(x)$, then $x = f^{-1}(y)$, and denote $x_0 = f^{-1}(0)$. For a linear translator a of function f, we have $a + x_0 = f^{-1}(f(a) - f(0)) = f^{-1}(d)$ because

$$f(a + x_0) = f(a) + f(x_0) - f(0) = f(a) - f(0) = d.$$

Thus, for any c in F, a is a linear translator of f

$\Longleftrightarrow$ $\quad f(x) + c(f(a) - f(0)) = f(x + ca) \qquad \forall x \in F^n$
$\Longleftrightarrow$ $\quad y + cd = f(x + ca) \qquad \forall x \in F^n$
$\Longleftrightarrow$ $\quad f^{-1}(y + cd) = x + ca = f^{-1}(y) + c[a + x_0 - x_0]$
$\qquad\qquad\qquad\qquad\quad = f^{-1}(y) + c[f^{-1}(d) - f^{-1}(0)] \quad \forall y \in F^n$
$\Longleftrightarrow$ $\quad d = f(a) - f(0)$ is a linear translator of f^{-1}.

Let $c = 1$ in the above proof, we obtain the proof for the case of additive translators.

To show the second part of the Theorem, let a and b be linear translators of f. Equation (4) and (5) implies that for any c_1, c_2 in F,

$$f(c_1 a + c_2 b) = c_1 f(a) - (c_1 - 1)f(0) + c_2 f(b) - (c_2 - 1)f(0) - f(0)$$
$$= c_1(f(a) - f(0)) + c_2(f(b) - f(0)) + f(0).$$

Thus,

$$f(c_1 a + c_2 b) = f(0) \quad \text{if and only if} \quad c_1(f(a) - f(0)) + c_2(f(b) - f(0)) = 0.$$

Because f is invertible, we see that

$$c_1 a + c_2 b = 0 \quad \text{if and only if} \quad c_1(f(a) - f(0)) + c_2(f(b) - f(0)) = 0.$$

Thus, we have shown that a and b are linearly independent linear translators of function f if and only if $f(a) - f(0)$ and $f(b) - f(0)$ are linearly independent linear translators of the inverse function f^{-1}. Therefore, L_f and $L_{f^{-1}}$ have the same dimension. $\quad\square$

Similar to Theorem 3, we consider the "normal" form of an invertible function having linear structures. A function $f : F^n \mapsto F^n$ will be expressed as

$$y = f(x) \quad \Longleftrightarrow \quad \begin{bmatrix} y_1 \\ y_2 \\ \vdots \\ y_n \end{bmatrix} = \begin{bmatrix} f_1(x_1, x_2, \ldots, x_n) \\ f_2(x_1, x_2, \ldots, x_n) \\ \vdots \\ f_n(x_1, x_2, \ldots, x_n) \end{bmatrix},$$

that is, f maps row vector $(x_1, \ldots, x_n)$ to column vector $(y_1, \ldots, y_n)$.

Theorem 8.   *Let $f: F^n \mapsto F^n$ be an invertible function, then f has r linearly independent linear translators if and only if there exist $n \times n$ invertible matrices A and M over F such that*

$$g(x) = Mf(xA) = \begin{bmatrix} x_1 + g_1^*(x_{r+1}, \ldots, x_n) \\ \vdots \\ x_r + g_r^*(x_{r+1}, \ldots, x_n) \\ g_{r+1}^*(x_{r+1}, \ldots, x_n) \\ \vdots \\ g_n^*(x_{r+1}, \ldots, x_n) \end{bmatrix}. \qquad (11)$$

Proof.   Let $a_i, i = 1, \ldots, r$ be r linearly independent linear translators of f. By Theorem 7, $f(a_i) - f(0)$, $i = 1, \ldots, r$, are r linear translators of f^{-1}, and they are linearly independent because f is invertible. Let A be an invertible matrix such that $e_i A = a_i$, by Theorem 3 function $f(xA)$ will have the form shown in (6):

$$f((x_1, \ldots, x_n)A) = x_1 v_1 + \cdots + x_r v_r + g'(x_{r+1}, \ldots, x_n) \qquad (12)$$

where $v_i = f(a_i) - f(0)$. Now let M be an invertible matrix such that $M(f(a_i) - f(0))^T = e_i^T$, then the function $g(x) = Mf(xA)$ will have the form in (11).

Conversely, suppose (11) holds, then Theorem 3 implies that the vectors $a_i = e_i A$, $i = 1, \ldots, r$, are r linearly independent linear translators of function f.   □

In cryptographic practice, the methods of *confusion* and *diffusion*, introduced by Shannon [7], are fundamental. A function will be said to achieve *complete diffusion* if each of its output variable depends on every input variable. The following result is a direct consequence of Theorem 8. It shows a relationship between the complete diffusion and the nonlinearity with respect to linear structures for an invertible function.

Corollary 9.   *Let $f: F^n \mapsto F^n$, $n \geq 3$, be an invertible function such that, for any $n \times n$ invertible matrices A and M, the function $g(x) = Mf(xA)$ achieves complete diffusion, then the function f has no linear structure.*

A remark on additive translators and differential cryptanalysis

The basic concept used in differential cryptanalysis [8] is that of 'differentials' and their probabilities [8, 9]. A differential of a function f can be defined as a couple (a,b) such that, if a pair of inputs of f has difference $\Delta x = x_1 - x = a$, then b is a possible value of the difference of the pair of outputs of f: $\Delta y = f(x_1) - f(x) = f(x + a) - f(x)$. It then follows from (3) that *a is an additive translator of f if and only if $(a, f(a) - f(0))$ is a differential of f with probability one.*

6. Summary

In this paper we have considered additive and linear structures of functions over a finite field in terms of their additive and linear translators. The main result was that an n-variable function has r linearly independent linear translators if and only if there is a non-singular linear transformation on the variables of the function such that the resulting function is the sum of a linear function of r variables and a function of the other $n - r$ variables.

Acknowledgement

The author would like to thank J.L. Massey for his encouraging criticism on this work when the author was at the ETH Zürich. The author would also like to thank K. Nyberg for her valuable suggestions (especially for using the term additive translators) which greatly improved this work.

References

1. D. Chaum, J.H. Evertse, Cryptanalysis of DES with a reduced number of rounds, *Advances in Cryptology - CRYPTO'85, Proceedings*, pp. 192–211, Springer-Verlag, 1986.
2. J.H. Evertse, Linear structures in block ciphers, *Advances in Cryptology - EURO-CRYPT'87, Proceedings*, pp. 249–266, Springer-Verlag, 1988.
3. M. Hellman, R. Merkle, R. Schroeppel, L. Washington, W. Diffie, S. Pohlig, P. Schweitzer, Results of an initial attempt to cryptanalyze the NBS Data Encryption Standard, Information System Lab. report SEL 76-042, Stanford University, 1976.
4. W. Meier, O. Staffelbach, Nonlinearity criteria for cryptographic functions, *Advances in Cryptology - EUROCRYPT'89, Proceedings*, pp. 549–562, Springer-Verlag, 1990.
5. K. Nyberg, On the construction of highly nonlinear permutations *Advances in Cryptology - EUROCRYPT'92, Proceedings*, pp. 92–98, Springer-Verlag, 1993.
6. J.A. Reeds, J.L. Manferdeli, DES has no per round linear factors, *Advances in Cryptology - CRYPTO'84, Proceedings*, pp. 377–389, Springer-Verlag, 1985.
7. C. E. Shannon, "Communication Theory of Secrecy Systems", Bell. System Technical Journal, Vol. 28, pp. 656-715, Oct. 1949.
8. E. Biham and A. Shamir, *Differential Cryptanalysis of the Data Encryption Standard*, Springer-Verlag, 1993.
9. X. Lai, J. L. Massey and S. Murphy, "Markov Ciphers and Differential Cryptanalysis", *Advances in Cryptology – EUROCRYPT'91, Proceedings*, LNCS 547, pp. 17-38, Springer-Verlag, Berlin, 1991.

The RC5 Encryption Algorithm*

Ronald L. Rivest

MIT Laboratory for Computer Science
545 Technology Square, Cambridge, Mass. 02139
rivest@theory.lcs.mit.edu

Abstract. This document describes the RC5 encryption algorithm, a fast symmetric block cipher suitable for hardware or software implementations. A novel feature of RC5 is the heavy use of *data-dependent rotations*. RC5 has a variable word size, a variable number of rounds, and a variable-length secret key. The encryption and decryption algorithms are exceptionally simple.

1 Introduction

RC5 was designed with the following objectives in mind.

- RC5 should be a *symmetric block cipher*. The same secret cryptographic key is used for encryption and for decryption. The plaintext and ciphertext are fixed-length bit sequences (blocks).
- RC5 should be *suitable for hardware or software*. This means that RC5 should use only computational primitive operations commonly found on typical microprocessors.
- RC5 should be *fast*. This more-or-less implies that RC5 be *word-oriented*: the basic computational operations should be operators that work on full words of data at a time.
- RC5 should be *adaptable to processors of different word-lengths*. For example, as 64-bit processors become available, it should be possible for RC5 to exploit their longer word length. Therefore, the number w of bits in a word is a *parameter* of RC5; different choices of this parameter result in different RC5 algorithms.
- RC5 should be iterative in structure, with a *variable number of rounds*. The user can explicitly manipulate the trade-off between higher speed and higher security. The number of rounds r is a second parameter of RC5.
- RC5 should have *a variable-length cryptographic key*. The user can choose the level of security appropriate for his application, or as required by external considerations such as export restrictions. The key length b (in bytes) is thus a third parameter of RC5.

---

* RC5 is a trademark of RSA Data Security. Patent pending.

- RC5 should be *simple*. It should be easy to implement. More importantly, a simpler structure is perhaps more interesting to analyze and evaluate, so that the cryptographic strength of RC5 can be more rapidly determined.
- RC5 should have a *low memory requirement*, so that it may be easily implemented on smart cards or other devices with restricted memory.
- (Last but not least!) RC5 should provide *high security* when suitable parameter values are chosen.

In addition, during the development of RC5, we began to focus our attention on a intriguing new cryptographic primitive: *data-dependent rotations*, in which one word of intermediate results is cyclically rotated by an amount determined by the low-order bits of another intermediate result. We thus developed an additional goal.

- RC5 should *highlight the use of data-dependent rotations*, and encourage the assessment of the cryptographic strength data-dependent rotations can provide.

The RC5 encryption algorithm presented here hopefully meets all of the above goals. Our use of "hopefully" refers of course to the fact that this is still a new proposal, and the cryptographic strength of RC5 is still being determined.

2   A Parameterized Family of Encryption Algorithms

In this section we discuss in somewhat greater detail the parameters of RC5, and the tradeoffs involved in choosing various parameters.

As noted above, RC5 is *word*-oriented: all of the basic computational operations have w-bit words as inputs and outputs. RC5 is a block-cipher with a two-word input (plaintext) block size and a two-word (ciphertext) output block size. The nominal choice for w is 32 bits, for which RC5 has 64-bit plaintext and ciphertext block sizes. RC5 is well-defined for any $w > 0$, although for simplicity it is proposed here that only the values 16, 32, and 64 be "allowable."

The number r of rounds is the second parameter of RC5. Choosing a larger number of rounds presumably provides an increased level of security. We note here that RC5 uses an "expanded key table," S, that is derived from the user's supplied secret key. The size t of table S also depends on the number r of rounds: S has $t = 2(r + 1)$ words. Choosing a larger number of rounds therefore also implies a need for somewhat more memory.

There are thus several distinct "RC5" algorithms, depending on the choice of parameters w and r. We summarize these parameters below:

w This is the *word size*, in bits; each word contains $u = (w/8)$ 8-bit bytes. The nominal value of w is 32 bits; allowable values of w are 16, 32, and 64. RC5 encrypts two-word blocks: plaintext and ciphertext blocks are each $2w$ bits long.

r This is the number of rounds. Also, the expanded key table S contains $t = 2(r + 1)$ words. Allowable values of r are 0, 1, ..., 255.

In addition to w and r, RC5 has a variable-length secret cryptographic key, specified by parameters b and K:

b The number of bytes in the secret key K. Allowable values of b are 0, 1, ..., 255.

K The b-byte secret key: $K[0]$, $K[1]$, ..., $K[b-1]$.

For notational convenience, we designate a particular (parameterized) RC5 algorithm as $RC5\text{-}w/r/b$. For example, RC5-32/16/10 has 32-bit words, 16 rounds, a 10-byte (80-bit) secret key variable, and an expanded key table of $2(16+1) = 34$ words. Parameters may be dropped, from last to first, to talk about RC5 with the dropped parameters unspecified. For example, one may ask: How many rounds should one use in RC5-32?

We propose RC5-32/12/16 as providing a "nominal" choice of parameters. That is, the nominal values of the parameters provide for $w = 32$ bit words, 12 rounds, and 16 bytes of key. Further analysis is needed to analyze the security of this choice. For RC5-64, we suggest increasing the number of rounds to $r = 16$.

We suggest that in an implementation, all of the parameters given above may be packaged together to form an *RC5 control block*, containing the following fields:

v 1 byte version number; 10 (hex) for version 1.0 here.

w 1 byte.

r 1 byte.

b 1 byte.

K b bytes.

A control block is thus represented using $b+4$ bytes. For example, the control block

```
10 20 0C 0A 20 33 7D 83 05 5F 62 51 BB 09 (in hexadecimal)
```

specifies an RC5 algorithm (version 1.0) with 32-bit words, 12 rounds, and a 10-byte (80-bit) key "20 33 ... 09". RC5 "key-management" schemes would then typically manage and transmit entire RC5 control blocks, containing all of the relevant parameters in addition to the usual secret cryptographic key variable.

2.1   Discussion of Parameterization

In this section we discuss the extensive parameterization that RC5 provides.

We should first note that it is not intended that RC5 be secure for all possible parameter values. For example, $r = 0$ provides essentially no encryption, and $r = 1$ is easily broken. And choosing $b = 0$ clearly gives no security.

On the other hand, choosing the maximum allowable parameter values would be overkill for most applications.

We allow a range of parameter values so that users may select an encryption algorithm whose security and speed are optimized for their application, while providing an evolutionary path for adjusting their parameters as necessary in the future.

As an example, consider the problem of replacing DES with an "equivalent" RC5 algorithm. One might reasonable choose RC5-32/16/7 as such a replacement. The input/output blocks are $2w = 64$ bits long, just as in DES. The number of rounds is also the same, although each RC5 round is more like two DES rounds since all data registers, rather than just half of them, are updated in one RC5 round. Finally, DES and RC5-32/16/7 each have 56-bit (7-byte) secret keys.

Unlike DES, which has no parameterization and hence no flexibility, RC5 permits upgrades as necessary. For example, one can upgrade the above choice for a DES replacement to an 80-bit key by moving to RC5-32/16/10. As technology improves, and as the true strength of RC5 algorithms becomes better understood through analysis, the most appropriate parameter values can be chosen.

The choice of r affects both encryption speed and security. For some applications, high speed may be the most critical requirement—one wishes for the best security obtainable within a given encryption time requirement. Choosing a small value of r (say $r = 6$) may provide some security, albeit modest, within the given speed constraint.

In other applications, such as key management, security is the primary concern, and speed is relatively unimportant. Choosing $r = 32$ rounds might be appropriate for such applications. Since RC5 is a new design, further study is required to determine the security provided by various values of r; RC5 users may wish to adjust the values of r they use based on the results of such studies.

Similarly, the word size w also affects speed and security. For example, choosing a value of w larger than the register size of the CPU can degrade encryption speed. The word size $w = 16$ is primarily for researchers who wish to examine the security properties of a natural "scaled-down" RC5. As 64-bit processors become common, one can move to RC5-64 as a natural extension of RC5-32. It may also be convenient to specify $w = 64$ (or larger) if RC5 is to be used as the basis for a hash function, in order to have 128-bit (or larger) input/output blocks.

It may be considered unusual and risky to specify an encryption algorithm that permits insecure parameter choices. We have two responses to this criticism:

1. A fixed set of parameters may be at least as dangerous, since the parameters can not be increased when necessary. Consider the problem DES has now: its key size is too short, and there is no easy way to increase it.
2. It is expected that implementors will provide implementations that ensure that suitably large parameters are chosen. While unsafe choices might be usable in principle, they would be forbidden in practice.

It is not expected that a typical RC5 implementation will work with any RC5 control block. Rather, it may only work for certain fixed parameter values, or parameters in a certain range. The parameters w, r, and b in a received or transmitted RC5 control block are then merely used for *type-checking*—values other than those supported by the implementation will be disallowed. The flexibility of RC5 is thus utilized at the system design stage, when the appropriate parameters are chosen, rather than at run time, when unsuitable parameters might be chosen by an unwary user.

Finally, we note that RC5 might be used in some applications that do not require cryptographic security. For example, one might consider using RC5-32/8/0 (with no secret key) applied to inputs 0, 1, 2, ... to generate a sequence of pseudo-random numbers to be used in a randomized computation.

3   Notation and RC5 Primitive Operations

We use $\lg(x)$ to denote the base-two logarithm of x.

RC5 uses only the following three primitive operations (and their inverses).

1. Two's complement addition of words, denoted by "+". This is modulo-2^w addition. The inverse operation, subtraction, is denoted "$-$".
2. Bit-wise exclusive-OR of words, denoted by $\oplus$.
3. A left-rotation (or "left-spin") of words: the cyclic rotation of word x left by y bits is denoted $x \lll y$. Here y is interpreted modulo w, so that when w is a power of two, only the $\lg(w)$ low-order bits of y are used to determine the rotation amount. The inverse operation, right-rotation, is denoted $x \ggg y$.

These operations are directly and efficiently supported by most processors.

A distinguishing feature of RC5 is that the rotations are rotations by "variable" (plaintext-dependent) amounts. We note that on modern microprocessors, a variable-rotation $x \lll y$ takes *constant time*: the time is independent of the rotation amount y. We also note that rotations are the only non-linear operator in RC5; there are no nonlinear substitution tables or other nonlinear operators. The strength of RC5 depends heavily on the cryptographic properties of data-dependent rotations.

4   The RC5 Algorithm

In this section we describe the RC5 algorithm, which consists of three components: a *key expansion* algorithm, an *encryption* algorithm, and a *decryption* algorithm. We present the encryption and decryption algorithms first.

Recall that the plaintext input to RC5 consists of two w-bit words, which we denote A and B. Recall also that RC5 uses an *expanded key table*, $S[0...t-1]$, consisting of $t = 2(r+1)$ w-bit words. The key-expansion algorithm initializes S from the user's given secret key parameter K. (We note that the S table in RC5 encryption is not an "S-box" such as is used by DES; RC5 uses the entries in S sequentially, one at a time.)

We assume standard *little-endian* conventions for packing bytes into input/output blocks: the first byte occupies the low-order bit positions of register A, and so on, so that the fourth byte occupies the high-order bit positions in A, the fifth byte occupies the low-order bit positions in B, and the eighth (last) byte occupies the high-order bit positions in B.

4.1 Encryption

We assume that the input block is given in two w-bit registers A and B. We also assume that key-expansion has already been performed, so that the array $S[0...t-1]$ has been computed. Here is the encryption algorithm in pseudo-code:

$A = A + S[0];$
$B = B + S[1];$
for $i = 1$ **to** r **do**
$\quad A = ((A \oplus B) \lll B) + S[2 * i];$
$\quad B = ((B \oplus A) \lll A) + S[2 * i + 1];$

The output is in the registers A and B.

We note the exceptional simplicity of this 5-line algorithm.

We also note that each RC5 round updates *both* registers A and B, whereas a "round" in DES updates only half of its registers. An RC5 "half-round" (one of the assignment statements updating A or B in the body of the loop above) is thus perhaps more analogous to a DES round.

4.2 Decryption

The decryption routine is easily derived from the encryption routine.

for $i = r$ **downto** 1 **do**
$\quad B = ((B - S[2 * i + 1]) \ggg A) \oplus A;$
$\quad A = ((A - S[2 * i]) \ggg B) \oplus B;$
$B = B - S[1];$
$A = A - S[0];$

4.3 Key Expansion

The key-expansion routine expands the user's secret key K to fill the expanded key array S, so that S resembles an array of $t = 2(r + 1)$ random binary words determined by K. The key expansion algorithm uses two "magic constants," and consists of three simple algorithmic parts.

Definition of the Magic Constants The key-expansion algorithm uses two word-sized binary constants P_w and Q_w. They are defined for arbitrary w as follows:

$$P_w = \text{Odd}((e - 2)2^w) \qquad (1)$$
$$Q_w = \text{Odd}((\phi - 1)2^w) \qquad (2)$$

where

$e = 2.718281828459...$ (base of natural logarithms)
$\phi = 1.618033988749...$ (golden ratio) ,

and where $\text{Odd}(x)$ is the odd integer nearest to x (rounded up if x is an even integer, although this won't happen here). For $w = 16$, 32, and 64, these constants are given below in binary and in hexadecimal.

```
P16 = 1011011111100001 = b7e1
Q16 = 1001111000110111 = 9e37

P32 = 10110111111000010101000101100011 = b7e15163
Q32 = 10011110001101110111100110111001 = 9e3779b9

P64 = 1011011111100001010100010111000101000101011101101001010101001101011
    = b7e151628aed2a6b
Q64 = 1001111000110111011110011011101110010111111110100010100111110000010101
    = 9e3779b97f4a7c15
```

Converting the Secret Key from Bytes to Words. The first algorithmic step of key expansion is to copy the secret key $K[0...b-1]$ into an array $L[0...c-1]$ of $c = \lceil b/u \rceil$ words, where $u = w/8$ is the number of bytes/word. This operation is done in a natural manner, using u consecutive key bytes of K to fill up each successive word in L, low-order byte to high-order byte. Any unfilled byte positions of L are zeroed.

On "little-endian" machines such as an Intel '486, the above task can be accomplished merely by zeroing the array L, and then copying the string K directly into the memory positions representing L. The following pseudo-code achieves the same effect, assuming that all bytes are "unsigned" and that array L is initially zeroed.

> **for** $i = b - 1$ **downto** 0 **do**
> $$L[i/u] = (L[i/u] \lll 8) + K[i];$$

Initializing the Array S. The second algorithmic step of key expansion is to initialize array S to a particular fixed (key-independent) pseudo-random bit pattern, using an arithmetic progression modulo 2^w determined by the "magic constants" P_w and Q_w. Since Q_w is odd, the arithmetic progression has period 2^w.

> $S[0] = P_w;$
> **for** $i = 1$ **to** $t - 1$ **do**
> $$S[i] = S[i-1] + Q_w;$$

Mixing in the Secret Key. The third algorithmic step of key expansion is to mix in the user's secret key in three passes over the arrays S and L. More precisely, due to the potentially different sizes of S and L, the larger array will be processed three times, and the other may be handled more times.

$$i = j = 0;$$
$$A = B = 0;$$
do $3 * \max(t, c)$ **times:**
$$A = S[i] = (S[i] + A + B) \lll 3;$$
$$B = L[j] = (L[j] + A + B) \lll (A + B);$$
$$i = (i + 1) \bmod(t);$$
$$j = (j + 1) \bmod(c);$$

The key-expansion function has a certain amount of "one-wayness": it is not so easy to determine K from S.

5   Discussion

A distinguishing feature of RC5 is its heavy use of *data-dependent rotations*— the amount of rotation performed is dependent on the input data, and is not predetermined.

The encryption/decryption routines are very simple. While other operations (such as substitution operations) could have been included in the basic round operations, our objective is to focus on the data-dependent rotations as a source of cryptographic strength.

Some of the expanded key table S is initially added to the plaintext, and each round ends by adding expanded key from S to the intermediate values just computed. This assures that each round acts in a potentially different manner, in terms of the rotation amounts used.

The xor operations back and forth between A and B provide some avalanche properties, causing a single-bit change in an input block to cause multiple-bit changes in following rounds.

6   Implementation

The encryption algorithm is very compact, and can be coded efficiently in assembly language on most processors. The table S is both quite small and accessed sequentially, minimizing issues of cache size.

A reference implementation of RC5-32/12/16, together with some sample input/output pairs, is provided in the Appendix.

This (non-optimized) reference implementation encrypts 100K bytes/second on a 50Mhz '486 laptop (16-bit Borland C++ compiler), and 2.4M bytes/second on a Sparc 5 (gcc compiler). These speeds can certainly be improved. In assembly language the rotation operator is directly accessible: an assembly-language routine for the '486 can perform each half-round with just four instructions. An initial assembly-language implementation runs at 1.2M bytes/sec on a 50MHz '486 SLC. A Pentium should be able to encrypt at several megabytes/second.

7 Analysis

This section contains some preliminary results on the strength of RC5. Much more work remains to be done. Here we report the results of two experiments studying how changing the number of rounds affects properties of RC5.

The first test involved examining the uniformity of correlation between input and output bits. We found that four rounds sufficed to get very uniform correlations between individual input and output bits in RC5-32.

The second test checked to see if the data-dependent rotation amounts depended on every plaintext bit, in 100 million trials with random plaintext and keys. That is, we checked whether flipping a plaintext bit caused some intermediate rotation to be a rotation by a different amount. We found that eight rounds in RC5-32 were sufficient to cause each message bit to affect some rotation amount.

The number of rounds chosen in practice should always be at least as great (if not substantially greater) than these simple tests would suggest. As noted above, we suggest 12 rounds as a "nominal" choice for RC5-32, and 16 rounds for RC5-64.

RC5's data-dependent rotations may help frustrate differential cryptanalysis (Biham/Shamir [1]) and linear cryptanalysis (Matsui [3]), since bits are rotated to "random" positions in each round.

There is no obvious way in which an RC5 key can be "weak," other than by being too short.

I invite the reader to help determine the strength of RC5.

8 Acknowledgements

I'd like to thank Burt Kaliski, Yiqun Lisa Yin, Paul Kocher, and everyone else at RSA Laboratories for their comments and constructive criticisms.
(Note added in press: I'd also like to thank Karl A. Siil for bringing to my attention a cipher due to W. E. Madryga [2] that also uses data-dependent rotations, albeit in a rather different manner.)

References

1. E. Biham and A. Shamir. *A Differential Cryptanalysis of the Data Encryption Standard*. Springer-Verlag, 1993.
2. W. E. Madryga. A high performance encryption algorithm. In *Computer Security: A Global Challenge*, pages 557–570. North Holland: Elsevier Science Publishers, 1984.
3. Mitsuru Matsui. The first experimental cryptanalysis of the Data Encryption Standard. In Yvo G. Desmedt, editor, *Proceedings CRYPTO 94*, pages 1–11. Springer, 1994. Lecture Notes in Computer Science No. 839.

9 Appendix

```
/* RC5REF.C -- Reference implementation of RC5-32/12/16 in C.        */
/* Copyright (C) 1995 RSA Data Security, Inc.                        */
#include <stdio.h>
typedef unsigned long int WORD; /* Should be 32-bit = 4 bytes        */
#define w       32              /* word size in bits                 */
#define r       12              /* number of rounds                  */
#define b       16              /* number of bytes in key            */
#define c        4              /* number  words in key = ceil(8*b/w)*/
#define t       26              /* size of table S = 2*(r+1) words   */
WORD S[t];                      /* expanded key table                */
WORD P = 0xb7e15163, Q = 0x9e3779b9;  /* magic constants             */
/* Rotation operators. x must be unsigned, to get logical right shift*/
#define ROTL(x,y) (((x)<<(y&(w-1))) | ((x)>>(w-(y&(w-1)))))
#define ROTR(x,y) (((x)>>(y&(w-1))) | ((x)<<(w-(y&(w-1)))))

void RC5_ENCRYPT(WORD *pt, WORD *ct) /* 2 WORD input pt/output ct    */
{ WORD i, A=pt[0]+S[0], B=pt[1]+S[1];
  for (i=1; i<=r; i++)
    { A = ROTL(A^B,B)+S[2*i];
      B = ROTL(B^A,A)+S[2*i+1];
    }
  ct[0] = A; ct[1] = B;
}

void RC5_DECRYPT(WORD *ct, WORD *pt) /* 2 WORD input ct/output pt    */
{ WORD i, B=ct[1], A=ct[0];
  for (i=r; i>0; i--)
    { B = ROTR(B-S[2*i+1],A)^A;
      A = ROTR(A-S[2*i],B)^B;
    }
  pt[1] = B-S[1]; pt[0] = A-S[0];
}

void RC5_SETUP(unsigned char *K) /* secret input key K[0...b-1]      */
{ WORD i, j, k, u=w/8, A, B, L[c];
  /* Initialize L, then S, then mix key into S */
  for (i=b-1,L[c-1]=0; i!=-1; i--) L[i/u] = (L[i/u]<<8)+K[i];
  for (S[0]=P,i=1; i<t; i++) S[i] = S[i-1]+Q;
  for (A=B=i=j=k=0; k<3*t; k++,i=(i+1)%t,j=(j+1)%c)   /* 3*t > 3*c */
    { A = S[i] = ROTL(S[i]+(A+B),3);
      B = L[j] = ROTL(L[j]+(A+B),(A+B));
    }
}
```

```
void main()
{ WORD i, j, pt1[2], pt2[2], ct[2] = {0,0};
  unsigned char key[b];
  if (sizeof(WORD)!=4)
    printf("RC5 error: WORD has %d bytes.\n",sizeof(WORD));
  printf("RC5-32/12/16 examples:\n");
  for (i=1;i<6;i++)
    { /* Initialize pt1 and key pseudorandomly based on previous ct */
      pt1[0]=ct[0]; pt1[1]=ct[1];
      for (j=0;j<b;j++) key[j] = ct[0]%(255-j);
      /* Setup, encrypt, and decrypt */
      RC5_SETUP(key);
      RC5_ENCRYPT(pt1,ct);
      RC5_DECRYPT(ct,pt2);
      /* Print out results, checking for decryption failure */
      printf("\n%d. key = ",i);
      for (j=0; j<b; j++) printf("%.2X ",key[j]);
      printf("\n    plaintext %.81X %.81X --->  ciphertext %.81X %.81X \n",
             pt1[0], pt1[1], ct[0], ct[1]);
      if (pt1[0]!=pt2[0] || pt1[1]!=pt2[1])
        printf("Decryption Error!");
    }
}
```

RC5-32/12/16 examples:

1. key = 00 00 00 00 00 00 00 00 00 00 00 00 00 00 00 00
   plaintext 00000000 00000000 ---> ciphertext EEDBA521 6D8F4B15

2. key = 91 5F 46 19 BE 41 B2 51 63 55 A5 01 10 A9 CE 91
   plaintext EEDBA521 6D8F4B15 ---> ciphertext AC13C0F7 52892B5B

3. key = 78 33 48 E7 5A EB 0F 2F D7 B1 69 BB 8D C1 67 87
   plaintext AC13C0F7 52892B5B ---> ciphertext B7B3422F 92FC6903

4. key = DC 49 DB 13 75 A5 58 4F 64 85 B4 13 B5 F1 2B AF
   plaintext B7B3422F 92FC6903 ---> ciphertext B278C165 CC97D184

5. key = 52 69 F1 49 D4 1B A0 15 24 97 57 4D 7F 15 31 25
   plaintext B278C165 CC97D184 ---> ciphertext 15E444EB 249831DA

The MacGuffin Block Cipher Algorithm

Matt Blaze[1] and Bruce Schneier[2]

[1] AT&T Bell Laboratories
101 Crawfords Corner Road, Holmdel, NJ 07733 USA
mab@research.att.com
[2] Counterpane Systems
730 Fair Oaks Avenue, Oak Park, IL 70302 USA
schneier@chinet.com

Abstract. This paper introduces MacGuffin, a 64 bit "codebook" block cipher. Many of its characteristics (block size, application domain, performance and implementation structure) are similar to those of the U.S. Data Encryption Standard (DES). It is based on a Feistel network, in which the cleartext is split into two sides with one side repeatedly modified according to a keyed function of the other. Previous block ciphers of this design, such as DES, operate on equal length sides. MacGuffin is unusual in that it is based on a *generalized unbalanced Feistel network (GUFN)* in which each round of the cipher modifies only 16 bits according to a function of the other 48. We describe the general characteristics of MacGuffin architecture and implementation and give a complete specification for the 32-round, 128-bit key version of the cipher.

1 Introduction

Feistel ciphers [1] operate by alternately encrypting the bits in one "side" of their input based on a keyed non-linear function of the bits in the other. This is done repeatedly, for a fixed number of "rounds". It is believed that, when iterated over sufficiently many rounds, even relatively simple non-linear functions can provide high security. Traditionally, such ciphers split their input block evenly about the middle; a 64 bit cipher would operate on two 32 bit internal blocks, swapping the "left" (the *target block*) and "right" (the *control block*) sides with each round. Several important block ciphers, including DES [3], are built upon this structure. We say these ciphers are based on *balanced Feistel networks (BFNs)*, since both sides are of equal length.

This paper describes a block cipher, called *MacGuffin*, that is based on a variant of this structure, the *generalized unbalanced Feistel network (GUFN)*, in which the target and control blocks need not be of equal length[3]. GUFNs, especially those in which the target block is smaller than the control block,

---

[3] Several cryptographic hash functions, such as MD5 [6] and SHA [5], employ an unbalanced structure similar in some respects to a GUFN.

appear to have a number of attractive properties for cipher design, particularly with respect to the design of the non-linear function. The principles underlying GUFNs are discussed in [7].

As its name suggests, MacGuffin is intended primarily as a catalyst for discussion and analysis. We believe it may also prove a practical, high security block cipher suitable for general use as an alternative to DES. It operates on 64 bit blocks of data, with an internal structure containing a 16 bit target block and a 48 bit control block ("48 on 16", in the notation of [7]). In principle, almost any length key and any number of rounds may be used, although we specify 32 rounds and a 128 bit key as "standard".

2   Architecture

We have been conservative in most aspects of MacGuffin's design, isolating most of its novel features to those parts of the design related to its unbalanced structure. As such, much of our design is adapted directly from DES. We hope that the many similarities between DES and MacGuffin will invite analysis of their differences.

Basically, the input cipherblock is partitioned into four 16 bit words, from left to right. In each round, the three rightmost words comprise the control block and are bitwise exclusive-ORed (XORed) with three words derived from the key. These 48 bits are then split eight ways according to a fixed permutation to provide input to eight functions of six bits (the "S-boxes"), each producing two bits of output. The 16 S-box output bits are then XORed, according to another fixed permutation, with the bits in the leftmost (target) word. Finally, the leftmost word is rotated into the rightmost position. The cipher can be reversed by a similar process, with the key derived bits applied in reverse order.

2.1   Design Principles

Because each round operates on only half as many bits as in a BFN (16 as opposed to 32), we use 32 rounds, twice as many as in DES, in our standard version. Because there are twice as many rounds, however, there are also a total of twice as many key bits XORed with the control blocks. These bits are obtained from the 128 bit key with the key expansion function described in the next section.

We adapt our S-boxes directly from those of DES. The eight DES S-boxes each produce four bits of output. Since we require only two bits from each (for a total of 16 bits), we use only the "outer" two output bits from each S-box.

In each round, each control block bit is XORed with one derived key bit and provides one input to exactly one S-box. There is no "expansion" permutation, since the number of control bits equals the number of S-box inputs. The control bits are mapped 1 : 1 to S-box inputs according to a fixed permutation. This permutation was designed so that each S-box receives two of its six inputs from each of the three registers in the control block.

S-box outputs are distributed across the 16 target bits. No S-box output goes to a bit position that is used as a direct input to itself in the next four rounds.

Observe that each of the three control registers contains bits produced in a different round of the cipher, and that each encrypted bit provides input to three different S-boxes (in the next three rounds), before it is encrypted again.

The cipher is designed for implementation in either hardware or software. Permutations were chosen to minimize the number of shift and mask operations and to allow time/memory optimizations in a software implementation.

3 Algorithm Description

3.1 Data Structures and Notation

We use the following notation:

$\oplus$ represents a 16 bit bitwise exclusive-OR (XOR) operation.

$\leftarrow$ is the conventional assignment operator, except as noted below.

$w, x, y, z \leftarrow i$ copies the data from 64 bit interface i, from lowest to highest bit position, into 16 bit registers w, x, y and z, respectively.

$i \leftarrow w, x, y, z$ copies the bits from 16 bit registers w, x, y and z, respectively into interface i, from lowest to highest bit position.

$s, t, u, v \leftarrow w, x, y, z$ copies w, x, y and z to s, t, u and v, respectively, in parallel (e.g., $x, y \leftarrow y, x$ swaps x and y).

$w \Leftarrow F(x, y, z)$ selects, according to a fixed permutation, bits from x, y and z as input to function F, storing the function output in bits of w, selected according to a fixed permutation.

The cipher employs the following internal structures:

$I_{0...63}, O_{0...63}$ are the 64 bit external input and output interfaces.

$left, a, b, c, t$ are 16 bit registers on which all cryptographic operations are performed. r_0 represents the least significant bit of r, r_{15} the most significant.

$k_{0...127}$ is a 128 bit secret key parameter.

$K[0...31, 0...2]$ is a 32×3 table of 16 bit words containing an expansion of k, as explained below.

3.2 S-boxes and Permutations

Nonlinearity in the encryption and key setup processes is provided primarily through eight functions, or "S-boxes", denoted $S_1...S_8$, each taking six bits of input selected from the a, b and c registers and producing two bits of output (which are XORed into the $left$ register).

Inputs to each S-box are selected uniquely from the a, b and c registers, as specified in Table 1. (In this table, input bit 0 is the least significant bit.) Outputs from each S-box are distributed across the 16 bit target block as specified in Table 2. Each S-box is defined as a 64×2 bit mapping of input values to outputs, as given in Table 3.

S-box	Input Bit					
	0	1	2	3	4	5
S_1	a_2	a_5	b_6	b_9	c_{11}	c_{13}
S_2	a_1	a_4	b_7	b_{10}	c_8	c_{14}
S_3	a_3	a_6	b_8	b_{13}	c_0	c_{15}
S_4	a_{12}	a_{14}	b_1	b_2	c_4	c_{10}
S_5	a_0	a_{10}	b_3	b_{14}	c_6	c_{12}
S_6	a_7	a_8	b_{12}	b_{15}	c_1	c_5
S_7	a_9	a_{15}	b_5	b_{11}	c_2	c_7
S_8	a_{11}	a_{13}	b_0	b_4	c_3	c_9

Table 1. S-Box Input Permutation

S-box	Output Bit	
	0	1
S_1	t_0	t_1
S_2	t_2	t_3
S_3	t_4	t_5
S_4	t_6	t_7
S_5	t_8	t_9
S_6	t_{10}	t_{11}
S_7	t_{12}	t_{13}
S_8	t_{14}	t_{15}

Table 2. S-Box Output Permutation

3.3 Key Setup

Each round of the cipher uses the secret key parameter to perturb the S-boxes by bitwise XOR against the S-box inputs. Each round thus requires 48 key bits. To convert the 128 bit k parameter to a sequence of 48 bit values for each round (the K table), MacGuffin uses an iterated version of its own block encryption function. See Figure 1.

3.4 Block Encryption

Block encryption is defined in Figure 2.

3.5 Block Decryption

Block decryption is similar to block encryption, and is defined in Figure 3.

$K \leftarrow 0$
$left, a, b, c \leftarrow k_{0...63}$
for $h = 0$ **to** 31 **do**
    for $i = 0$ **to** 31 **do**
        for $j = 1$ **to** 8 **do**
            $t \Leftarrow S_j(a \oplus K[i,0], b \oplus K[i,1], c \oplus K[i,2])$
        $left \leftarrow left \oplus t$
        $left, a, b, c \leftarrow a, b, c, left$
    $K[h,0] \leftarrow left$
    $K[h,1] \leftarrow a$
    $K[h,2] \leftarrow b$
$left, a, b, c \leftarrow k_{64...127}$
for $h = 0$ **to** 31 **do**
    for $i = 0$ **to** 31 **do**
        for $j = 1$ **to** 8 **do**
            $t \Leftarrow S_j(a \oplus K[i,0], b \oplus K[i,1], c \oplus K[i,2])$
        $left \leftarrow left \oplus t$
        $left, a, b, c \leftarrow a, b, c, left$
    $K[h,0] \leftarrow K[h,0] \oplus left$
    $K[h,1] \leftarrow K[h,1] \oplus a$
    $K[h,2] \leftarrow K[h,2] \oplus b$

Fig. 1. MacGuffin Key Setup

$left, a, b, c \leftarrow I$
for $i = 0$ **to** 31 **do**
    for $j = 1$ **to** 8 **do**
        $t \Leftarrow S_j(a \oplus K[i,0], b \oplus K[i,1], c \oplus K[i,2])$
    $left \leftarrow left \oplus t$
    $left, a, b, c \leftarrow a, b, c, left$
$O \leftarrow left, a, b, c$

Fig. 2. MacGuffin Block Encryption

$c, left, a, b \leftarrow I$
for $i = 31$ **downto** 0 **do**
    for $j = 1$ **to** 8 **do**
        $t \Leftarrow S_j(a \oplus K[i,0], b \oplus K[i,1], c \oplus K[i,2])$
    $left \leftarrow left \oplus t$
    $left, a, b, c \leftarrow c, left, a, b$
$O \leftarrow left, a, b, c$

Fig. 3. MacGuffin Block Decryption

$$S_1$$

```
2 0 0 3 3 1 1 0 0 2 3 0 3 3 2 1 1 2 2 0 0 2 2 3 1 3 3 1 0 1 1 2
0 3 1 2 2 2 2 0 3 0 0 3 0 1 3 1 3 1 2 3 3 1 1 2 1 2 2 0 1 0 0 3
```

$$S_2$$

```
3 1 1 3 2 0 2 1 0 3 3 0 1 2 0 2 3 2 1 0 0 1 3 2 2 0 0 3 1 3 2 1
0 3 2 2 1 2 3 1 2 1 0 3 3 0 1 0 1 3 2 0 2 1 0 2 3 0 1 1 0 2 3 3
```

$$S_3$$

```
2 3 0 1 3 0 2 3 0 1 1 0 3 0 1 2 1 0 3 2 2 1 1 2 3 2 0 3 0 3 2 1
3 1 0 2 0 3 3 0 2 0 3 3 1 2 0 1 3 0 1 3 0 2 2 1 1 3 2 1 2 0 1 2
```

$$S_4$$

```
1 3 3 2 2 3 1 1 0 0 0 3 3 0 2 1 1 0 0 1 2 0 1 2 3 1 2 2 0 2 3 3
2 1 0 3 3 0 0 0 2 2 3 1 1 3 3 2 3 3 1 0 1 1 2 3 1 2 0 1 2 0 0 2
```

$$S_5$$

```
0 2 2 3 0 0 1 2 1 0 2 1 3 3 0 1 2 1 1 0 1 3 3 2 3 1 0 3 2 2 3 0
0 3 0 2 1 2 3 1 2 1 3 2 1 0 2 3 3 0 3 3 2 0 1 3 0 2 1 0 0 1 2 1
```

$$S_6$$

```
2 2 1 3 2 0 3 0 3 1 0 2 0 3 2 1 0 0 3 1 1 3 0 2 2 0 1 3 1 1 3 2
3 0 2 1 3 0 1 2 0 3 2 1 2 3 1 2 1 3 0 2 0 1 2 1 1 0 3 0 3 2 0 3
```

$$S_7$$

```
0 3 3 0 0 3 2 1 3 0 0 3 2 1 3 2 1 2 2 1 3 1 1 2 1 0 2 3 0 2 1 0
1 0 0 3 3 3 3 2 2 1 1 0 1 2 2 1 2 3 3 1 0 0 2 3 0 2 1 0 3 1 0 2
```

$$S_8$$

```
3 1 0 3 2 3 0 2 0 2 3 1 3 1 1 0 2 2 3 1 1 0 2 3 1 0 0 2 2 3 1 0
1 0 3 1 0 2 1 1 3 0 2 2 2 2 0 3 0 3 0 2 2 3 3 0 3 1 1 1 1 0 2 3
```

Table 3. MacGuffin S-Boxes

4 Implementation, Performance and Applications

Feistel ciphers, with their many permutation operations and table lookups, are particularly well suited to hardware implementation. Because permutations in hardware are "free" (they are implemented with simple connections), and because S-box lookups can occur in parallel, each round can be implemented with conventional modern hardware in two clock cycles.

Software implementations of Feistel ciphers on general-purpose computers are typically much slower than their hardware counterparts, since the S-boxes must be evaluated in sequence and bit permutations must be simulated with shifts, ANDs, ORs and other operators. Depending on the specific permutations and S-box structures, however, many of these operations can be made faster with table lookups and by combining several operations into one.

The permutations in MacGuffin have been designed explicitly to permit software optimization. First, the six inputs to each S-box are from different bits from each of the a, b and c registers, allowing the three registers to be masked and ORed together (without individual shifting) for a single lookup for each S-box. Furthermore, for each S-box there is a unique "mate" S-box with which it shares no common inputs. This allows the eight S-boxes to be "paired off" and looked up two at a time with a single 2^{16} entry table containing the combined outputs of both S-boxes. (The pairs are S_1S_2, S_3S_4, S_5S_7 and S_6S_8).

An optimized software implementation (given in the Appendix) of 32 round MacGuffin runs at close to the speed of optimized 16 round DES in software. An implementation on a 486/66 processor has a bandwidth of about 1.5Mbps; a reasonable DES implementation [2] on the same processor runs at 2.1Mbps.

The MacGuffin interface is similar to that of DES (except for the larger keyspace). It can be used with the standard "FIPS-81" modes of operation[4]. Note that key setup is an explicitly time consuming process. This is intended to discourage exhaustive search of poorly chosen keys. In an implementation where rapid selection among many keys is required (such as a packet-based network security protocol) the 1536 bit expanded key may be passed directly as the cryptovariable.

Experiments with MacGuffin are detailed in [7].

While we believe the GUFN structure is superior to the conventional BFN cipher structure, much more discussion and analysis is required before we can recommend its use for protecting sensitive data. We encourage attacks against MacGuffin in particular and the GUFN structure in general.

References

1. H. Feistel. Cryptography and Computer Privacy. *Scientific American,* May 1973.
2. J. Lacy, D.P. Mitchell, and W.M. Schell. CryptoLib: Cryptography in Software. *Proceedings of USENIX Security Symposium IV,* October 1993.
3. National Bureau of Standards. Data Encryption Standard, *Federal Information Processing Standards Publication 46,* US Government Printing Office, Washington, D.C., 1977.
4. National Bureau of Standards. Data Encryption Standard Modes of Operation, *Federal Information Processing Standards Publication 81,* US Government Printing Office, Washington, D.C., 1980.
5. National Institute for Standards and Technology. Secure Hash Standard. *Federal Information Processing Standard Publication 180,* US Government Printing Office, April 1993.
6. R. Rivest. The MD5 Message Digest Algorithm. *RFC 1321,* IETF, April 1992.
7. B. Schneier and M. Blaze. Unbalanced Feistel Network Block Ciphers. *To appear,* 1994.

Appendix: Optimized C Language Implementation

```c
/*
 * MacGuffin Cipher
 * 10/3/94 - Matt Blaze
 * (fast, unrolled version)
 */

#define ROUNDS 32
#define KSIZE (ROUNDS*3)

/* expanded key structure */
typedef struct mcg_key {
  unsigned short val[KSIZE];
} mcg_key;

#define TSIZE (1<<16)

/* the 8 s-boxes, expanded to put the output bits in the right
 * places.  note that these are the des s-boxes (in left-right,
 * not canonical, order), but with only the "outer" two output
 * bits. */
unsigned short sboxes[8][64] = {
/* 0 (S1) */
  {0x0002, 0x0000, 0x0000, 0x0003, 0x0003, 0x0001, 0x0001, 0x0000,
   0x0000, 0x0002, 0x0003, 0x0000, 0x0003, 0x0003, 0x0002, 0x0001,
   0x0001, 0x0002, 0x0002, 0x0000, 0x0000, 0x0002, 0x0002, 0x0003,
   0x0001, 0x0003, 0x0003, 0x0001, 0x0000, 0x0001, 0x0001, 0x0002,
   0x0000, 0x0003, 0x0001, 0x0002, 0x0002, 0x0002, 0x0002, 0x0000,
   0x0003, 0x0000, 0x0000, 0x0003, 0x0000, 0x0001, 0x0003, 0x0001,
   0x0003, 0x0001, 0x0002, 0x0003, 0x0003, 0x0001, 0x0001, 0x0002,
   0x0001, 0x0002, 0x0002, 0x0000, 0x0001, 0x0000, 0x0000, 0x0003},
/* 1 (S2) */
  {0x000c, 0x0004, 0x0004, 0x000c, 0x0008, 0x0000, 0x0008, 0x0004,
   0x0000, 0x000c, 0x000c, 0x0000, 0x0004, 0x0008, 0x0000, 0x0008,
   0x000c, 0x0008, 0x0004, 0x0000, 0x0000, 0x0004, 0x000c, 0x0008,
   0x0008, 0x0000, 0x0000, 0x000c, 0x0004, 0x000c, 0x0008, 0x0004,
   0x0000, 0x000c, 0x0008, 0x0008, 0x0004, 0x0008, 0x000c, 0x0004,
   0x0008, 0x0004, 0x0000, 0x000c, 0x000c, 0x0000, 0x0004, 0x0000,
   0x0004, 0x000c, 0x0008, 0x0000, 0x0008, 0x0004, 0x0000, 0x0008,
   0x000c, 0x0000, 0x0004, 0x0004, 0x0000, 0x0008, 0x000c, 0x000c},
/* 2 (S3) */
  {0x0020, 0x0030, 0x0000, 0x0010, 0x0030, 0x0000, 0x0020, 0x0030,
   0x0000, 0x0010, 0x0010, 0x0000, 0x0030, 0x0000, 0x0010, 0x0020,
   0x0010, 0x0000, 0x0030, 0x0020, 0x0020, 0x0010, 0x0010, 0x0020,
   0x0030, 0x0020, 0x0000, 0x0030, 0x0000, 0x0030, 0x0020, 0x0010,
```

```
  0x0030, 0x0010, 0x0000, 0x0020, 0x0000, 0x0030, 0x0030, 0x0000,
  0x0020, 0x0000, 0x0030, 0x0030, 0x0010, 0x0020, 0x0000, 0x0010,
  0x0030, 0x0000, 0x0010, 0x0030, 0x0000, 0x0020, 0x0020, 0x0010,
  0x0010, 0x0030, 0x0020, 0x0010, 0x0020, 0x0000, 0x0010, 0x0020},
/* 3 (S4) */
 {0x0040, 0x00c0, 0x00c0, 0x0080, 0x0080, 0x00c0, 0x0040, 0x0040,
  0x0000, 0x0000, 0x0000, 0x00c0, 0x00c0, 0x0000, 0x0080, 0x0040,
  0x0040, 0x0000, 0x0000, 0x0040, 0x0080, 0x0000, 0x0040, 0x0080,
  0x00c0, 0x0040, 0x0080, 0x0080, 0x0000, 0x0080, 0x00c0, 0x00c0,
  0x0080, 0x0040, 0x0000, 0x00c0, 0x00c0, 0x0000, 0x0000, 0x0000,
  0x0080, 0x0080, 0x00c0, 0x0040, 0x0040, 0x00c0, 0x00c0, 0x0080,
  0x00c0, 0x00c0, 0x0040, 0x0000, 0x0040, 0x0040, 0x0080, 0x00c0,
  0x0040, 0x0080, 0x0000, 0x0040, 0x0080, 0x0000, 0x0000, 0x0080},
/* 4 (S5) */
 {0x0000, 0x0200, 0x0200, 0x0300, 0x0000, 0x0000, 0x0100, 0x0200,
  0x0100, 0x0000, 0x0200, 0x0100, 0x0300, 0x0300, 0x0000, 0x0100,
  0x0200, 0x0100, 0x0100, 0x0000, 0x0100, 0x0300, 0x0300, 0x0200,
  0x0300, 0x0100, 0x0000, 0x0300, 0x0200, 0x0200, 0x0300, 0x0000,
  0x0000, 0x0300, 0x0000, 0x0200, 0x0100, 0x0200, 0x0300, 0x0100,
  0x0200, 0x0100, 0x0300, 0x0200, 0x0100, 0x0000, 0x0200, 0x0300,
  0x0300, 0x0000, 0x0300, 0x0300, 0x0200, 0x0000, 0x0100, 0x0300,
  0x0000, 0x0200, 0x0100, 0x0000, 0x0000, 0x0100, 0x0200, 0x0100},
/* 5 (S6) */
 {0x0800, 0x0800, 0x0400, 0x0c00, 0x0800, 0x0000, 0x0c00, 0x0000,
  0x0c00, 0x0400, 0x0000, 0x0800, 0x0000, 0x0c00, 0x0800, 0x0400,
  0x0000, 0x0000, 0x0c00, 0x0400, 0x0400, 0x0c00, 0x0000, 0x0800,
  0x0800, 0x0000, 0x0400, 0x0c00, 0x0400, 0x0400, 0x0c00, 0x0800,
  0x0c00, 0x0000, 0x0800, 0x0400, 0x0c00, 0x0000, 0x0400, 0x0800,
  0x0000, 0x0c00, 0x0800, 0x0400, 0x0800, 0x0c00, 0x0400, 0x0800,
  0x0400, 0x0c00, 0x0000, 0x0800, 0x0000, 0x0400, 0x0800, 0x0400,
  0x0400, 0x0000, 0x0c00, 0x0000, 0x0c00, 0x0800, 0x0000, 0x0c00},
/* 6 (S7) */
 {0x0000, 0x3000, 0x3000, 0x0000, 0x0000, 0x3000, 0x2000, 0x1000,
  0x3000, 0x0000, 0x0000, 0x3000, 0x2000, 0x1000, 0x3000, 0x2000,
  0x1000, 0x2000, 0x2000, 0x1000, 0x3000, 0x1000, 0x1000, 0x2000,
  0x1000, 0x0000, 0x2000, 0x3000, 0x0000, 0x2000, 0x1000, 0x0000,
  0x1000, 0x0000, 0x0000, 0x3000, 0x3000, 0x3000, 0x3000, 0x2000,
  0x2000, 0x1000, 0x1000, 0x0000, 0x1000, 0x2000, 0x2000, 0x1000,
  0x2000, 0x3000, 0x3000, 0x1000, 0x0000, 0x0000, 0x2000, 0x3000,
  0x0000, 0x2000, 0x1000, 0x0000, 0x3000, 0x1000, 0x0000, 0x2000},
/* 7 (S8) */
 {0xc000, 0x4000, 0x0000, 0xc000, 0x8000, 0xc000, 0x0000, 0x8000,
  0x0000, 0x8000, 0xc000, 0x4000, 0xc000, 0x4000, 0x4000, 0x0000,
  0x8000, 0x8000, 0xc000, 0x4000, 0x4000, 0x0000, 0x8000, 0xc000,
  0x4000, 0x0000, 0x0000, 0x8000, 0x8000, 0xc000, 0x4000, 0x0000,
```

```
     0x4000, 0x0000, 0xc000, 0x4000, 0x0000, 0x8000, 0x4000, 0x4000,
     0xc000, 0x0000, 0x8000, 0x8000, 0x8000, 0x8000, 0x0000, 0xc000,
     0x0000, 0xc000, 0x0000, 0x8000, 0x8000, 0xc000, 0xc000, 0x0000,
     0xc000, 0x4000, 0x4000, 0x4000, 0x4000, 0x0000, 0x8000, 0xc000}
};

/* table of s-box outputs, expanded for 16 bit input.
 * this one table includes all 8 sboxes - just mask off
 * the output bits not in use. */
unsigned short stable[TSIZE];

/* we exploit two features of the s-box input & output perms -
 * first, each s-box uses as input two different bits from each
 * of the three registers in the right side, and, second,
 * for each s-box there is another-sbox with no common input bits
 * between them. therefore we can lookup two s-box outputs in one
 * probe of the table.  just mask off the approprate input bits
 * in the table below for each of the three registers and OR
 * together for the table lookup index.
 * these masks are also available below in #defines, for better
 * lookup speed in unrolled loops. */
unsigned short lookupmasks[4][3] = {
    /* a    ,    b    ,    c    */
    {0x0036, 0x06c0, 0x6900},  /* s1+s2 */
    {0x5048, 0x2106, 0x8411},  /* s3+s4 */
    {0x8601, 0x4828, 0x10c4},  /* s5+s7 */
    {0x2980, 0x9011, 0x022a}}; /* s6+s8 */

/* this table contains the corresponding output masks for the
 * lookup procedure mentioned above.
 * (similarly available below in #defines). */
unsigned short outputmasks[4] = {
    0x000f /*s1+s2*/, 0x00f0 /*s3+s4*/,
    0x3300 /*s5+s7*/, 0xcc00 /*s6+s8*/};

/* input and output lookup masks (see above) */
/* s1+s2 */
#define IN00   0x0036
#define IN01   0x06c0
#define IN02   0x6900
#define OUT0   0x000f
/* s3+s4 */
#define IN10   0x5048
#define IN11   0x2106
#define IN12   0x8411
```

```
#define OUT1   0x00f0
/* s5+s7 */
#define IN20   0x8601
#define IN21   0x4828
#define IN22   0x10c4
#define OUT2   0x3300
/* s6+s8 */
#define IN30   0x2980
#define IN31   0x9011
#define IN32   0x022a
#define OUT3   0xcc00


/*
 * initialize the macguffin s-box tables.
 * this takes a while, but is only done once.
 */
mcg_init()
{
  unsigned int i,j,k;
  int b;
  /*
   * input permutation for the 8 s-boxes.
   * each row entry is a bit position from
   * one of the three right hand registers,
   * as follows:
   *    a,a,b,b,c,c
   */
  static int sbits[8][6] = {
    {2,5,6,9,11,13},  {1,4,7,10,8,14},
    {3,6,8,13,0,15},  {12,14,1,2,4,10},
    {0,10,3,14,6,12}, {7,8,12,15,1,5},
    {9,15,5,11,2,7},  {11,13,0,4,3,9}};

  for (i=0; i<TSIZE; i++) {
    stable[i]=0;
    for (j=0; j<8; j++)
      stable[i] |=
        sboxes[j][((i>>sbits[j][0])&1)
            |(((i>>sbits[j][1])&1)<<1)
            |(((i>>sbits[j][2])&1)<<2)
            |(((i>>sbits[j][3])&1)<<3)
            |(((i>>sbits[j][4])&1)<<4)
            |(((i>>sbits[j][5])&1)<<5)];
  }
}
```

```
/*
 * expand key to ek
 */
mcg_keyset(key,ek)
     unsigned char *key;
     mcg_key *ek;
{
  int i,j;
  unsigned char k[2][8];

  mcg_init();
  bcopy(&key[0],k[0],8);
  bcopy(&key[8],k[1],8);
  for (i=0; i<KSIZE; i++)
    ek->val[i]=0;
  for (i=0; i<2; i++)
    for (j=0; j<32; j++) {
      mcg_block_encrypt(k[i],ek);
      ek->val[j*3]   ^= k[i][0] | (k[i][1]<<8);
      ek->val[j*3+1] ^= k[i][2] | (k[i][3]<<8);
      ek->val[j*3+2] ^= k[i][4] | (k[i][5]<<8);
    }
}

/*
 * codebook encrypt one block with given expanded key
 */
mcg_block_encrypt(blk,key)
     unsigned char *blk;
     mcg_key *key;
{
  unsigned short r0, r1, r2, r3, a, b, c;
  int i;
  unsigned short *ek;

  /* copy cleartext into local words */
  r0=blk[0]|(blk[1]<<8);
  r1=blk[2]|(blk[3]<<8);
  r2=blk[4]|(blk[5]<<8);
  r3=blk[6]|(blk[7]<<8);

  ek = &(key->val[0]);
  /* round loop, unrolled 4x */
  for (i=0; i<(ROUNDS/4); i++) {
```

```
      a = r1 ^ *(ek++);  b = r2 ^ *(ek++);  c = r3 ^ *(ek++);
      r0 ^=((OUT0 & stable[(a & IN00)|(b & IN01)|(c & IN02)])
          | (OUT1 & stable[(a & IN10)|(b & IN11)|(c & IN12)])
          | (OUT2 & stable[(a & IN20)|(b & IN21)|(c & IN22)])
          | (OUT3 & stable[(a & IN30)|(b & IN31)|(c & IN32)]));
      a = r2 ^ *(ek++);  b = r3 ^ *(ek++);  c = r0 ^ *(ek++);
      r1 ^=((OUT0 & stable[(a & IN00)|(b & IN01)|(c & IN02)])
          | (OUT1 & stable[(a & IN10)|(b & IN11)|(c & IN12)])
          | (OUT2 & stable[(a & IN20)|(b & IN21)|(c & IN22)])
          | (OUT3 & stable[(a & IN30)|(b & IN31)|(c & IN32)]));
      a = r3 ^ *(ek++);  b = r0 ^ *(ek++);  c = r1 ^ *(ek++);
      r2 ^=((OUT0 & stable[(a & IN00)|(b & IN01)|(c & IN02)])
          | (OUT1 & stable[(a & IN10)|(b & IN11)|(c & IN12)])
          | (OUT2 & stable[(a & IN20)|(b & IN21)|(c & IN22)])
          | (OUT3 & stable[(a & IN30)|(b & IN31)|(c & IN32)]));
      a = r0 ^ *(ek++);  b = r1 ^ *(ek++);  c = r2 ^ *(ek++);
      r3 ^=((OUT0 & stable[(a & IN00)|(b & IN01)|(c & IN02)])
          | (OUT1 & stable[(a & IN10)|(b & IN11)|(c & IN12)])
          | (OUT2 & stable[(a & IN20)|(b & IN21)|(c & IN22)])
          | (OUT3 & stable[(a & IN30)|(b & IN31)|(c & IN32)]));
   }
   /* copy 4 encrypted words back to output */
   blk[0] = r0;  blk[1] = r0>>8;
   blk[2] = r1;  blk[3] = r1>>8;
   blk[4] = r2;  blk[5] = r2>>8;
   blk[6] = r3;  blk[7] = r3>>8;
}


/*
 * codebook decrypt one block with given expanded key
 */
mcg_block_decrypt(blk,key)
     unsigned char *blk;
     mcg_key *key;
{
   unsigned short r0, r1, r2, r3, a, b, c;
   int i;
   unsigned short *ek;

   /* copy ciphertext to 4 local words */
   r0=blk[0]|(blk[1]<<8);
   r1=blk[2]|(blk[3]<<8);
   r2=blk[4]|(blk[5]<<8);
   r3=blk[6]|(blk[7]<<8);
```

```
ek = &(key->val[KSIZE]);
/* round loop, unrolled 4x */
for (i=0; i<(ROUNDS/4); ++i) {
  c = r2 ^ *(--ek);  b = r1 ^ *(--ek);  a = r0 ^ *(--ek);
  r3 ^=((OUT0 & stable[(a & IN00)|(b & IN01)|(c & IN02)])
     | (OUT1 & stable[(a & IN10)|(b & IN11)|(c & IN12)])
     | (OUT2 & stable[(a & IN20)|(b & IN21)|(c & IN22)])
     | (OUT3 & stable[(a & IN30)|(b & IN31)|(c & IN32)]));
  c = r1 ^ *(--ek);  b = r0 ^ *(--ek);  a = r3 ^ *(--ek);
  r2 ^=((OUT0 & stable[(a & IN00)|(b & IN01)|(c & IN02)])
     | (OUT1 & stable[(a & IN10)|(b & IN11)|(c & IN12)])
     | (OUT2 & stable[(a & IN20)|(b & IN21)|(c & IN22)])
     | (OUT3 & stable[(a & IN30)|(b & IN31)|(c & IN32)]));
  c = r0 ^ *(--ek);  b = r3 ^ *(--ek);  a = r2 ^ *(--ek);
  r1 ^=((OUT0 & stable[(a & IN00)|(b & IN01)|(c & IN02)])
     | (OUT1 & stable[(a & IN10)|(b & IN11)|(c & IN12)])
     | (OUT2 & stable[(a & IN20)|(b & IN21)|(c & IN22)])
     | (OUT3 & stable[(a & IN30)|(b & IN31)|(c & IN32)]));
  c = r3 ^ *(--ek);  b = r2 ^ *(--ek);  a = r1 ^ *(--ek);
  r0 ^=((OUT0 & stable[(a & IN00)|(b & IN01)|(c & IN02)])
     | (OUT1 & stable[(a & IN10)|(b & IN11)|(c & IN12)])
     | (OUT2 & stable[(a & IN20)|(b & IN21)|(c & IN22)])
     | (OUT3 & stable[(a & IN30)|(b & IN31)|(c & IN32)]));
}
/* copy decrypted bits back to output */
blk[0] = r0;  blk[1] = r0>>8;
blk[2] = r1;  blk[3] = r1>>8;
blk[4] = r2;  blk[5] = r2>>8;
blk[6] = r3;  blk[7] = r3>>8;
}
```

S-Boxes and Round Functions
with Controllable
Linearity and Differential Uniformity

Kaisa Nyberg [*]

Prinz Eugen-Straße 18/6, A-1040 Vienna, Austria

nyberg@ict.tuwien.ac.at

Abstract. In this contribution we consider the stability of linearity and differential uniformity of vector Boolean functions under certain constructions and modifications. These include compositions with affine surjections onto the input space and with affine surjections from the output space, inversions, adding coordinate functions, forming direct sums and restrictions to affine subspaces. As examples we consider some true round function and S-box constructions. More theoretical examples are offered by the bent and almost perfect nonlinear functions. We also include some facts about functions with partially bent components.

1 Introduction

Several methods of constructing S-boxes for an iterated block cipher have been previously presented. The most common methods are based on

- random generation,
- testing against a set of design criteria,
- algebraic constructions having certain good properties,
- or a combination of these.

The round functions typically consist of S-boxes combined in certain ways (e.g. parallel or summing up) and finally the whole cipher is formed by iterating (e.g. DES-like or SPN) certain number of rounds.

At each step of the design of a block cipher algebraic constructions and compositions are used. In this contribution we focus on algebraic properties that are necessary and, in some cases sufficient, to guarantee resistance against the differential and linear cryptanalysis.

For example, in ciphers using small parallel S-boxes the bit permutations between rounds play a crucial role in the security of the cipher (cf. DES and substitution-permutation networks [9]). On the other hand, proven security based

---

[*] Sponsored by the Matine Board, Finland.

only on the properties of the round functions can be achieved [18]. In both cases low differential uniformity and low linearity of S-boxes and round functions of iterated block ciphers seem to be necessary conditions and are accepted as useful design criteria of S-boxes and round functions.

Different algebraic methods to combine and modify S-boxes in the construction of a round function have been previously proposed. It is of essential importance to understand how well low differential uniformity and linearity are preserved under different combinations and modifications. We will consider the following:

1. composition with a linear (or affine) surjective mapping onto the input space,
2. composition with a linear (or affine) surjective mapping from the output space,
3. inversion,
4. adding coordinate functions,
5. restriction to a linear (or affine) subspace, and
6. sum of functions with independent inputs

This list is not meant to be exhaustive but represents the most common constructions. For example 4 and 6 are used in the CAST algorithm [2]. We will see that differential uniformity can be controlled under 4, but not linearity. A probabilistic method to overcome these problems was presented in [8].

Modification 2 contains as a special case the chopping of an S-box. It gives a controlled way to modify S-boxes and round functions. It was used in [18] and received special attention in [23]. We will give a simple general treatment of 2. For simplicity, it is assumed that the input and output spaces are linear spaces over $\mathbf{F} = GF(2)$, but the results can be generalized to any finite field.

We conclude that in general, differential uniformity and linearity behave in different ways under the modifications 1–6. As an application of the results on 1–6 we give constructions of round functions of iterated ciphers with proven resistance against differential cryptanalysis, but which can be trivially broken by linear cryptanalysis. Similarly, we show that a cipher can be secure against linear cryptanalysis but easily broken using the differential method.

2   Linearity and Nonlinearity

2.1   Boolean Functions

We denote by $\mathbf{F}$ the finite field $GF(2)$. Let $f : \mathbf{F}^n \to \mathbf{F}$ be a Boolean function. The *nonlinearity* of f is defined as follows [13].

$$\mathcal{NL}(f) = \min_{A \text{ aff.}} \#\{x \in \mathbf{F}^n \mid f(x) \neq A(x)\}$$
$$= \min_{L \text{ lin.}} \min\{\#\{x \in \mathbf{F}^n \mid f(x) = L(x)\}, 2^n - \#\{x \in \mathbf{F}^n \mid f(x) = L(x)\}\}$$
$$= 2^{n-1} - \frac{1}{2} \max_{L \text{ lin.}} |\#\{x \in \mathbf{F}^n \mid f(x) = L(x)\} - \#\{x \in \mathbf{F}^n \mid f(x) \neq L(x)\}|$$
$$= 2^{n-1} - 2^{n-1} \max_{L \text{ lin.}} |c(f, L)|$$

where, for $L(x) = L_a(x) = a \cdot x$,

$$c(f, L) = \mathrm{Pr}_X(f(X) = L(X)) - \mathrm{Pr}_X(f(X) \neq L(X))$$
$$= 2(\ \mathrm{Pr}_X(f(X) = a \cdot X)) - 1/2)$$
$$= 2^{-n}\widehat{F}(a)$$

measures the *correlation* between f and $L = L_a$, and $\widehat{F}$ denotes the Walsh transform of f,

$$\widehat{F}(w) = \sum_{x \in \mathbf{F}^n} (-1)^{f(x)+w \cdot x}, \ w \in \mathbf{F}^n.$$

Various measures of the *linearity* of a Boolean function have been previously used in the literature. In this contribution (see also [17]) we use the following.

Definition 1. The *linearity* of a Boolean function is

$$\mathcal{L}(f) = \max_{L \ \mathrm{lin.}} |c(f, L)|.$$

The relationships with the linearity measure Λ_f of Chabaud and Vaudenay [5] and with the linearity measure R_f of Dobbertin [7] are

$$\Lambda_f = 2^{n-1}\mathcal{L}(f),$$
$$R_f = 2^n\mathcal{L}(f).$$

The linearity and nonlinearity are related as follows

$$\mathcal{NL}(f) = 2^{n-1} - 2^{n-1}\mathcal{L}(f). \tag{1}$$

By Parseval's theorem

$$\sum_{L \ \mathrm{lin.}} c(f, L)^2 = 1$$

from where it follows that

$$2^{-n/2} \leq \mathcal{L}(f) \leq 1.$$

For n even, the lower bound of linearity is tight and is reached by the *bent functions*. For n odd this lower bound is not reached by any functions, and the general tight lower bound is unknown. For some n, at least for $n = 1, 3, 5$ and 7, the tight lower bound is $2^{-\frac{n-1}{2}}$. For $n = 15$, it was shown in [19] that there exist functions $f : \mathbf{F}^n \to \mathbf{F}$ with $2^{-\frac{n}{2}} < \mathcal{L}(f) = \frac{27}{32}2^{-\frac{n-1}{2}}$. Let $f : \mathbf{F}^n \to \mathbf{F}$ be a function with linearity $\mathcal{L}(f)$. Then the function $g : \mathbf{F}^n \times \mathbf{F}^2 \to \mathbf{F}$, $g(x, y, z) = f(x) + yz$, $x \in \mathbf{F}^n$, $y, z \in \mathbf{F}$, has linearity $\mathcal{L}(g) = 2\mathcal{L}(f)$. Hence for all odd n, $n \geq 15$, there exist Boolean functions f with $2^{-\frac{n}{2}} < \mathcal{L}(f) \leq \frac{27}{32}2^{-\frac{n-1}{2}}$. An important conjecture [7] is that the lower bound is asymptotically tight.

Since bent functions are not balanced, the minimal linearity is not reached by balanced Boolean functions. In fact, the tight lower bound is not known for the balanced Boolean functions. Upper bounds of the minimal linearity of balanced Boolean functions in even dimension can be found in [7] and [21].

2.2 Vector Boolean Functions

From now on we consider a vector Boolean function $f : \mathbf{F}^n \to \mathbf{F}^m$. Let $b \in \mathbf{F}^m$ be a nonzero element, $b = (b_1, \ldots, b_m)$. We denote by $b \cdot f$ the Boolean function, which is the linear combination $b_1 f_1 + \ldots + b_m f_m$ of the coordinate functions $f_1, \ldots, f_m$ of f. Against the usual convention, which is to use the term *component* as a synonyme of *coordinate function*, we will, throughout this paper, call the *nonzero linear combinations $b \cdot f$ of the coordinate functions the components of f*. In [15] the notion of nonlinearity was extended to vector functions as follows. The *nonlinearity of a vector Boolean function f* is $\mathcal{NL}(f) = \min_{b \neq 0} \mathcal{NL}(b \cdot f)$. The following definition is then in full accordance with (1) and extends (1) and the relationship with the measure of Chabaud and Vaudenay to hold also for vector Boolean functions.

Definition 2. The *linearity of a vector Boolean function f* is

$$\mathcal{L}(f) = \max_{b \neq 0} \mathcal{L}(b \cdot f).$$

It follows immediately from the absolute lower bound of linearity of Boolean functions that $\mathcal{L}(f) \geq 2^{-\frac{n}{2}}$. It was proven in [14] that this lower bound is tight if and only if $n \geq 2m$ and n is even. The functions reaching this minimum linearity are called *bent*. In [5] Chabaud and Vaudenay proved the following lower bound of linearity of a vector Boolean function $f : \mathbf{F}^n \to \mathbf{F}^m$.

Theorem 3. [5]

$$\mathcal{L}(f) \geq \frac{1}{2^n}(3 \cdot 2^n - 2 - 2\frac{(2^n - 1)(2^{n-1} - 1)}{2^m - 1})^{1/2} = C(n, m). \tag{2}$$

Observe that $C(n, m)$ is negative if $m = 1$, except for $n = 2$, and

$$C(n, m) < 2^{-\frac{n}{2}}, \text{ if } 1 < m < n - 1$$
$$C(n, m) = 2^{-\frac{n}{2}}, \text{ if } m = n - 1$$
$$C(n, m) = 2^{-\frac{n-1}{2}}, \text{ if } m = n$$
$$C(n, m) > 2^{-\frac{n-1}{2}}, \text{ if } m > n$$

Hence the lower bound $C(n, m)$ cannot be reached if $m < n$ (except for $n = 2$). Neither is it tight for $m > n$ [5]. Indeed, for $\frac{n}{2} < m < n$ and $m > n$ the minimum linearity is unknown.

On the other hand, it is known (see e.g. [15] and [16]) that for $m = n$, functions (even bijective) $f : \mathbf{F}^n \to \mathbf{F}^n$ exist with $\mathcal{L}(f) = 2^{-\frac{n-1}{2}}$. Such functions are called *almost bent* [5]. Almost bent functions exist only, if n is odd, and are characterized by the property that their components have an almost flat correlation spectrum. More precisely, $|c(b \cdot f, L)| = 0$ or $2^{-\frac{n-1}{2}}$ for all $b \in \mathbf{F}^n$, if and only if $f : \mathbf{F}^n \to \mathbf{F}^n$ is almost bent. For example, the power functions $f(x) = x^{2^k+1}$ and $f(x) = x^{2^{2k}-2^k+1}$ in $GF(2^n)$, n odd and $\gcd(n, k) = 1$, have this property [10] and are almost bent.

3   Differential Uniformity

In [16] a function $f : \mathbf{F}^n \rightarrow \mathbf{F}^n$ is called *differentially δ-uniform* if

$$\#\{x \in \mathbf{F}^n \mid f(x + a) + f(x) = b\} \leq \delta \text{ for all } a \in \mathbf{F}^n, b \in \mathbf{F}^m, a \neq 0.$$

Hence the following definition is natural (see also [22]).

Definition 4. *Differential uniformity* $\Delta(f)$ of a function $f : \mathbf{F}^n \rightarrow \mathbf{F}^m$ is

$$\Delta(f) = \max_{a \neq 0, b} \#\{x \in \mathbf{F}^n \mid f(x + a) + f(x) = b\}.$$

Clearly $\Delta(f) \geq \max\{2, 2^{n-m}\}$. It was shown in [14] that for $m < n$ the minimum differential uniformity 2^{n-m} is reached if and only if $2m \leq n$ and n is even. Such functions are called *perfect nonlinear* and they are the same as the bent functions. For $\frac{n}{2} < m < n$ the minimum differential uniformity is unknown. If $m \geq n$ the minimum differential uniformity is 2. A function which reaches this bound is called *almost perfect nonlinear (APN)* in [18], where examples of such functions are given in the case where m and n are equal and odd.

For $m = n$ even, the minimum differential uniformity is unknown. It was shown in [22] that, for $m = n$ even, there is no APN quadratic permutation. In the next section we generalize this result by repeating the approach of [18] and we show that, for $m = n$ even, there is no APN permutation with partially bent components. Let us mention that no examples of differentially 2-uniform functions are known for $m = n$ even and $n > 2$.

For $m > n$ the minimum differential uniformity is 2, and can be reached by simple modifications of APN functions, as we will see below. Such functions may even have linear components.

We may conclude, that for $m \geq n$ differential uniformity is a weaker notion than linearity. In [5] Chabaud and Vaudenay show that almost bentness implies almost perfect nonlinearity. If $m = n$ odd, the permutation $f : GF(2^n) \rightarrow GF(2^n)$, $f(x) = x^{-1}$, $f(0) = 0$, is differentially 2-uniform without having the minimum nonlinearity. An interesting open question is that of what is the maximum linearity of an APN function when $m = n$.

4   Functions with Partially Bent Components

4.1   Partially Bent Boolean Functions

Functions with quadratic components were studied in [15] and [18] and later in [23]. In this section we adopt the techniques from [5] and generalize the approach of [18] to functions with partially bent components. Such functions have a simple and clear structure and therefore they are useful as illustrative examples of linearity properties.

Definition 5. [4] A Boolean function $f : \mathbf{F}^n \rightarrow \mathbf{F}$ is *partially bent*, if there exists a linear subspace U of $\mathbf{F}^n$ such that the restriction of f to U is affine and the restriction of f to any complementary subspace V of U, $V \oplus U = \mathbf{F}^n$, is bent, and f can be represented as a direct sum of the restricted functions, i.e., $f(y + z) = f(y) + f(z)$, for all $z \in U$ and $y \in V$.

The space U is formed by the *linear structures* of f, that is, vectors $\alpha \in \mathbf{F}^n$ such that $f(x + \alpha) + f(x)$ is constant. The dimension ℓ of U is called *linearity dimension* of f [15]. Bent functions exist only in even dimension, hence $n - \ell$ is even.

Let us briefly outline some properties of the autocorrelation function and the Walsh transform of partially bent functions. For more details and other properties we refer to [4].

Let us first consider the autocorrelation function $\hat{r}$ of a partially bent function f. Let $\alpha \in U$ and $\beta \in V$. Then

$$\hat{r}(\alpha + \beta) = \sum_{x \in \mathbf{F}^n} (-1)^{f(x+\alpha+\beta)+f(x)} = \sum_{y \in V} \sum_{z \in U} (-1)^{f(y+z+\alpha+\beta)+f(y+z)}$$

$$= 2^\ell \sum_{y \in V} (-1)^{f(y+\alpha+\beta)+f(y)} = 2^\ell (-1)^{f(\alpha)+f(0)} \sum_{y \in V} (-1)^{f(y+\beta)+f(y)}$$

$$= \begin{cases} 0, \text{if } \beta \neq 0, \\ (-1)^{f(\alpha)+f(0)} 2^n, \text{if } \beta = 0. \end{cases}$$

Hence the autocorrelation function of a partially bent function has the following values

$$\hat{r}(s) = \begin{cases} 0, \text{if } s \notin U, \\ (-1)^{f(s)+f(0)} 2^n, \text{if } s \in U, \end{cases} \tag{3}$$

where U is as in Definition 5. Conversely, if the autocorrelation function of a Boolean function $f : \mathbf{F}^n \to \mathbf{F}$ has only values 0 and $\pm 2^n$, then f is partially bent, which can be seen as follows. The vectors $\alpha \in \mathbf{F}^n$, for which $|\hat{r}(\alpha)| = 2^n$ are exactly those, for which $f(x + \alpha) + f(x)$ is constant. Clearly, they form a linear subspace U of $\mathbf{F}^n$ and the restriction of f to U is linear. Let V be any complementary subspace of U. Since

$$f(x + z) = f(x) + f(z) + f(0)$$

for all $x \in \mathbf{F}^n$ and $z \in U$, this holds particularly for all $y \in V$ and $z \in U$. It remains to show that the restriction of f to V is bent. Let $\beta \in V$ be not equal to zero. Then $\beta \notin U$ and thus $\hat{r}(\beta) = 0$. Consequently,

$$0 = \sum_{x \in \mathbf{F}^n} (-1)^{f(x+\beta)+f(x)} = \sum_{y \in V} \sum_{z \in U} (-1)^{f(y+z+\beta)+f(y+z)}$$

$$= \sum_{z \in U} \sum_{y \in V} (-1)^{f(y+\beta)+f(y)} = 2^\ell \sum_{y \in V} (-1)^{f(y+\beta)+f(y)}.$$

Hence $\sum_{y \in V} (-1)^{f(y+\beta)+f(y)} = 0$, for all $\beta \in V$, $\beta \neq 0$, and therefore the restriction of f to V is bent.

A quadratic Boolean function is partially bent. This follows from the fact that then the difference $f(x + \alpha) + f(x)$ is an affine function of x, for all α, and hence either constant or balanced. Therefore the autocorrelation function of a quadratic function takes only values $\pm 2^n$ and 0.

Let us now calculate the values of the Walsh transform $\widehat{F}$ of a partially bent function $f : \mathbf{F}^n \to \mathbf{F}$. By the Wiener-Khintchin theorem we get

$$\widehat{F}(w)^2 = \sum_{s \in \mathbf{F}^n} \widehat{r}(s)(-1)^{w \cdot s} = \sum_{\alpha \in U} \sum_{\beta \in V} \widehat{r}(\alpha + \beta)(-1)^{w \cdot (\alpha + \beta)}$$

$$= \sum_{\alpha \in U} (-1)^{w \cdot \alpha} = 2^n \sum_{\alpha \in U} (-1)^{f(\alpha) + f(0) + w \cdot \alpha}.$$

Recall that the restriction of f to the ℓ-dimensional linear subspace U is affine. Hence we have

$$\widehat{F}(w)^2 = \begin{cases} 2^{n+\ell}, & \text{if } f(x) + w \cdot x \text{ is constant on } U, \\ 0, & \text{if } f(x) + w \cdot x \text{ is not constant on } U. \end{cases}$$

It follows that the linearity of a partially bent function $f : \mathbf{F}^n \to \mathbf{F}$ is

$$\mathcal{L}(f) = 2^{\frac{\ell - n}{2}},$$

where ℓ is the linearity dimension of f. We also see that f is balanced, i.e. $\widehat{F}(0) = 0$, if and only if the restriction of f is a nonconstant affine function on U, or equivalently, if and only if f has a linear structure $\alpha \in \mathbf{F}^n$ such that $f(x + \alpha) + f(x) = 1$ for all $x \in \mathbf{F}^n$.

4.2 Functions with Partially Bent Components

The purpose of this section is to discuss some basic properties of functions with partially bent components. We also precisize and improve some results from [18], [22] and [23] and simplify their proofs. Examples of functions with partially bent components are offered by the power functions $f(x) = x^{2^k+1}$, $x \in GF(2^n)$, the components of which are quadratic [18].

For a function $f : \mathbf{F}^n \to \mathbf{F}^m$ and vectors $a \in \mathbf{F}^n$, $a \neq 0$, and $b \in \mathbf{F}^m$, we make the following notation.

$$\delta_f(a, b) = \#\{x \in \mathbf{F}^n \mid f(x + a) + f(x) = b\}$$
$$\widehat{r}_f(a, b) = \sum_{x \in \mathbf{F}^n} (-1)^{b \cdot f(x+a) + b \cdot f(x)}.$$

Then (see also [6])

$$\sum_{c \in \mathbf{F}^m} \widehat{r}_f(a, c)(-1)^{c \cdot b} = \sum_{x \in \mathbf{F}^n} \sum_{c \in \mathbf{F}^m} (-1)^{c \cdot f(x+a) + c \cdot f(x) + c \cdot b}$$
$$= 2^m \#\{x \in \mathbf{F}^n \mid f(x + a) + f(x) + b = 0\} = 2^m \delta_f(a, b).$$

Applying the inverse Walsh-Hadamard transform we get

$$\widehat{r}_f(a, c) = \sum_{b \in \mathbf{F}^m} \delta_f(a, b)(-1)^{b \cdot c},$$

and further,

$$\sum_{c \in \mathbf{F}^m} \widehat{r}_f(a, c)^2 = 2^m \sum_{b \in \mathbf{F}^m} \delta_f(a, b)^2. \tag{4}$$

Let us now focus on a special case where $\delta_f(a, b)$ takes at most two values, and the second value (if it exists) is zero. This property is a generalization of perfect nonlinearity and almost perfect nonlinearity, and was introduced and studied in [23].

Lemma 6. Let $f : \mathbf{F}^n \to \mathbf{F}^m$ be a function and $a \in \mathbf{F}^n$, $a \neq 0$. Let us assume that there is a $\delta > 0$ such that, for all $b \in \mathbf{F}^m$, $\delta_f(a, b) = 0$ or δ. Then

$$\sum_{b \in \mathbf{F}^m} \widehat{r}_f(a, b)^2 = \delta 2^{n+m}.$$

Proof. Since $\sum_{b \in \mathbf{F}^m} \delta_f(a, b) = 2^n$, the claim follows directly from (4).   □

Lemma 7. Let $f : \mathbf{F}^n \to \mathbf{F}^m$ be a function with partially bent components and $a \neq 0$. Let us assume that there is $\delta > 0$ such that for all $b \in \mathbf{F}^m$, $\delta_f(a, b) = 0$ or δ. Then δ is a power of 2, say $\delta = 2^{n-m+t_a}$, $n - m + t_a \geq 1$, and

$$\sum_{b \in \mathbf{F}^m} \widehat{r}_f(a, b)^2 = 2^{2n+t_a}. \tag{5}$$

Moreover, a is a linear structure of exactly $2^{t_a} - 1$ components of f.

Proof. The vectors $c \in \mathbf{F}^m$ such that $c \cdot f(x + a) + c \cdot f(x)$ is constant form a linear subspace of $\mathbf{F}^m$. Let t_a be the dimension of this subspace. Then by (3) and Lemma 6

$$\sum_{b \in \mathbf{F}^m} \widehat{r}_f(a, b)^2 = 2^{n-m+t_a} 2^{n+m} = 2^{2n+t_a}.$$

□

Theorem 8. Let $f : \mathbf{F}^n \to \mathbf{F}^m$ have partially bent components and $\ell_b \geq 0$ be the linearity dimension of the component $b \cdot f$. If there is a $\delta > 0$ such that $\delta_f(a, b) = 0$ or δ, for all $a \in \mathbf{F}^n$, $a \neq 0$, and for all $b \in \mathbf{F}^m$, then there is $t \geq 0$ such that $\delta = 2^{n-m+t}$ and

$$\sum_{b \neq 0} (2^{\ell_b} - 1) = (2^n - 1)(2^t - 1). \tag{6}$$

Proof. It follows from the assumption and Lemma 7, that (5) holds with $t_a = t$, for all $a \neq 0$. Then by (5)

$$\sum_{b \neq 0} \widehat{r}_f(a, b)^2 = 2^{2n+t} - 2^{2n} = 2^n(2^t - 1),$$

for all $a \neq 0$. By summing this up over $a \neq 0$ we get

$$2^{2n} \sum_{b \neq 0} (2^{\ell_b} - 1) = \sum_{b \neq 0} \sum_{a \neq 0} \widehat{r}_f(a, b)^2$$

$$= \sum_{a \neq 0} \sum_{b \neq 0} \widehat{r}_f(a, b)^2 = 2^{2n}(2^t - 1)(2^n - 1).$$

□

We are using the same notation for n and t as in [23], Section 2.2., and our m corresponds to their s. Hence we can see from (6) that the corresponding unproven formula in [23] is incorrect. Consequently, Theorem 2 of [23] remains unproven. From Theorem 8 we get the following corollary.

Corollary 9. *Assume that n is odd. If there exists a function $f : \mathbf{F}^n \to \mathbf{F}^m$ with $\delta_f(a, b) = 0$ or 2^{n-m+t}, for all $a \neq 0$ and b, then $t \geq 1$, and t and m have the same parity.*

Proof. It follows from (6) that

$$\sum_{b \neq 0} (2^{\ell_b} + 1) = (2^n - 1)(2^t - 1) + 2(2^m - 1).$$

Since n is odd, all ℓ_b are odd, and the left hand side is divisible by 3 while $2^n - 1$ is not. Consequently, 3 divides $2^t - 1$ if and only if 3 divides $2^m - 1$.   □

In the case of odd n and $t = 1$ the equation (6) has exactly one solution, that is, $\ell_b = 1$, for all $b \neq 0$. From this and (4) we get the following result.

Theorem 10. *Let $f : \mathbf{F}^n \to \mathbf{F}^n$ be an almost perfect nonlinear function with partially bent components and n odd. Then each component of f has exactly one nonzero linear structure and the nonzero linear structures of different components are distinct.*

Conversely, if each $a \neq 0$ is a linear structure of exactly one component of a function f with partially bent components, and $m = n$, then by (4)

$$\sum_{b \in \mathbf{F}^m} \delta_f(a, b)^2 = 2^{n+1},$$

for all $a \neq 0$. Since $\sum_{b \in \mathbf{F}^n} \delta_f(a, b) = 2^n$ it follows that $\delta_f(a, b) = 0$ or 2 for all $a \neq 0$ and for all b, that is, f is almost perfect nonlinear.

The case $n - m + t = 1$ was considered in [23] for functions with balanced quadratic components. Based on Theorem 8 we can replace 'quadratic' by 'partially bent'. Particularly, we see that there is no APN permutation with partially bent components in even dimension.

In general, it is not known whether APN functions $f : \mathbf{F}^n \to \mathbf{F}^n$ with partially bent components exist in even dimension, except for $n = 2$, where for example, $f = (f_1, f_2)$,

$$f_1(x_1, x_2) = x_1 x_2$$
$$f_2(x_1, x_2) = x_1$$

is clearly an APN function.

For $n = m$ even, and $t = 1$, we get from (6) that the number of bent components is at least $\frac{2}{3}(2^n - 1)$. One solution of (6) is that $\ell_b = 0$ for $\frac{2}{3}(2^n - 1)$ components and $\ell_b = 2$ for $\frac{1}{3}(2^n - 1)$ components. Do such functions exist remains an open problem for $n \geq 4$. The other extreme solution of (6) is that $\ell_b = n$ for one component, which is linear, and $\ell_b = 0$ for all other components. The existence of such functions would imply the existence of a bent function from $\mathbf{F}^n$ to $\mathbf{F}^{n-1}$ which is not possible for $n > 2$.

5   Affine Surjections onto the Input Space

Let $A = L + a : \mathbf{F}^s \to \mathbf{F}^n$ be an affine surjection, where L is a linear surjection. When composed with a function $f : \mathbf{F}^n \to \mathbf{F}^m$ the linearity and differential uniformity are as follows.

Theorem 11.

1. $\mathcal{L}(f \circ A) = \mathcal{L}(f)$
2. $\Delta(f \circ A) = \begin{cases} 2^s, & \text{if } s > n, \text{and} \\ \Delta(f), & \text{if } s = n. \end{cases}$

Proof.

1. It suffices to prove the claim for the components of f. Hence we may assume that $m = 1$. We denote the zero space of L by $\mathrm{Ker}L$. Let V be an n- dimensional linear subspace of $\mathbf{F}^s$ such that $\mathbf{F}^s = V \oplus \mathrm{Ker}L$. Then the restriction of A to V is an affine bijection and every $x \in \mathbf{F}^s$ has a unique representation in the form $x = y + z$, where $y \in V$ and $z \in \mathrm{Ker}L$. Let us denote the Walsh transform of f and $f \circ A$ by $\widehat{F}$ and $\widehat{G}$, respectively. Let $w \in \mathbf{F}^s$ be arbitrary. Then

$$\widehat{G}(w) = \sum_{x \in \mathbf{F}^s} (-1)^{f(Lx+a)+w \cdot x} = \sum_{y \in V} (-1)^{f(Ly+a)+w \cdot y} \sum_{z \in \mathrm{Ker}L} (-1)^{w \cdot z}$$

$$= \begin{cases} 2^{s-n} \sum_{y \in V} (-1)^{f(Ly+a)+w \cdot y}, & \text{if } w \cdot z = 0, \text{ for all } z \in \mathrm{Ker}L, \\ 0, & \text{otherwise.} \end{cases}$$

Hence if $\widehat{G}(w) \neq 0$, then $w \cdot z = 0$, for all $z \in \mathrm{Ker}L$. In this case there is a unique $u \in \mathbf{F}^n$ such that $w \cdot x = u \cdot Lx = L^t \cdot x$, for all $x \in \mathbf{F}^s$, where we denote the transpose of L by L^t. This means that w has a unique representation in the form $L^t u$, where $u \in \mathbf{F}^n$, and we have

$$\widehat{G}(w) = 2^{s-n}(-1)^{u \cdot a} \sum_{y \in V} (-1)^{f(Ly+a)+u \cdot (Ly+a)} = 2^{s-n}(-1)^{a \cdot u} \widehat{F}(u).$$

So we have shown that either $\widehat{G}(w) = 0$, or $w = L^t u$ and $|\widehat{G}(w)| = 2^{s-n}|\widehat{F}(u)|$. This proves the first claim.

2. If $s > n$, then there exists an $\alpha \in \mathbf{F}^s$, $\alpha \neq 0$, such that $A(x + \alpha) = A(x)$, for all $x \in \mathbf{F}^s$. Hence $\Delta(f \circ A) = 2^s$. If $s = n$, then for all $\alpha \neq 0$ and for all β

$$
\begin{aligned}
&\#\{x \in \mathbf{F}^s \mid f(A(x + \alpha)) + f(A(x)) = \beta\} \\
&= \#\{x \in \mathbf{F}^s \mid f(A(x) + A(\alpha) + a) + f(A(x)) = \beta\} \\
&\leq \Delta(f),
\end{aligned}
$$

and the equality is achieved with a suitable choice of α, since in this case, A is bijective. $\qquad\square$

The pitfalls in S-box construction presented in Sections 6.1. and 6.4 of [21] are special cases of the preceding theorem. As far as known to this author affine enlargements of the input space have never been used in the design of S-boxes.

6   Affine Surjections from the Output Space

Let $A = L + a : \mathbf{F}^m \to \mathbf{F}^s$ be an affine surjection, where L is a linear surjection, $a \in \mathbf{F}^s$ and $s \leq m$. When composed with a function $f : \mathbf{F}^n \to \mathbf{F}^m$ the linearity and differential uniformity are as follows.

Theorem 12.

1. $\mathcal{L}(A \circ f) \leq \mathcal{L}(f)$, *with equality if* $s = m$.
2. $\Delta(f) \leq \Delta(A \circ f) \leq 2^{m-s} \Delta(f)$.

Proof.
1. The components of $A \circ f$ form a subset of the components of f plus some constants. More precisely,

$$
b \cdot (A \circ f) = (L^t b) \cdot f + b \cdot a.
$$

Hence the claim is true and holds with equality, if L is bijective.

2. For $\beta \in \mathbf{F}^s$, let $B = A^{-1}\{\beta + a\}$ denote the preimage set of $\beta + a$ under A. Then

$$
\begin{aligned}
&\#\{x \in \mathbf{F}^n \mid A(f(x + \alpha)) + A(f(x)) = \beta\} \\
&= \sum_{b \in B} \#\{x \in \mathbf{F}^n \mid f(x + \alpha) + f(x) = b\}.
\end{aligned}
$$

Since the cardinality of B equals 2^{m-s}, the claim follows. $\qquad\square$

6.1   The S-Boxes of MacGuffin

Deletion of output bits of an S-box is a special case of a surjection applied to the output space of a substitution box. By the preceeding theorem the linearity may decrease while the differential uniformity may increase when output bits are deleted. A recent example of this phenomenon is offered by the MacGuffin block cipher algorithm [3], which makes use of the S-boxes of DES, chopped to the half, i.e., from the original four output bits two are deleted. An analysis of this cipher performed by Rijmen and Preneel [20] shows that linear cryptanalysis of the MacGuffin cipher is about as hard as it is for the DES, but the MacGuffin cipher is slightly weaker against differential cryptanalysis.

However, chopping S-boxes does not always result in a decreased security level. In [18] an example of a DES-like cipher construction is given where the nonlinear substitution function constitutes of one large substitution box, constructed from an almost perfect nonlinear permutation in odd dimension, say 33, by deleting one output bit. If such a cipher has independent round keys and sufficiently many rounds, so that differentials over at least four rounds need to be considered in differential cryptanalysis, then the differential attack is proven to be in average as hard as exhaustive key search [18].

6.2   Chopping of Bent and APN Functions

Let us first consider a bent function $f : \mathbf{F}^n \to \mathbf{F}^m$, $2m \le n$, n even. It follows immediately from the definition of bent functions that chopping t, $0 \le t < m$ output coordinates results in increase of differential uniformity by a factor of exactly 2^t, that is, the upperbound of the theorem is reached. However, the chopped function remains perfect nonlinear and bent .

A second example of a function, that preserves linearity and increases differential uniformity by a factor of 2^t if t output bits are deleted, is an APN function with partially bent components in odd dimension.

Theorem 13. *Let $f : \mathbf{F}^n \to \mathbf{F}^n$ be an almost perfect nonlinear function and n odd. Then all components of f are partially bent if and only if $\Delta(A \circ f) = 2^{n-s}\Delta(f)$ for all affine surjections $A : \mathbf{F}^n \to \mathbf{F}^s$ and for all s, $1 \le s \le n$.*

Proof. Let us assume first, that all components of f are partially bent. The components of $A \circ f$ form a subset of $2^s - 1$ components of f plus 0 or 1.

Let $\alpha \in \mathbf{F}^n$ be an arbitrary nonzero vector. If α is not a linear structure of any of these components, then $b \cdot ((A \circ f)(x + \alpha)(A \circ f)(x))$ is balanced for all $b \in \mathbf{F}^s$, $b \ne 0$. Therefore (see Appendix)

$$\#\{x \in \mathbf{F}^n \,|\, (A \circ f)(x + \alpha) + (A \circ f)(x) = \beta\} = 2^{n-s}$$

for all $\beta \in \mathbf{F}^s$. Note that by Theorem 10 there are $2^n - 2^s$ such α. The other $2^s - 1$ vectors are the linear structures of the components of $A \circ f$.

If α is a linear structure of a component, say g_1 of $A \circ f$, then there are $s - 1$ components $g_2, \dots, g_s$ of $A \circ f$ such that the vector equation

$$(A \circ f)(x + \alpha) + (A \circ f)(x) = \beta$$

is a linear transformation of the system

$$g_i(x + \alpha) + g_i(x) = \gamma_i, \quad i = 1, 2, \dots, s. \tag{7}$$

By Theorem 10 α is not a linear structure of any of $g_2, \ldots, g_s$. Hence $g_i(x + \alpha) + g_i(x)$ is a balanced function of x for all $i = 2, 3, \ldots, s$. Consequently (see Appendix), the number of solutions of (7) is either 2^{n-s+1} or zero. So we have proved that $\Delta(A \circ f) = 2^{n-s+1} = 2^{n-s} \Delta(f)$.

To prove the converse, let us assume that $\Delta(A \circ f) = 2^{n-s} \Delta(f)$ for all affine surjections $A : \mathbf{F}^n \to \mathbf{F}^s$ and for all s, $1 \leq s \leq n$. In fact, we need this only for $s = 1$ and $s = 2$. Applying the assumption in the case where the dimension of the output space of A is one, we get that $\Delta(A \circ f) = 2^n$, which means that every component of f has a linear structure. By Lemma 6 the linear structure is unique for each component.

Let f_0 be an arbitrary component of f. It suffices to show that $f_0(x+\alpha)+f_0(x)$ is a balanced function of x if α is not a linear structure of f_0. Let f_α be the component of f whose linear structure is α. Then by the assumption the system

$$f_0(x + \alpha) + f_0(x) = \beta_0$$
$$f_\alpha(x + \alpha) + f_\alpha(x) = \beta_\alpha$$

has at most 2^{n-1} solutions. Since the second equation holds either never or always, depending only on the value of β_α, it follows that the first equation has always 2^{n-1} solutions, that is, $f_0(x + \alpha) + f_0(x)$ is balanced. Therefore, f_0 is partially bent. $\square$

In the view of this theorem we might extend the definition of APN function to the case $m < n$ saying that a function $f : \mathbf{F}^n \to \mathbf{F}^m$, $m \leq n$, is almost perfect nonlinear, if $\Delta(f) \leq 2^{n-m+1}$.

7   Affine Bijections to the Input Space and from the Output Space

As a corollary of Theorems 11 and 12 we get the following (see also [15] and [16]).

Corollary 14. *Let $f : \mathbf{F}^n \to \mathbf{F}^m$ be a function and let $A : \mathbf{F}^m \to \mathbf{F}^m$ and $B : \mathbf{F}^n \to \mathbf{F}^n$ be linear (or affine) bijections. Then*
  1. $\mathcal{L}(A \circ f \circ B) = \mathcal{L}(f)$,
  2. $\Delta(A \circ f \circ B) = \Delta(f)$.

8   Inverted Function

The following results were given in [15] and [16] but the proofs were omitted. We take this opportunity to present the simple proofs.

Theorem 15. *If a function $f : \mathbf{F}^n \to \mathbf{F}^n$ is invertible then*
  1. $\mathcal{L}(f^{-1}) = \mathcal{L}(f)$,
  2. $\Delta(f^{-1}) = \Delta(f)$.

Proof.

1. Let $\widehat{F}_b$ and $\widehat{G}_c$ be the Walsh transforms of $b \cdot f$, $b \neq 0$, and $c \cdot f^{-1}$, $c \neq 0$, respectively. Since f is bijective, we have $\widehat{F}_b(0) = \widehat{G}_c(0) = 0$ for all $b \neq 0$ and $c \neq 0$. Hence it suffices to consider the values of $\widehat{F}_b$ and $\widehat{G}_c$ outside the zero. Let b and c be nonzero vectors in $\mathbf{F}^n$. The claim follows from the following equality.

$$\widehat{F}_b(c) = \sum_{x \in \mathbf{F}^n} (-1)^{b \cdot f(x) + c \cdot x} = \sum_{y \in \mathbf{F}^n} (-1)^{b \cdot y + c \cdot f^{-1}(y)} = \widehat{G}_c(b).$$

2. Since f is a permutation, we have $\#\{x \in \mathbf{F}^n \mid f(x+\alpha) + f(x) = 0\} = 0$, for all $\alpha \in \mathbf{F}^n$, $\alpha \neq 0$. Further, $f(x+\alpha) + f(x) = \beta$ if and only if $f(f^{-1}(y)+\alpha) = y+\beta$, or what is equivalent, $f^{-1}(y + \beta) + f^{-1}(y) = \alpha$, for all $\alpha \neq 0$ and $\beta \neq 0$. This proves the second claim.   □

9   Adding Coordinate Functions

Given a function $f : \mathbf{F}^n \rightarrow \mathbf{F}^m$ with coordinate functions $f_1, \ldots, f_m$ and a function $g : \mathbf{F}^n \rightarrow \mathbf{F}$, we set $\tilde{f} = (f_1, \ldots, f_m, g)$. As a corollary of Theorem 12 we get the following.

Theorem 16.

  1. $\mathcal{L}(\tilde{f}) \geq max \{\mathcal{L}(f), \mathcal{L}(g)\} \geq \mathcal{L}(f)$
  2. $\Delta(f) \geq \Delta(\tilde{f}) \geq \dfrac{1}{2}\Delta(f)$.

This method has been previously used in the CAST algorithm [2], in which S-boxes of 8 input bits and 32 output bits are constructed by selecting 32 bent Boolean functions in $\mathbf{F}^8$ as coordinate functions. It is exactly as hard for the designer to prove upperbounds to the linearity of such S-box as it is to the cryptanalyst to find the best linear approximation. In [8] the probability that the linearity of such an $m \times n$ S-box is below a given bound is estimated under the assumption *"that the 2^n functions determined from all linear combinations of the n output functions of an S-box may be considered independently in an analysis of their nonlinearities and the probability distribution of the nonlinearity of each function is the same as that of a randomly generated function"*. The estimated probabilities are encouragingly large, but the relevance of the assumption about independence remains an open problem.

10   Direct Sum of Functions

The full substitution function of the CAST algorithm takes 32 input bits and outputs 32 bits, and is constructed by forming the direct sum of four 8×32 S-boxes. The design method of the S-boxes was discussed in the previous section. In this section we discuss the linearity and differential uniformity of direct sums of functions.

Let $f_1 : \mathbf{F}^{n_1} \to \mathbf{F}^m$ and $f_2 : \mathbf{F}^{n_2} \to \mathbf{F}^m$ be two functions. The *direct sum* of f_1 and f_2 is the function $f : \mathbf{F}^{n_1} \times \mathbf{F}^{n_2} \to \mathbf{F}^m$, $f(x, y) = f_1(x) + f_2(y)$. We denote $f = f_1 + f_2$.

Theorem 17.

   1. $\mathcal{L}(f_1 + f_2) \leq \mathcal{L}(f_1)\mathcal{L}(f_2)$,

   2. $\Delta(f_1 + f_2) \leq \min \{2^{n_2}\Delta(f_1), 2^{n_1}\Delta(f_2), 2^m \Delta(f_1)\Delta(f_2)\}$.

Note that if $m = 1$ then 1. is satisfied with equality and is the same as what sometimes is called the "piling-up lemma".
Proof.
1. Let $b \in \mathbf{F}^m$ be nonzero and let us denote the Walsh transform of $b \cdot (f_1 + f_2)$, $b \cdot f_1$ and $b \cdot f_2$ by $\widehat{F_b}$, $\widehat{G_b}$ and $\widehat{G'_b}$, respectively. It is well known and easy to check that $\widehat{F_b}(u, v) = \widehat{G_b}(u)\widehat{G'_b}(v)$, for all $u \in \mathbf{F}^{n_1}$ and $v \in \mathbf{F}^{n_2}$.
2. Let $\beta \in \mathbf{F}^{n_1}$ and $\gamma \in \mathbf{F}^{n_2}$ be nonzero, and let $\delta \in \mathbf{F}^m$. Then

$$\#\{(y, z) \mid f_1(y + \beta) + f_1(y) + f_2(z + \gamma) + f_2(z) = \delta\}$$
$$= \sum_{z \in \mathbf{F}^{n_2}} \#\{y \in \mathbf{F}^{n_1} \mid f_1(y + \beta) + f_1(y) = f_2(z + \gamma) + f_2(z) + \delta\}$$
$$\leq \sum_{z \in \mathbf{F}^{n_2}} \Delta(f_1) = 2^{n_2}\Delta(f_1).$$

This gives the first upper bound. The second is obtained from this by changing the roles of f_1 and f_2. We get the third upperbound as follows.

$$\#\{(y, z) \mid f_1(y + \beta) + f_1(y) + f_2(z + \gamma) + f_2(z) = \delta\}$$
$$= \sum_{b \in \mathbf{F}^m} \#\{y \in \mathbf{F}^{n_1} \mid f_1(y + \beta) + f_1(y) = b\}$$
$$\times \#\{z \in \mathbf{F}^{n_2} \mid f_2(z + \gamma) + f_2(z) = \delta + b\}$$
$$\leq 2^m \Delta(f_1)\Delta(f_2).$$

$\square$

If f_1 and f_2 are bent functions, then their direct sum is a bent function and $\mathcal{L}(f_1 + f_2) = \mathcal{L}(f_1)\mathcal{L}(f_2)$. Morever, all three upperbounds in 2. are reached and are hence equal. Note that in this case m is small compared to n_1 and n_2.

With the CAST algorithm the situation is different. The round function is a direct sum of four S-boxes $f_i : \mathbf{F}^{n_i} \to \mathbf{F}^m$, $i = 1, 2, 3, 4$. Since $m = n_1 + n_2 + n_3 + n_4$, the third upperbound can never be reached. Therefore

$$\Delta(f_1 + f_2 + f_3 + f_4) \leq \min_i 2^{\sum_{j \neq i} n_j} \Delta(f_i).$$

Hence the upperbound only depends of the S-box with least differential uniformity. With the parameters $n_i = 8$ and $m = 32$ this gives the upperbound of 2^{-7} to the probability of the most likely one round differential (characteristic), assuming that the best S-box is differentially 2-uniform.

However, it does not seem likely that the differential uniformity of the round function is as high as indicated by Theorem 17. The reason is the large number of zero entries on the rows of the difference distribution tables. From the proof of the theorem we see that the upperbound is reached if there is a permutation of $\mathbf{F}^m$ originating from a translation (by the vector δ, see above) such that after permuting the columns of the difference distribution table of one function there is a row in this difference distribution table which has nonzero entries exactly at the same locations as some row in the difference distribution table of the second function. This kind of coincidence may be rare. More generally, it might be possible to estimate the expected differential uniformity of the CAST f-function by estimating the expected number of coincidences of locations of nonzero entries.

11   Restrictions to Linear (or Affine) Subspaces

Given a function $f : \mathbf{F}^n \to \mathbf{F}^m$ let $g : \mathbf{F}^s \to \mathbf{F}^m$ be the restriction of f to an s-dimensional affine subspace $a + V$.

Let first $m = 1$. By the linearity $\mathcal{L}(g)$ of the restriction g of a Boolean function f to $a + V$ we mean the maximum value taken over all $w \in \mathbf{F}^n$ of

$$2^{-s} |\#\{x \in a + V \mid f(x) = w \cdot x\} - \#\{x \in a + V \mid f(x) \neq w \cdot x\}|$$

$$= 2^{-s} | \sum_{x \in a+V} (-1)^{f(x)+w \cdot x}| = 2|\mathrm{Pr}_{X \in a+V}\,(f(X) = w \cdot X) - \frac{1}{2}|.$$

For $m > 1$ we set $\mathcal{L}(g) = \max_{b \neq 0} \mathcal{L}(b \cdot g)$.

Theorem 18.

1. $\mathcal{L}(g) \leq 2^{n-s} \mathcal{L}(f)\}$,
2. $\Delta(g) \leq \Delta(f)$.

Proof.
1. Since the components of g are restrictions of the components of f, it suffices to prove the claim for Boolean functions. Let $\widehat{F}$ be the Walsh transform of f. Then taking the Walsh-Hadamerd transform of $\widehat{F}$ we have

$$(-1)^{f(x)} = 2^{-n} \sum_{t \in \mathbf{F}^n} \widehat{F}(t)(-1)^{t \cdot x}.$$

Using this we get

$$\sum_{x \in a+V} (-1)^{f(x)+w \cdot x} = \sum_{x \in V}(-1)^{f(x+a)+w \cdot x+w \cdot a}$$

$$= 2^{-n} \sum_{x \in V} \sum_{t \in \mathbf{F}^n} \widehat{F}(t)(-1)^{t \cdot (x+a)}(-1)^{w \cdot x+w \cdot a}$$

$$= 2^{-n}(-1)^{w \cdot a} \sum_{t \in \mathbf{F}^n} \widehat{F}(t)(-1)^{t \cdot a} \sum_{x \in V}(-1)^{(t+w) \cdot x}$$

$$= 2^{s-n}(-1)^{w \cdot a} \sum_{t \in w+V^\perp} \widehat{F}(t)(-1)^{t \cdot a}$$

where $V^\perp$ is the orthogonal subspace of V formed by $v \in \mathbf{F}^n$ such that $v \cdot x = 0$ for all $x \in V$. Then the dimension of $V^\perp$ is $n - s$. Consequently,

$$| \sum_{x \in a+V} (-1)^{f(x)+w \cdot x}| \le 2^{s-n} \sum_{t \in w+V^\perp} |\widehat{F}(t)| \le \max_{t \in \mathbf{F}^n} |\widehat{F}(t)|,$$

which proves the claim.

2. The proof of the second claim follows directly from the definition of differential uniformity. □

Again, bent functions offer examples of functions satisfying the equality. Let us consider the bent function

$$f(x_1, \ldots, x_{2s}) = x_1 x_{s+1} + \ldots + x_s x_{2s}.$$

Then $\mathcal{L}(f) = 2^{-s}$, and the linearity of the restricted function to the s-dimensional subspace $x_1 = x_2 = \ldots = x_s = 0$ is equal to $1 = 2^{2s-s}\mathcal{L}(f)$.

Restricted functions occur in DES-like ciphers, where the input data to the round is first expanded, then added to the round key, and then taken as input to the substitution function. Let $f : \mathbf{F}^n \to \mathbf{F}^m$ be the substitution function of a DES- like cipher with nonlinearity $\mathcal{NL}(f)$. Let $E : \mathbf{F}^m \to \mathbf{F}^n$ be a linear expansion mapping. Let k be a fixed round key, and we denote by V_k the affine subspace of $\mathbf{F}^n$ consisting of elements of the form $E(x) + k$, $x \in \mathbf{F}^m$ and by $f|_k$ the restriction of f to V_k. Then

$$| \Pr_X \{b \cdot f(E(X) + k) = a \cdot X\} - \frac{1}{2}| \le \frac{1}{2}\mathcal{L}(f|_k) \le \frac{1}{2}2^{n-m}\mathcal{L}(f)$$
$$= \frac{2^{n-1} - \mathcal{NL}(f)}{2^m},$$

to replace an unproven formula in [11], page 152, by a correct one.

12   Examples

Applying the results discussed above let us first show that for all $n < m$ there exists a differentially 2-uniform function $f : \mathbf{F}^n \to \mathbf{F}^m$. If n is odd we can take any APN function from $\mathbf{F}^n$ to $\mathbf{F}^n$ and add sufficiently many new coordinate functions. Then the differential uniformity can only decrease, even if the new coordinates were linear or the same as old components. If n is even, we start with an APN function from $\mathbf{F}^{n+1}$ to $\mathbf{F}^{n+1}$, restrict it to $\mathbf{F}^n$, and then add new coordinate functions if necessary.

The second example is a function $g : \mathbf{F}^n \to \mathbf{F}^n$ such that $\mathcal{L}(g)$ is low but g has a linear structure. Let us start with any function $f : \mathbf{F}^n \to \mathbf{F}^n$ such that $\mathcal{L}(f)$ is small. We denote by $\tilde{f}$ a modification of f which is obtained by deleting one input coordinate. Then $\mathcal{L}(\tilde{f}) \leq 2\mathcal{L}(f)$. By composing $\tilde{f}$ with a linear projection $L : \mathbf{F}^n \to \mathbf{F}^{n-1}$ we get a function $g = \tilde{f} \circ L$, such that

$$\mathcal{L}(g) = \mathcal{L}(\tilde{f} \circ L) \leq \mathcal{L}(\tilde{f}) \leq 2\mathcal{L}(f)$$
$$\Delta(g) = 2^n.$$

Since $\mathcal{L}(g)$ is small, a round function of a DES-like cipher can be based on g to guarantee proven resistance against linear attacks [17]. But as it is easy to see, such a cipher has an iterative characteristic with probability 1, that is, a linear structure over the whole cipher, which can be exploited to reduce the complexity of exhaustive key search by a factor of 2.

It is not any harder to give an opposite example, that is, a function $g : \mathbf{F}^n \to \mathbf{F}^n$ such that $\mathcal{L}(g) = 1$ and $\Delta(g) = 4$. Let us start with any function $f : \mathbf{F}^n \to \mathbf{F}^n$ such that $\Delta(f)$ is small. We denote by $\tilde{f}$ a modification of f which is obtained by deleting one output coordinate. Then $\Delta(\tilde{f}) \leq 2\Delta(f)$. By replacing the deleted component by the all zero Boolean function, we get a function g such that

$$\mathcal{L}(g) = 1 \text{ and } \Delta(g) \leq 2\Delta(f).$$

Since $\Delta(g)$ is small, a round function of a DES-like cipher can be based on g to guarantee proven resistance against differential attacks [18]. But as it is easy to see, such a cipher has an iterative linear approximation over all rounds of the cipher with probability 1, which can be exploited to determine one bit of the unknown key.

Without going into the details let us mention that it is possible to modify the first example in such a way that the linearity does not increase significantly while the probability of the one-round differential to be iterated is strictly less than one, but is still large enough to give a substantial differential over all but the last round. Then the differential cryptanalysis method can be exploited to search for the last round key exhaustively. Note that if the last round differential holds with probability 1, then there is no way to make distinction between wrong and correct candidates for the last round key.

A similar modification of the second example gives a round function of a DES-like cipher, which is resistant against differential cryptanalysis, but where the last round key can be determined by the linear cryptanalysis method.

References

1. C. Adams and S. E. Tavares, *The structured design of cryptographically good S-boxes*, Journal of Cryptology **3**, 1, 1990, pp. 27-42.
2. C. Adams and S. E. Tavares, *Designing S-boxes for ciphers resistant to differential cryptanalysis*, Proceedings of SPRC'93, Fondazione Ugo Bòrdoni, 1993.
3. M. Blaze and B. Schneier, *The MacGuffin block cipher algorithm*, these proceedings, pp. 97–110.
4. C. Carlet, *Partially-bent functions*, Advances in Cryptology – CRYPTO'92, Lecture Notes in Computer Science, Springer-Verlag, 1993.
5. F. Chabaud and S. Vaudenay, *Links between differential and linear cryptanalysis*, Proceedings of Eurocrypt'94 (to appear).

6. J. Daemen, *Correlation matrices*, these proceedings, pp. 275–285.
7. H. Dobbertin, *Construction of bent functions and balanced Boolean functions with high nonlinearity*, these proceedings, pp. 61–74.
8. H. M. Heys and S. E. Tavares, *On the security of the CAST encryption algorithm*, to appear in the proceedings of Canadian Conference on Computer and Electrical Engineering, Halifax, September 1994.
9. H. M. Heys and S. E. Tavares, *Substitution-permutation networks resistant to differential and linear cryptanalysis*, 2nd ACM CCCS, Fairfax, Virginia, November 1994.
10. T. Kasami, *Weight enumerators for several classes of the 2nd order binary Reed-Muller codes*, Information and Control 18, 1971.
11. L. Ramkilde Knudsen, *Block ciphers – analysis, design and applications*, Ph.D. thesis, DAIMI PB – 485, November 1994.
12. R. Lidl and H. Niederreiter, "Finite Fields", Encyclopedia of Mathematics and its applications **20**, Addison-Wesley, Reading, Massachusetts, 1983.
13. W. Meier and O. Staffelbach, *Nonlinearity criteria for cryptographic functions*, Advances in Cryptology – EUROCRYPT'89, Lecture Notes in Computer Science, Springer-Verlag (1990), pp. 549–562.
14. K. Nyberg, *Perfect nonlinear S-boxes*, Advances in Cryptology – EUROCRYPT '91, Lecture Notes in Computer Science 547, Springer-Verlag (1991), pp. 378–385.
15. K. Nyberg, *On the construction of highly nonlinear permutations*, Advances in Cryptology – EUROCRYPT '92, Lecture Notes in Computer Science 658, Springer-Verlag (1993), pp. 92–98.
16. K. Nyberg, *Differentially uniform mappings for cryptography*, Advances in Cryptology – EUROCRYPT '93, Lecture Notes in Computer Science 765, Springer-Verlag (1994), pp. 55–64.
17. K. Nyberg, *Linear approximation of block ciphers*, Proceedings of Eurocrypt'94 (to appear).
18. K. Nyberg and L. R. Knudsen, *Provable security against a differential attack*, to appear in J. Crypt. 8, No. 1, 1995 (preliminary version: *Proven security against differential cryptanalysis*, Proceedings of Crypto'92).
19. N. J. Patterson and D. H. Wiedemann, *The covering radius of the* $(2^{15}, 16)$ *Reed-Müller code it at least 16276*, IEEE Trans. on Information Theory **29** (1983), pp. 354–356.
20. V. Rijmen and B. Preneel, *Cryptanalysis of MacGuffin*, these proceedings, pp. 353–358.
21. J. Seberry, X.-M. Zhang and Y. Zheng, *Nonlinearity and propagation characteristics of balanced Boolean functions*, Information and Computation (to appear).
22. J. Seberry, X.-M. Zhang and Y. Zheng, *Nonlinearity characteristics of quadratic substitution boxes*, Proceedings of the Workshop on Selected Areas in Cryptography (SAC '94), May 5-6, 1994, Kingston, Canada. To appear under the title *Relationships among nonlinearity criteria* in the proceedings of EUROCRYPT'94.
23. J. Seberry, X.-M. Zhang and Y. Zheng, *Pitfalls in designing substitution boxes*, Advances in Cryptology – CRYPTO'94, Lecture Notes in Computer Science 839, Springer-Verlag (1994), pp. 383–396.

Appendix: On the Distribution of Values of a Vector Boolean Function

It is well known that a function $f = (f_1, \ldots, f_m) : \mathbf{F}^n \to \mathbf{F}^m$, $1 \leq m \leq n$, over any finite field $\mathbf{F}$ of order q takes all values in $\mathbf{F}^m$ equally many, i.e., q^{n-m} times, if and only if each component $b \cdot f$, $b \neq 0$, takes each value in $\mathbf{F}$ equally many times (see e.g. [12]). In this appendix we give a short proof of this fact in the special case of $\mathbf{F} = GF(2)$. For "concrete" proofs in the case of $\mathbf{F} = GF(2)$ and $m = n$ we refer to [1], and the appendix of the Eurocrypt version of [22].

Let $f : \mathbf{F}^n \to \mathbf{F}^m$ be a function and $\mathbf{F} = GF(2)$. According to [5] we denote by θ_f the characteristic function of f,

$$\theta_f(x, y) = \begin{cases} 1, & \text{if } y = f(x), \\ 0, & \text{otherwise.} \end{cases}$$

Let $b \in \mathbf{F}^m$ and $\widehat{F}_b$ be the Walsh transform of $b \cdot f$. Then

$$\sum_{x,y} \theta_f(x,y)(-1)^{a \cdot x + b \cdot y} = \sum_{x \in \mathbf{F}^n} (-1)^{a \cdot x + b \cdot f(x)} = \widehat{F}_b(a). \tag{8}$$

Applying the inverse Walsh-Hadamard transform with respect to the second variable in $\mathbf{F}^m$, we get

$$\sum_{x \in \mathbf{F}^n} \theta_f(x,y)(-1)^{a \cdot x} = 2^{-m} \sum_{b \in \mathbf{F}^m} \widehat{F}_b(a)(-1)^{b \cdot y}. \tag{9}$$

As an easy application of (8) and (9) we get the proof of the result about uniform distribution of values:

A function $f : \mathbf{F}^n \to \mathbf{F}^m$, $1 \leq m \leq n$, *takes each value in* $\mathbf{F}^m$ *equally many times if and only if each component of* f *is balanced.*

Proof. First, let us observe that by (9) we have for all $y \in \mathbf{F}^m$

$$\#\{x \in \mathbf{F}^n \mid f(x) = y\} = \sum_{x \in \mathbf{F}^n} \theta_f(x,y) = 2^{-m} \sum_{b \in \mathbf{F}^m} \widehat{F}_b(0)(-1)^{b \cdot y}.$$

If each component is balanced, then $\widehat{F}_b(0) = 0$, for all $b \neq 0$, and we get

$$\#\{x \in \mathbf{F}^n \mid f(x) = y\} = 2^{-m} \widehat{F}_0(0) = 2^{n-m},$$

for all $y \in \mathbf{F}^m$.

To prove the converse, let us assume that

$$\#\{x \in \mathbf{F}^n \mid f(x) = y\} = \sum_{x \in \mathbf{F}^n} \theta_f(x,y) = 2^{n-m},$$

for all $y \in \mathbf{F}^m$. Then by (8)

$$\widehat{F}_b(0) = \sum_{y \in \mathbf{F}^m} (-1)^{b \cdot y} \sum_{x \in \mathbf{F}^n} \theta_f(x,y) = \begin{cases} 0, & \text{if } b \neq 0, \\ 2^n, & \text{if } b = 0. \end{cases}$$

$\square$

Properties of Linear Approximation Tables

Luke O'Connor [*,1,2]

[1] Distributed Systems Technology Centre (DSTC), Brisbane, Australia
[2] Information Security Research Centre, Queensland University of Technology
GPO Box 2434, Brisbane Q 4001, Australia

Email: oconnor@dstc.edu.au

Abstract. Linear cryptanalysis is an attack that derives a linear approximation between bits of the plaintext, ciphertext and key. This global approximation is constructed from the linear approximation tables of the nonlinear mappings used by the cipher, usually the S-boxes, as in the case of DES. In this paper we will describe the distribution of these tables for bijective mappings (permutations), concentrating on the expected value of the largest entry, and use our results to construct Feistel ciphers provably resistant to linear cryptanalysis.

1 Introduction

Linear cryptanalysis [11, 10] is a recently proposed attack due to M. Matsui. When successful, the attack recovers information about the secret key K used by approximating several nonlinear components of the cipher. For DES, the S-boxes can be approximated by deriving linear relations between the inputs and outputs to each S-box, where each relation is true with some probability p_i. We will call each such approximation a *linearization*. Matsui has shown that it is possible to derive linearizations to the S-boxes of DES at various rounds such that when these linearizations are added modulo 2, the remaining linearization involves bits of the key, plaintext and ciphertext only. The probability p of this 'global' linearization being correct is determined directly from the probabilities p_i of the 'local' linearizations when it is assumed that the subkeys are independent. One bit of information concerning the key can be recovered using maximum likelihood estimation when approximately $|p - \frac{1}{2}|^{-2}$ plaintext-ciphertext pairs are known. The quantity $|p - \frac{1}{2}|^{-2}$ is referred to as the (data) complexity of the attack. The attack can be modified to obtain more information about the key.

There have been several responses to the introduction of linear cryptanalysis. Kim *et al.* [8] have given a list of conditions for DES-like S-boxes to satisfy where

---

[*] The work reported in this paper has been funded in part by the Cooperative Research Centres program through the Department of the Prime Minister and Cabinet of Australia.

the probability p of the best global approximation based on these S-boxes will have the property that $|p - \frac{1}{2}| > 2^{28}$, implying that the complexity of the attack exceeds the cost of exhaustive key search. More general ciphers resistant to linear cryptanalysis have been proposed by Knudsen [9] and Heys and Tavares [7]. Also, several similarities between linear cryptanalysis and the more familiar differential cryptanalysis [2] have been noted [1, 3, 9].

The main result of this paper is to derive an upper bound on the largest entry in the linear approximation table of a bijective mapping, which when combined with a bound on the number of rounds that must be linearized, yields a lower bound on the complexity of linear cryptanalysis.

2 Linear approximation tables

Let $\pi : Z_2^n \to Z_2^n$ be a bijective n-bit mapping, and let S_{2^n} be the set of all such mappings, known as the symmetric group. For an n-bit vector $X \in Z_2^n$ let $X[i]$ denote the ith bit of X. The linear approximation table for π, denoted LAT_π, is a $2^n \times 2^n$ table such that

$$LAT_\pi(\alpha, \beta) \stackrel{\text{def}}{=} \# \left\{ X \mid X \in Z_2^n, \; \bigoplus_{i=1}^n X[i] \cdot \alpha[i] = \bigoplus_{i=1}^n \pi(X)[i] \cdot \beta[i] \right\} \quad (1)$$

where $\alpha, \beta \in Z_2^n$ and '$\cdot$' denotes bitwise logical AND. Thus $LAT_\pi(\alpha, \beta)$ gives the number of equal parity checks between a linear combination of the input bits (specified by α) and a linear combination of the output bits (specified by β).

Theorem 1. Let $\lambda(\alpha, \beta)$ be a random variable describing $LAT_\pi(\alpha, \beta)$ when π is selected uniformly from S_{2^n}, and α, β are nonzero. Then $\lambda(\alpha, \beta)$ only assumes even values and

$$\Pr(\lambda(\alpha, \beta) = 2k) = \frac{(2^{n-1}!)^2}{2^n!} \cdot \binom{2^{n-1}}{k}^2. \quad (2)$$

for $0 \leq k \leq 2^{n-1}$. $\square$

In linear cryptanalysis we are interested in those entries in the linear approximation table that differ from 2^{n-1} as this represents the correlation between linear combinations of the inputs and outputs. Matsui [10] calls this the *effectiveness* of the linearization. For this reason we define the 'normalized' linear approximation table, denoted by LAT_π^*, as

$$LAT_\pi^*(\alpha, \beta) = |LAT_\pi(\alpha, \beta) - 2^{n-1}|. \quad (3)$$

The distribution of $LAT_\pi^*(\alpha, \beta)$ follows directly from Theorem 1.

Corollary 2.1 Let $\lambda^*(\alpha, \beta)$ be a random variable describing $LAT_\pi^*(\alpha, \beta)$ when π is selected uniformly from S_{2^n}, and α, β are nonzero. Then $\lambda^*(\alpha, \beta)$ only assumes even values and

$$\Pr(\lambda^*(\alpha, \beta) = 2k) = \frac{(2 - [k = 0]) \cdot (2^{n-1}!)^2}{2^n!} \cdot \binom{2^{n-1}}{2^{n-2} + k}^2. \quad (4)$$

for $0 \leq k \leq 2^{n-2}$, where $[k = 0]$ evaluates to zero or one.   □

Using Stirling's approximation, it can be shown that the expected number of zero entries in the table is $4/\sqrt{2\pi \cdot 2^n}$, which is tending to zero with n. However, our main goal is to determine a bound on the largest entry in LAT_π^* which is useful in the design of ciphers that are to be resistant to linear cryptanalysis. To this end let $\lambda(\pi)$ be the largest entry in LAT_π^* for the mapping π taken over all nontrivial α, β,

$$\lambda(\pi) \stackrel{\text{def}}{=} \max_{\alpha, \beta \neq 0} LAT_\pi^*(\alpha, \beta). \tag{5}$$

In the next section we derive an upper bound on $\lambda(\pi)$.

3   An upper bound

Let $\mathbf{E}[\lambda(\pi, 2k)]$ denote the expected number of entries in LAT_π^* of size $2k$. Consider the following bound on $\lambda(\pi)$,

$$\Pr(\lambda(\pi) = 2k) \leq \Pr(LAT_\pi^* \text{ has at least one entry of size } 2k) < \mathbf{E}[\lambda(\pi, 2k)] \tag{6}$$

which is valid since

$$\mathbf{E}[\lambda(\pi, 2k)] = \sum_{t \geq 0} t \cdot \Pr(LAT_\pi^* \text{ has } t \text{ entries of size } 2k). \tag{7}$$

If $\mathbf{E}[\lambda(\pi, 2k)]$ is tending rapidly to zero as a function of k, then we are likely to obtain a useful bound on $\pi(\lambda)$. From Corollary 2.1, we can derive $\mathbf{E}[\lambda(\pi, 2k)]$, $k > 0$ as

$$\mathbf{E}[\lambda(\pi, 2k)] = \frac{2 \cdot (2^n - 1)^2 \cdot (2^{n-1}!)^2}{2^n!} \cdot \binom{2^{n-1}}{2^{n-2} + k}^2. \tag{8}$$

In the next theorem we derive an approximation for the tail of the $\mathbf{E}[\lambda(\pi, 2k)]$ distribution using a well-known bound on the sum of consecutive binomial coefficients.

Theorem 2. For $t > 0$,

$$\sum_{k=2^{n-2}+t}^{2^{n-1}} \mathbf{E}[\lambda(\pi, 2k)] < \sqrt{\frac{\pi}{2}} \cdot 2^{2^n H(m) - 2^n + 5n/2} \tag{9}$$

where $m = (2^{n-2} - t)/2^{n-1}$ and $H(x) = -x \log x - (1 - x) \log(1 - x)$.   □

From Theorem 2 we are now able to derive a bound on the probability that $\lambda(\pi)$ is at most $2t$ since

$$\Pr(\lambda(\pi) \leq 2t) > 1 - \sum_{k=2^{n-2}+t}^{2^{n-1}} \mathbf{E}[\lambda(\pi, 2k)]. \tag{10}$$

In particular, we will determine the smallest value of t (according to our bounds) for which $\Pr(\lambda(\pi) \leq 2t) > \frac{1}{2}$, which we will denote as t_n. Here t_n is referred to as the median of the distribution [6]. The practical implication of t_n is that when a bijective n-bit mapping π is selected uniformly, with odds better than 50/50 the mappings will have $\lambda(\pi) \leq 2t_n$.

Our results are shown in Table 1. The second column shows the actual tail computation for $\mathbf{E}[\lambda(\pi, 2k)]$ using (8), while the third column shows the tail computation using the approximation of (9). The last column shows the results of generating random bijective mappings and determining t_n from this sample. That is, if there were k mappings generated, t_n is determined to be the minimum value for which at least half the mappings had the largest entry bounded by t_n. The sample k actually used was relatively small (several hundred) because of the time required to determine t_n. For example, it requires half an hour of clock time to determine t_{10} for one 10-bit mapping.

n	t_n via (8)	t_n via (9)	t_n via experiment
6	9	10	8
7	13	15	12
8	19	22	18
9	28	33	26
10	41	49	40
11	61	72	59
12	90	106	88

Table 1. Upper bounds on t_n where $\Pr(\lambda(\pi) \leq 2t_n) > \frac{1}{2}$.

Note that the bound on t_n derived from (9) is consistently higher than the bound derived from (8). Of course this is expected as various approximations are used to derive the bound in (9). The advantage however is that the expression in (9) is easy to evaluate and can give useful information about the best approximation for large mappings. On the other hand, the expression in (8) can only be evaluated for relatively small n. For example, using (9), we have determined that the largest entry in the linear approximation table of a 64-bit mapping is at most 16057555882 for more than half the possible such mappings, giving a best possible linear approximation of

$$|p - \frac{1}{2}| \leq \frac{16057555882}{2^{64}} = 0.87048 \times 10^{-9} \approx 2^{-30}. \tag{11}$$

We may then assume that the probability of the best (global) linear approximations for 64-bit mappings such as DES and FEAL would be close to 2^{-30} *if the mappings were random and not derived from an iterative process*. However, using the fact that the nonlinear components of the round function are fixed, Matsui has derived an approximation for 16-round DES that has $|p - \frac{1}{2}| = 1.19 \times 2^{-21}$.

4 Conclusion

The success of linear cryptanalysis depends on the nonlinearity of the round function **F**, which for ciphers such as DES reduces to the nonlinearity of the S-boxes. It is clear that if this nonlinearity can be made sufficiently large then the complexity of the attack will exceed the cost of exhaustive key search. Our approach has been to bound the largest value in the linear approximation table of a bijective mapping, which can now be used to construct ciphers resistant to linear cryptanalysis using similar methods to those employed by Knudsen [9] and Heys and Tavares [7]. Similar observations have been used by O'Connor [13] to show that sufficiently large S-boxes will also defeat differential cryptanalysis, as the largest value in the XOR table can be bounded asymptotically.

Acknowledgements

I would like to thank Kaisa Nyberg for for her comments on the original version of this manuscript.

References

1. E. Biham. On Matsui's Linear Cryptanalysis. *to appear, proceedings of EUROCRYPT 94, Perugia, Italy*, 1994.
2. E. Biham and A. Shamir. Differential cryptanalysis of DES-like cryptosystems. *Journal of Cryptology*, 4(1):3–72, 1991.
3. F. Chabaud and S. Vandenay. Links between differential and linear cryptanalysis. *to appear, proceedings of EUROCRYPT 94, Perugia, Italy*, 1994.
4. H. Feistel. Cryptography and computer privacy. *Scientific American*, 228(5):15–23, 1973.
5. H. Feistel, W. A. Notz, and J. Lynn Smith. Some cryptographic techniques for machine-to-machine data communications. *proceedings of the IEEE*, 63(11):1545–1554, 1975.
6. W. Feller. *An Introduction to Probability Theory and its Applications*. New York: Wiley, 3rd edition, Volume 1, 1968.
7. H. M. Heys and S. E. Tavares. Substitution-permutation networks resistent to differential and linear cryptanalysis. submitted to the Journal of Crytology.
8. K. Kim, S. Lee, S. Park, and D. Lee. DES can be immune to linear cryptanalysis. *proceedings of the Workshop on Selected Areas in Cryptography, Kingston, Canada, May 1994*, pages 70–81, 1994.
9. L. R. Knudsen. Practically secure Feistel ciphers. *proceedings of Fast Software Encryption, Cambridge Security Workshop, Lecture Notes in Conputer Science, vol. 809, 1994*, pages 211–221, 1994.
10. M. Matsui. Linear cryptanalysis of DES cipher (I). (version 1.03) private communication.
11. M. Matsui. Linear cryptanalysis method for DES cipher. *Advances in Cryptology, EUROCRYPT 93, Lecture Notes in Computer Science, vol. 65, T. Helleseth ed., Springer-Verlag*, pages 386–397, 1994.

12. W. Meier and O. Staffelbach. Nonlinearity criteria for cryptographic functions. *Advances in Cryptology, EUROCRYPT 89, Lecture Notes in. Computer Science, vol. 434, J.-J. Quisquater, J. Vandewalle eds., Springer-Verlag*, pages 549–562, 1990.
13. L. J. O'Connor. On the distribution of characteristics in bijective mappings. *Advances in Cryptology, EUROCRYPT 93, Lecture Notes in Computer Science, vol. 765, T. Helleseth ed., Springer-Verlag*, pages 360–370, 1994.
14. J. Pieprzyk, C. Charnes, and Seberry J. Linear approximation versus nonlinearity. *proceedings of the Workshop on Selected Areas in Cryptography, Kingston, Canada, May 1994*, pages 82–89, 1994.

Searching for the Optimum Correlation Attack

Ross Anderson

Computer Laboratory, Pembroke Street, Cambridge CB2 3QG
Email: rja14@cl.cam.ac.uk

Abstract. We present some new ideas on attacking stream ciphers based on regularly clocked shift registers. The nonlinear filter functions used in such systems may leak information if they interact with shifted copies of themselves, and this gives us a systematic way to search for correlations between a keystream and the underlying shift register sequence.

1 Introduction

A number of cipher systems use a nonlinear filter generator to expand a short key into a long keystream. This generator is based on a linear feedback shift register, some of whose state bits are filtered through a nonlinear function to provide the keystream (figure 1):

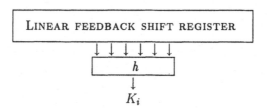

Fig. 1 - The nonlinear filter generator

Typical systems have shift registers of between 61 and 127 bits in length, and nonlinear filter functions of varying complexity [MFB] [KBS] [CSh]. Some variants use several functions simultaneously to generate a number of keystream bits in parallel [M1].

The conventional attack on the filter generator [S2] [S3] [MS1] proceeds in two stages. Firstly, we find a function of the keystream which is correlated with the underlying shift register sequence; it can be shown that such a function always exists [XM], even if the combining function possesses memory [G]. The keystream is viewed as a noisy version of the shift register sequence, and is reconstructed by various techniques (figure 2):

138   R. Anderson

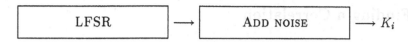

Fig. 2 - The standard model

This 'standard model', which has been the focus of most of the published work on the subject, was first proposed by Siegenthaler [S4]. His original attack involved an exhaustive search through all phases of the shift register to find the highest correlation [S3]; Meier and Staffelbach later showed that iterative reconstruction techniques were much faster, and especially so if the shift register's connection polynomial $f(x)$ is of low weight [MS1], while Mihaljević and Golić proved conditions under which these fast correlation attacks converge [MG].

Where $f(x)$ is not sparse, one can look for a decimation of the sequence whose polynomial is sparse [A1], or more generally a sparse multiple of f (i.e., a low weight parity check) [CS]. Meier and Staffelbach pointed out that low weight checks can be found by meet-in-the-middle techniques [MS1]; and if $f(x)$ has degree n, this will take about $\frac{n}{2}2^{(\frac{n}{2})}$ operations. Recent work by van Oorschot and Wiener has shown that it is feasible to construct special-purpose collision search hardware for n up to 128 or so [VW]. Thus the security of the nonlinear filter generators under consideration boils down to finding good correlations between the keystream and the underlying shift register.

However, the problem of finding an actual correlation tends to have been dismissed with an existence proof. Our principle that robust security depends on explicitness [A2] made us suspicious of this, and inspired us to look for a construction. We would ideally like to have an algorithm which will find the maximum correlation which an attacker can obtain; we can then use this together with the convergence bounds found in [MG] and [M2] to establish whether a given design is vulnerable to a fast correlation attack. Our model is therefore

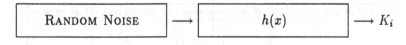

Fig. 3 - Our model

This is essentially the dual of the problem studied in the standard model; the goal is to find out how much information about an arbitrary signal leaks through $h(x)$ to K_i. Its solution depends on the nonlinear structure of h: if $K_i = S_i + S_{i+1}$, then knowing K_i tells us nothing about S_i, while if $K_i = S_i S_{i+1}$, then whenever $K_i = 1$ we know that $S_i = 1$ too. Note that when attacking a filter generator we can always discount a linear function by moving to a different phase of the underlying shift register, so we are really interested in finding the maximum leakage of $h(l(x))$ over all linear functions $l(x)$.

2   Finding a Correlation

Let us take a concrete example. Suppose that the nonlinear combining function h is given by

$$h(x_1, x_2, x_3, x_4, x_5) = x_1 + x_2 + (x_1 + x_3)(x_2 + x_4 + x_5) + (x_1 + x_4)(x_2 + x_3)x_5$$

This function is used as a primitive in [K] and appears to have been used in other designs too [KBS]; it is distinguished by being as small a function as one can get which is both balanced and correlation immune of degree two.

As already noted, the standard attack on such a cipher would be to look for linear functions of the keystream and of the underlying shift register sequence which are correlated; a variant is the 'best affine approximation' attack of Ding, Xiao and Shan [DXS]. However, both these attacks throw away a lot of information about the nonlinear structure of h, and our goal is to try and identify - and if possible use - all the information which h leaks.

If $K_i = f(S_{i-2}, S_{i-1}, S_i, S_{i+1}, S_{i+2})$, the keystream bits $K_{i-2}, \cdots, K_{i+2}$ all depend on S_i, and if we want to approximate S_i we need a function of the form $S_i = g(K_{i-2}, K_{i-1}, K_i, K_{i+1}, K_{i+2})$. However, the bits $K_{i-2}, \cdots, K_{i+2}$ depend on the nine bits $S_{i-4}, \cdots, S_{i+4}$; and so it is natural to look at the set of input 9-tuples from the shift register sequence which give rise to each possible 5-bit keystream output. We will call the 9 bit to 5 bit function the *augmented function* of h and write it as $\overline{h}$.

When we count the outputs of $\overline{h}$ for each of the 512 possible inputs, we find:

output	# inputs	output	# inputs	output	# inputs	output	# inputs
0	18	8	11	16	16	24	23
1	16	9	17	17	18	25	17
2	14	10	12	18	16	26	10
3	20	11	12	19	18	27	18
4	16	12	23	20	12	28	17
5	14	13	13	21	10	29	15
6	21	14	13	22	15	30	19
7	17	15	19	23	15	31	17

We might have hoped that a good system would have each output generated by sixteen inputs, but the actual table is irregular, and this may give us a way in. For example, there are two outputs (21 and 26) generated by only ten inputs, and if we look at the inputs which generate 26, we find that they are:

```
001010101
001110001
001110010
100110001
100110010
101001011
101110001
101110010
110110001
110110010
```

Casting our eye down these columns, we see that there is only a single zero in the fifth column, and a single one in the sixth and seventh. In other words, if $K_i, ..., K_{i+4} = 11010$, then $S_{i+2} = 1$, $S_{i+3} = 0$ and $S_{i+4} = 0$ with probability 0.9 in each case. The other columns give us correlations of 0.7, 0.8, and so on. Now the fact that the correlation between a nonlinearly filtered sequence and the underlying shift register is uneven was first pointed out by Forré [F], but she did not investigate the matter further. At last we have explained this irregularity — it is simply a matter of counting the input/output stability of the augmented function $\bar{h}$.

Of course, we do not just get information from those inputs which give rise to the least common outputs. For example, when we consider the 17 inputs which generate the output 9, a delightful discovery awaits us: these 17 inputs are all zero in the fourth bit. So whenever we see that $K_i, ..., K_{i+4} = 01001$, we know that $S_{i+1} = 0$.

Now, one might at first think that this yields an optimal correlation attack. After all, if h is an m-bit to 1-bit function, then each shift register bit will contribute to precisely m keystream bits, and we might expect that all the correlation information could be found by examining the augmented function which generates these bits. If we are lucky, we will find correlations of one and be able to solve the cipher outright using linear algebra; otherwise, we should still get lots of correlations of the order of 0.8 or 0.9, with which a probabilistic reconstruction becomes fairly straightforward [M2].

However, we can get more than just correlations between the K_i and the S_i. On looking more closely at the above table, we notice that columns 5 and 6 are inverses of each other. Thus whenever $K_i, ..., K_{i+4} = 11010$, then $S_{i+2} = 1 + S_{i+3}$. Similar relations are to be found in the other output sets; for example, if K_i through K_{i+4} are all equal to zero, then $S_{i+3} = S_{i+4} = S_{i+5}$. Every such equation halves the keyspace which we still have to search.

How well does our technique work against other nonlinear combiners? In recent years, a lot of attention has been paid to bent functions [MS2]. We looked at a typical bent function, $h(x_1, ..., x_6) = x_1 + x_2 + x_3 + x_1 x_4 + x_2 x_5 + x_3 x_6$, and found it to be significantly worse than the correlation-immune function discussed above. In fact, its augmented function never attains nine possible output values (12, 15, 30, 31, 47, 60, 61, 62, 63), while the zero output is attained 100 times. Its information leakage is heavy; for example, the output value 17 is attained 12 times, with all twelve inputs having $S_1 = S_2 = 1$ and $S_4 = S_5 = S_{11} = 0$.

The use of almost bent functions has also been suggested; these are bent functions which have been made balanced by changing a few output values. When we changed the above bent function's output from 0 to 1 for input values of 7, 11, 23, and 27, we found that the columnar behaviour was somewhat less marked, but seven of the nine missing output values were now attained by small numbers of inputs (and the output 31 was only generated by the single input 00010111011). We concluded that attacking a filter generator using a bent or almost bent function would be easy.

The device reported in [M1] and [MFB] uses as its filter what we might call a De Bruijn function — a function on k input bits whose value at the input n is the n-th bit of a 2^k bit De Bruijn sequence. This filter has the interesting property that its augmented function is balanced, with every possible output attained exactly 2^k times; it also has the property that for each output, the last k bits of each of the inputs are identical. For example, with one of the De Bruijn functions implemented in [MFB], the output 21 is attained for 32 inputs of the form ******10110. In other words, the information leakage is total.

One also notices from looking at a few candidates that most randomly chosen balanced filter functions appear to leak rather badly. This suggests that just as it is a bad idea to replace the DES S-boxes with random ones [BS], it is also a bad idea to use randomly chosen filters; and of course a knowledgeable designer can easily place a trapdoor in the filter.

Many further implications remain to be worked out. For example, even if h has too many inputs for exhaustive analysis, it may still have some structure which we can use. It might have a tractable mathematical definition as in [CSh]; if it has a regular tree structure which is key dependent [K] [A3], then these keybits might be deduced by observing which patterns are most common in decimations of the keystream; where an unknown permutation is introduced in the tree structure, the ideas of [MDO] may be useful; and even where none of these tricks can be used, statistical sampling of the augmented function may still give information to the opponent.

Some systems use a number of nonlinear filters in parallel to generate more than one bit of keystream at a time [M1]; but these functions could well interact in a way which facilitates an attack. In fact, in a recent NSA patent on a device for generating simultaneous keystreams, the underlying generator is run at high speed to ensure that the keystreams are linearly independent [S1].

Another implication is that when doing a correlation attack, the 'lumpiness' of the correlation will mean that we have little information about some bits in the shift register sequence, but will know others with high probability. Existing shift register reconstruction algorithms will no longer work, as the problem is now that that of finding S such that given m bits Z and an n by m matrix $\mathbf{A}$, $|\mathbf{A}S - Z|$ is minimised. This problem is tackled in another paper in this volume [M2], which was motivated in part by this work and which extends the techniques pioneered by Meier and Staffelbach.

Clearly, when designing keystream generators, we must write down all the keystream bits K_i to which an arbitrary shift register bit S_i contributes, and then all the equations whereby these bits in turn are generated. We should then consider all the occurrences of S_i in these equations and check for information leakage. On considering a few examples, it appears that using either multiple shift registers or multiple filter functions makes the security harder to evaluate.

3   Conclusions

We have shown a practical method for searching for the best possible local correlations nonlinearly filtered shift register sequences. The key is to look at how the filter function reacts with shifted copies of itself. Many functions react badly — including both bent functions and De Bruijn functions — and if a number of different filter functions are used simultaneously, then their interactions must also be taken into account.

On the theoretical side, we have given students of Boolean functions new problems to investigate, namely what functions have low leakage (defined as the maximum imbalance in any input variable for any given output of the augmented function), and, in general, what properties a set of functions must have in order not to interact in a harmful way.

References

[A1]   RJ Anderson, "Derived Sequence Attacks on Stream Ciphers", presented at the rump session of Crypto 93

[A2]   RJ Anderson, "Why Cryptosystems Fail", in *Communications of the ACM* v **37** no 11 (November 1994) pp 32–40

[A3]   RJ Anderson, "Tree Functions and Cipher Systems", in *Cryptologia* v **XV** no 3 (July 1991) pp 194–202

[BS]   E Biham, A Shamir, '*Differential Cryptanalysis of the Data Encryption Standard*', Springer 1993

[CS]   V Chepyzhov, B Smeets, " On a Fast Correlation Attack on Certain Stream Ciphers", in *Advances in Cryptology — Eurocrypt 91*, Springer LNCS v **547** pp 176–185

[CSh]   TR Cain, AT Sherman, "How to Break Gifford's Cipher", in *Proceedings of the 2nd ACM Conference on Computer and Communications Security* (ACM, Nov 94) pp 198–209

[DXS]   C Ding, G Xiao, W Shan, *'The Stability Theory of Stream Ciphers'*, Springer LNCS v **561** (1991)

[F]   R Forré, "A Fast Correlation Attack on Nonlinearly Feedforward Filtered Shift-register Sequences", in *Advances in Cryptology — Eurocrypt 89*, Springer LNCS v **434**, pp 586–562

[G]   JD Golić, "Correlation via Linear Sequential Circuit Approximation of Combiners with Memory", in *Advances in Cryptology — Eurocrypt 92*, Springer LNCS v **658**, pp 113–123

[K]   GJ Kühn, "Algorithms for Self-Synchronising Ciphers", in *Proc COMSIG 88*

[KBS]   G Kühn, F Bruwer, W Smit, " 'n Vinnige Veeldoelige Enkripsievlokkie", supplementary paper to *Proceedings of Infosec 1990*

[M1]   G Mayhew, "A Low Cost, High Speed Encryption System and Method", in *Proc 1994 IEEE Computer Society Symposium on Research in Security and Privacy* (IEEE, 1994) pp 147–154

[M2]   DJC MacKay, "A Free Energy Minimization Framework for Inference Problems in Modulo 2 Arithmetic" *in this volume* pp 179–195

[MDO]   W Millan, EP Dawson, LJ O'Connor, "Fast Attacks on Tree-structured Ciphers", in *Proceedings of Workshop in Selected Areas in Cryptography* (Queen's University, 1994) pp 146–158

[MFB]   G Mayhew, R Frazee, M Bianco, "The Kinetic Protection Device", in *Proceedings of the 15th National Computer Security Conference* (NIST, 1992) pp 310–318

[MG]   MJ Mihaljević, JD Golić, "Convergence of a Bayesian Iterative Error-correction Procedure on a Noisy Shift Register Sequence", in *Advances in Cryptology — Eurocrypt 92*, Springer LNCS v **658**, pp 124–137

[MS1]   W Meier, O Staffelbach, "Fast correlation attacks on certain stream ciphers", in *Journal of Cryptology* v 1 1989 pp 159–176

[MS2]   W Meier, O Staffelbach, "Nonlinearity criteria for cryptographic functions", in *Advances in Cryptology — Eurocrypt 89*, Springer LNCS v **434** pp 549–562

[S1]   B Snow, *'Multiple Independent Binary Bit Stream Generator'*, US Patent 5,237,615 (17 August 1993)

[S2]   T Siegenthaler, " Correlation Immunity of Nonlinear Combining Functions for Cryptographic Applications", in *IEEE Transactions on Information Theory* **IT-30** no 5 (Sep 1984) pp 776–780

[S3]   T Siegenthaler, "Decrypting a Class of Stream Ciphers Using Ciphertext Only", in *IEEE Transactions on Computers* **C-34** no 1 (Jan 1985) pp 81–85

[S4]   T Siegenthaler, "Cryptanalysts' Representation of Nonlinearly Filtered m-Sequences", in *Advances in Cryptology — Eurocrypt 85*, Springer LNCS v **219** pp 103–110

[VW]   PC van Oorschot, MJ Wiener, "Parallel Collision Search with Application to Hash Functions and Discrete Logarithms", in *Proceedings of the 2nd ACM Conference on Computer and Communications Security* (ACM, Nov 94) pp 210–218

[XM]   GZ Xiao, JL Massey, "A spectral characterisation of correlation-immune combining functions", in *IEEE Transactions on Information Theory* **IT-34** (May 1988) pp 569–571

A Known Plaintext Attack on the PKZIP Stream Cipher

Eli Biham*   Paul C. Kocher**

Abstract. The PKZIP program is one of the more widely used archive/ compression programs on personal computers. It also has many compatible variants on other computers, and is used by most BBS's and ftp sites to compress their archives. PKZIP provides a stream cipher which allows users to scramble files with variable length keys (passwords).

In this paper we describe a known plaintext attack on this cipher, which can find the internal representation of the key within a few hours on a personal computer using a few hundred bytes of known plaintext. In many cases, the actual user keys can also be found from the internal representation. We conclude that the PKZIP cipher is weak, and should not be used to protect valuable data.

1 Introduction

The PKZIP program is one of the more widely used archive/compression programs on personal computers. It also has many compatible variants on other computers (such as Infozip's zip/unzip), and is used by most BBS's and ftp sites to compress their archives. PKZIP provides a stream cipher which allows users to scramble the archived files under variable length keys (passwords). This stream cipher was designed by Roger Schlafly.

In this paper we describe a known plaintext attack on the PKZIP stream cipher which takes a few hours on a personal computer and requires about 13–40 (compressed) known plaintext bytes, or the first 30–200 uncompressed bytes, when the file is compressed. The attack primarily finds the 96-bit internal representation of the key, which suffices to decrypt the whole file and any other file encrypted under the same key. Later, the original key can be constructed. This attack was used to find the key of the PKZIP contest.

The analysis in this paper holds to both versions of PKZIP: version 1.10 and version 2.04g. The ciphers used in the two versions differ in minor details, which does not affect the analysis.

The structure of this paper is as follows: Section 2 describes PKZIP and the PKZIP stream cipher. The attack is described in Section 3, and a summary of the results is given in Section 4.

---

* Computer Science Department, Technion - Israel Institute of Technology, Haifa 32000, Israel

** Independent cryptographic consultant, 7700 N.W. Ridgewood Dr., Corvallis, OR 97330, USA

2   The PKZIP Stream Cipher

PKZIP manages a ZIP file[1] which is an archive containing many files in a compressed form, along with file headers describing (for each file) the file name, the compression method, whether the file is encrypted, the CRC-32 value, the original and compressed sizes of the file, and other auxiliary information.

The files are kept in the zip-file in the shortest form possible of several compression methods. In case that the compression methods do not shrink the size of the file, the files are stored without compression. If encryption is required, 12 bytes (called the *encryption header*) are prepended to the compressed form, and the encrypted form of the result is kept in the zip-file. The 12 prepended bytes are used for randomization, but also include header dependent data to allow identification of wrong keys when decrypting. In particular, in PKZIP 1.10 the last two bytes of these 12 bytes are derived from the CRC-32 field of the header, and many of the other prepended bytes are constant or can be predicted from other values in the file header. In PKZIP 2.04g, only the last byte of these 12 bytes is derived from the CRC-32 field. The file headers are not encrypted in both versions.

The cipher is byte-oriented, encrypting under variable length keys. It has a 96-bit internal memory, divided into three 32-bit words called key0, key1 and key2. An 8-bit variable key3 (not part of the internal memory) is derived from key2. The key initializes the memory: each key has an equivalent internal representation as three 32-bit words. Two keys are equivalent if their internal representations are the same. The plaintext bytes update the memory during encryption.

The main function of the cipher is called update_keys, and is used to update the internal memory and to derive the variable key3, for each given input (usually plaintext) byte:

update_keys$_i$(char) :
   local unsigned short temp
   $key0_{i+1} \leftarrow crc32(key0_i, char)$
   $key1_{i+1} \leftarrow (key1_i + LSB(key0_{i+1})) * 134775813 + 1 \quad (mod\ 2^{32})$
   $key2_{i+1} \leftarrow crc32(key2_i, MSB(key1_{i+1}))$
   $temp_{i+1} \leftarrow key2_{i+1} \mid 3 \quad$ (16 LS bits)
   $key3_{i+1} \leftarrow LSB((temp_{i+1} * (temp_{i+1} \oplus 1)) \gg 8)$
end update_keys

where | is the binary inclusive-or operator, and $\gg$ denotes the right shift operator (as in the C programming language). LSB and MSB denote the least significant byte and the most significant byte of the operands, respectively. Note that the indices are used only for future references and are not part of the algorithm, and that the results of key3 using inclusive-or with 3 in the calculation of temp are the same as with the original inclusive-or with 2 used in the original algorithm. We prefer this notation in order to reduce one bit of uncertainty about temp in the following discussion.

Before encrypting, the key (password) is processed to update the initial value of the internal memory by:

```
process_keys(key) :
    key0_{1-l} ← 0x12345678
    key1_{1-l} ← 0x23456789
    key2_{1-l} ← 0x34567890
    loop for i ← 1 to l
        update_keys_{i-l}(key_i)
    end loop
end process_keys
```

where l is the length of the key (in bytes) and hexadecimal numbers are prefixed by 0x (as in the C programming language). After executing this procedure, the internal memory contains the internal representation of the key, which is denoted by $key0_1$, $key1_1$ and $key2_1$.

The encryption algorithm itself processes the bytes of the compressed form along with the prepended 12 bytes by:

Encryption	**Decryption**
prepend $P_1, \ldots, P_{12}$	
loop for $i \leftarrow 1$ to n	loop for $i \leftarrow 1$ to n
$\quad C_i \leftarrow P_i \oplus key3_i$	$\quad P_i \leftarrow C_i \oplus key3_i$
$\quad$ update_keys$_i(P_i)$	$\quad$ update_keys$_i(P_i)$
end loop	end loop
	discard $P_1, \ldots, P_{12}$

The decryption process is similar, except that it discards the 12 prepended bytes.

The crc32 operation takes a previous 32-bit value and a byte, XORs them and calculates the next 32-bit value by the crc polynomial denoted by 0xEDB88320. In practice, a table crctab can be precomputed, and the crc32 calculation becomes:

$$crc32 = crc32(pval, char) = (pval \gg 8) \oplus crctab[LSB(pval) \oplus char]$$

The crc32 equation is invertible in the following sense:

$$pval = crc32^{-1}(crc32, char) = (crc32 \ll 8) \oplus crcinvtab[MSB(crc32)] \oplus char$$

crctab and crcinvtab are precomputed as:

```
init_crc() :
    local unsigned long temp
    loop for i ← 0 to 255
        temp ← crc32(0, i)
        crctab[i] ← temp
        crcinvtab[temp ≫ 24] ← (temp ≪ 8) ⊕ i
    end loop
end init_crc
```

in which crc32 refers to the original definition of crc32:

```
crc32(temp, i) :
  temp ← temp ⊕ i
  loop for j ← 0 to 7
    if odd(temp) then
      temp ← temp ≫ 1 ⊕ 0xEDB88320
    else
      temp ← temp ≫ 1
    endif
  end loop
  return temp
end crc32
```

3   The Attack

The attack we describe works even if the known plaintext bytes are not the first bytes (if the file is compressed, it needs the compressed bytes, rather than the uncompressed bytes). In the following discussion the subscripts of the n known plaintext bytes are denoted by $1,\ldots,n$, even if the known bytes are not the first bytes. We ignore the subscripts when the meaning is clear and the discussion holds for all the indices.

Under a known plaintext attack, both the plaintext and the ciphertext are known. In the PKZIP cipher, given a plaintext byte and the corresponding ciphertext byte, the value of the variable key3 can be calculated by

$$\text{key3}_i = P_i \oplus C_i.$$

Given $P_1,\ldots,P_n$ and $C_1,\ldots,C_n$, we receive the values of $\text{key3}_1,\ldots,\text{key3}_n$. The known plaintext bytes are the inputs of the update_keys function, and the derived key3's are the outputs. Therefore, in order to break the cipher, it suffices to solve the set of equations derived from update_keys, and find the initial values of key0, key1 and key2.

In the following subsections we describe how we find many possible values for key2, then how we extend these possible values to possible values of key1, and key0, and how we discard all the wrong values. Then, we remain with only the right values which correspond to the internal representation of the key.

3.1   key2

The value of key3 depends only on the 14 bits of key2 that participate in temp. Any value of key3 is suggested by exactly 64 possible values of temp (and thus 64 possible values of the 14 bits of key2). The two least significant bits of key2 and the 16 most significant bits do not affect key3 (neither temp).

Given the 64 possibilities of temp in one location of the encrypted text, we complete the 16 most significant bits of key2 with all the 2^{16} possible values,

Side	Term	Bits Number	Bits Position	Number of Values
Left	$key2_i$	14	2–15	64
Right	$key2_{i+1} \ll 8$	22	10–31	1
	$crcinvtab[MSB(key2_{i+1})]$	32	0–31	1
	$MSB(key1_{i+1})$	24	8–31	1
Total Left Hand Side		14	2–15	64
Total Right Hand Side		22	10–31	1
Common bits		6	10–15	
Total bits		30	2–31	

Table 1. The Known Bits in Equation 1

and get 2^{22} possible values for the 30 most significant bits of key2. $key2_{i+1}$ is calculated by $key2_{i+1} \leftarrow crc32(key2_i, MSB(key1_{i+1}))$. Thus,

$$
\begin{aligned}
key2_i &= crc32^{-1}(key2_{i+1}, MSB(key1_{i+1})) \\
&= (key2_{i+1} \ll 8) \oplus crcinvtab[MSB(key2_{i+1})] \oplus MSB(key1_{i+1}).
\end{aligned}
\tag{1}
$$

Given any particular value of $key2_{i+1}$, for each term of this equation we can calculate the value of the 22 most significant bits of the right hand side of the equation, and we know 64 possibilities for the value of 14 bits of the left hand side, as described in Table 1. From the table, we can see that six bits are common to the right hand side and the left hand side. Only about 2^{-6} of the possible values of the 14 bits of $key2_i$ have the same value of the common bits as in the right hand side, and thus, we remain with only one possible value of the 14 bits of $key2_i$ in average, for each possible value of the 30 bits of $key2_{i+1}$. When this equation holds, we can complete additional bits of the right and the left sides, up to the total of the 30 bits known in at least one of the sides. Thus, we can deduce the 30 most significant bits of $key2_i$. We get in average one value for these 30 most significant bits of $key2_i$, for each value of the 30 most significant bits of $key2_{i+1}$. Therefore, we are now just in the same situation with $key2_i$ as we were before with $key2_{i+1}$, and we can now find values of 30 bits of $key2_{i-1}$, $key2_{i-2}$, ..., $key2_1$. Given this list of 30-bit values, we can complete the 32-bit values of $key2_n$, $key2_{n-1}$, ..., $key2_2$ (excluding $key2_1$) using the same equation. We remain with about 2^{22} lists of possible values ($key2_n$, $key2_{n-1}$, ..., $key2_2$), of which one must be the list actually calculated during encryption.

3.2   Reducing the number of possible values of key2

The total complexity of our attack is (as described later) 2^{16} times the number of possible lists of key2's. If we remain with 2^{22} lists, the total complexity becomes 2^{38}. This complexity can be reduced if we can reduce the number of lists of key2's without discarding the right list.

bytes	key2 list entries
1 2^{22} =	4194304
2	3473408
3	2152448
4	1789183
5	1521084
10	798169
15	538746
20	409011
25	332606
30	283930
40	213751
50	174471
100	88248
200	43796
300	31088
500	16822
1000	7785
2000	5196
4000	3976
6000	3000
8000	1296
10000	1857
12000	243
12289	801

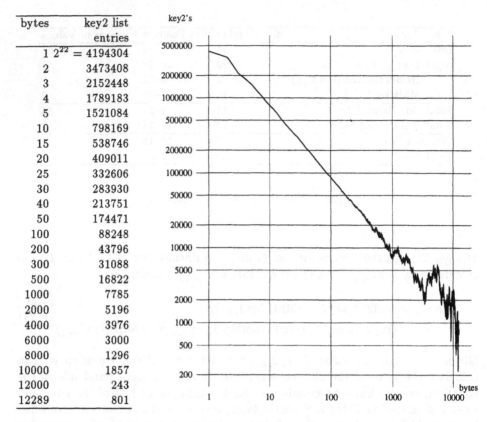

Fig. 1. Decrease in the number of key2 candidates using varying amounts of known plaintext. These results are for the PKZIP contest file and are fairly typical, though the entry 12000 is unusually low. (logarithmic scaling).

We observed that the attack requires only 12–13 known plaintext bytes (as we describe later). Our idea is to use longer known plaintext streams, and to reduce the number of lists based on the additional plaintext. In particular, we are interested only in the values of $key2_{13}$, and not in all the list of $key2_i, i = 13, \ldots, n$. $key2_{13}$ is then used in the attack as is described above.

We start with the 2^{22} possible values of $key2_n$, and calculate the possible values of $key2_{n-1}$, $key2_{n-2}$, etc. using Equation 1. The number of possible values of $key2_i$ ($i = n-1, n-2$, etc.) remains about 2^{22}. However, some of the values are duplicates of other values. When these duplicates are discarded, the number of possible values of $key2_i$ is substantially decreased. To speed things up, we calculate all the possible values of $key2_{n-1}$, and remove the duplicates. Then we calculate all the possible values of $key2_{n-2}$, and remove the duplicates, and so on. When the duplicates fraction becomes smaller, we can remove the duplicates only every several bytes, to save overhead. Figure 1 shows the number of remaining values for any given size of known plaintext participating in the reduction, as was measured on the PKZIP contest file (which is typical). We observed that using about 40 known plaintext bytes (28 of them are used for the reduction and 12 for the attack), the number of possible values of $key2_{13}$ is reduced to about 2^{18}, and the complexity of the attack is 2^{34}. Using 10000-byte known plaintext, the complexity of our attack is reduced to 2^{24}–2^{27}.

3.3   key1

From the list of $(key2_n, key2_{n-1}, \ldots, key2_2)$ we can calculate the values of the most significant bytes the key1's by

$$\mathrm{MSB}(key1_{i+1}) = (key2_{i+1} \ll 8) \oplus \mathrm{crcinvtab}[\mathrm{MSB}(key2_{i+1})] \oplus key2_i.$$

We receive the list $(\mathrm{MSB}(key1_n), \mathrm{MSB}(key1_{n-1}), \ldots, \mathrm{MSB}(key1_3))$ (excluding $\mathrm{MSB}(key1_2)$).

Given $\mathrm{MSB}(key1_n)$ and $\mathrm{MSB}(key1_{n-1})$, we can calculate about 2^{16} values for the full values of $key1_n$ and $key1_{n-1} + \mathrm{LSB}(key0_n)$. This calculation can be done efficiently using lookup tables of size 256–1024. Note that

$$key1_{n-1} + \mathrm{LSB}(key0_n) = (key1_n - 1) \cdot 134775813^{-1} \pmod{2^{32}}$$

and that $\mathrm{LSB}(key0_n)$ is in the range $0, \ldots, 255$. At this point we have about $2^{11} \cdot 2^{16} = 2^{27}$ (or $2^{22} \cdot 2^{16} = 2^{38}$) possible lists of key2's and $key1_n$. Note that in the remainder of the attack no additional complexity is added, and all the additional operations contain a fixed number of instructions for each of the already existing list.

The values of $key1_{n-1} + \mathrm{LSB}(key0_n)$ are very close to the values of $key1_{n-1}$ (since we lack only the 8-bit value $\mathrm{LSB}(key0_n)$). Thus, an average of only $256 \cdot 2^{-8} = 1$ possible value of $key1_{n-1}$ that leads to the most significant byte of $key1_{n-2}$ from the list. This value can be found efficiently using the same lookup tables used for finding $key1_n$, with only a few operations. Then, we remain with a similar situation, in which $key1_{n-1}$ is known and we lack only eight bits of $key1_{n-2}$. We find $key1_{n-2}$ with the same algorithm, and then find the rest of $key1_{n-3}$, $key1_{n-4}$, and so on with the same algorithm. We result with about 2^{27} possible lists, each containing the values of $(key2_n, key2_{n-1}, \ldots, key2_2,$ and $key1_n, key1_{n-1}, \ldots, key1_4)$ (again, $key1_3$ cannot be fully recovered since two successive values of $\mathrm{MSB}(key1)$ are required to find each value of key1).

3.4   key0

Given a list of $(key1_n, key1_{n-1}, \ldots, key1_4)$, we can easily calculate the values of the least significant bytes of $(key0_n, key0_{n-1}, \ldots, key0_5)$ by

$$\mathrm{LSB}(key0_{i+1}) = ((key1_{i+1} - 1) \cdot 134775813^{-1}) - key1_i \pmod{2^{32}}.$$

$key0_{i+1}$ is calculated by

$$key0_{i+1} \leftarrow \mathrm{crc32}(key0_i, P_i)$$
$$= (key0_i \gg 8) \oplus \mathrm{crctab}[\mathrm{LSB}(key0_i) \oplus P_i].$$

Crc32 is a linear function, and from any four consecutive LSB(key0) values, together with the corresponding known plaintext bytes it is possible to recover the full four key0's. Moreover, given one full key0, it is possible to reconstruct all the other key0's by calculating forward and backward, when the plaintext bytes are given. Thus, we can now receive $key0_n, \ldots, key0_1$ (this time including $key0_1$). We can now compare the values of the least significant bytes of $key0_{n-4}$, $\ldots, key0_{n-7}$ to the corresponding values from the lists. Only a fraction of 2^{-32} of the lists satisfy the equality. Since we have only about 2^{27} possible lists, it is expected that only one list remain. This list must have the right values of the key0's, key1's, and key2's, and in particular the right values of $key0_n$, $key1_n$ and $key2_n$. In total we need 12 known plaintext bytes for this analysis (except for reducing the number of key2 lists) since in the lists the values of LSB($key0_i$) start with $i = 5$, and $n - 7 = 5 \Rightarrow n = 12$.

If no reduction of the number of key2 lists is performed, 2^{38} lists of (key0, key1, key2) remain at this point, rather than 2^{27}. Thus, we need to compare five bytes $key0_{n-4}, \ldots, key0_{n-8}$ in order to remain with only one list. In this case, 13 known plaintext bytes are required for the whole attack, and the complexity of analysis is 2^{38}.

3.5   The Internal Representation of the Key

Given $key0_n$, $key1_n$ and $key2_n$, it is possible to construct $key0_i$, $key1_i$ and $key2_i$ for any $i < n$ using only the ciphertext bytes, without using the known plaintext, and even if the known plaintext starts in the middle of the encrypted file this construction works and provides also the unknown plaintext and the 12 prepended bytes. In particular it can find the internal representation of the key, denoted by $key0_1$, $key1_1$ and $key2_1$ (where the index denotes again the index in the encrypted text, rather than in the known plaintext). The calculation is as follows:

$$
\begin{aligned}
key2_i &= crc32^{-1}(key2_{i+1}, MSB(key1_{i+1})) \\
key1_i &= ((key1_{i+1} - 1) * 134775813^{-1}) - LSB(key0_{i+1}) \pmod{2^{32}} \\
temp_i &= key2_i \mid 3 \\
key3_i &= LSB((temp_i * (temp_i \oplus 1)) \gg 8) \\
P_i &= C_i \oplus key3_i \\
key0_i &= crc32(key0_{i+1}, P_i)
\end{aligned}
\tag{2}
$$

The resulting value of ($key0_1$, $key1_1$, $key2_1$) is the internal representation of the key. It is independent of the plaintext and the prepended bytes, and depends only on the key. With this internal representation of the key we can decrypt any ciphertext encrypted under the same key. The two bytes of crc32 (one byte in version 2.04g) which are included in the 12 prepended bytes allow further verification that the file is really encrypted under the found internal representation of the key.

Key length	1-6	7	8	9	10	11	12	13
Complexity	1	2^8	2^{16}	2^{24}	2^{32}	2^{40}	2^{48}	2^{56}

Table 2. Complexity of finding the key itself

3.6   The Key (Password)

The internal representation of the key suffices to break the cipher. However, we can go even further and find the key itself from this internal representation with the complexities summarized in Table 2. The algorithm tries all key lengths 0, 1, 2, ..., up to some maximal length; for each key length it does as described in the following paragraphs.

For $l \leq 4$ it knows $key0_{1-l}$ and $key0_1$. Only $l \leq 4$ key bytes are entered to the crc32 calculations that update $key0_{1-l}$ into $key0_1$. Crc32 is a linear function, and these $l \leq 4$ key bytes can be recovered, just as $key0_n$, ..., $key0_{n-3}$ recovered above. Given the l key bytes, we reconstruct the internal representation, and verify that we get $key1_1$ and $key2_1$ as expected ($key0_1$ must be as expected, due to the construction). If the verification succeeds, we found the key (or an equivalent key). Otherwise, we try the next key length.

For $5 \leq l \leq 6$ we can calculate $key1_0$, $key2_0$ and $key2_{-1}$, as in Equation 2. Then, $key2_{2-l}$, ..., $key2_{-2}$ can be recovered since they are also calculated with crc32, and depend on $l - 2 \leq 4$ unknown bytes (of $key1$'s). These unknown bytes $MSB(key1_{2-l})$, ..., $MSB(key1_{-1})$ are also recovered at the same time. $key1_{1-l}$ is known. Thus, we can receive an average of one possible value for $key1_{2-l}$ and for $key1_{3-l}$, together with $LSB(key0_{2-l})$ and $LSB(key0_{3-l})$, using the same lookup tables used in the attack. From $LSB(key0_{2-l})$ and $LSB(key0_{3-l})$ and $key0_{1-l}$, we can complete $key0_{2-l}$ and $key0_{3-l}$ and get key_1 and key_2. The remaining $l - 2$ key bytes are found by solving the $l - 2 \leq 4$ missing bytes of the crc32 as is done for the case of $l \leq 4$. Finally, we verify that the received key has the expected internal representation. If so, we have found the key (or an equivalent key). Otherwise, we try the next key length.

For $l > 6$, we try all the possible values of key_1, ..., key_{l-6}, calculating $key0_{-5}$, $key1_{-5}$ and $key2_{-5}$. Then we used the $l = 6$ algorithm to find the remaining six key bytes. In total we try about $2^{8 \cdot (l-6)}$ keys. Only a fraction of 2^{-64} of them pass the verification (2^{-32} due to each of $key1$ and $key2$). Thus, we expect to remain with only the right key (or an equivalent) in trials of up to 13-byte keys. Note that keys longer than 12 bytes will almost always have equivalents with up to 12 (or sometimes 13) bytes, since the internal representation is only 12-bytes long.

4   Summary

In this paper we describe a new attack on the PKZIP stream cipher which finds the internal representation of the key, which suffices to decrypt the whole file

Bytes	13	40	110	310	510	1000	4000	10000
Complexity	2^{38}	2^{34}	2^{32}	2^{31}	2^{30}	2^{29}	2^{28}	2^{27}

Table 3. Complexity of the attack by the size of the known plaintext

and any other file which is encrypted by the same key. This known plaintext attack breaks the cipher using 40 (compressed) known plaintext bytes, or about the 200 first uncompressed bytes (if the file is compressed), with complexity 2^{34}. Using about 10000 known plaintext bytes, the complexity is reduced to about 2^{27}. Table 3 describes the complexity of the attack for various sizes of known plaintext. The original key (password) can be constructed from the internal representation. An implementation of this attack in software was applied against the PKZIP cipher contest. It found the key "f7 30 69 89 77 b1 20" (in hexadecimal) within a few hours on a personal computer.

A variant of the attack requires only 13 known plaintext bytes, in price of a higher complexity 2^{38}. Since the last two bytes (one in version 2.04g) of the 12 prepended bytes are always known, if the known plaintext portion of the file is in its beginning, the attack requires only 11 (12) known plaintext bytes of the compressed file. (In version 1.10 several additional prepended bytes might be predictable, thus the attack might actually require even fewer known plaintext bytes.)

We conclude that the PKZIP cipher is weak and that it should not be used to protect valuable information.

References

1. PKWARE, Inc., *General Format of a ZIP File*, technical note, included in PKZIP 1.10 distribution (pkz110.exe: file appnote.txt).

Linear Cryptanalysis of Stream Ciphers

Jovan Dj. Golić *

Information Security Research Centre, Queensland University of Technology
GPO Box 2434, Brisbane Q 4001, Australia
School of Electrical Engineering, University of Belgrade
Email: golic@fit.qut.edu.au

Abstract. Starting from recent results on a linear statistical weakness of keystream generators and on linear correlation properties of combiners with memory, linear cryptanalysis of stream ciphers based on the linear sequential circuit approximation of finite-state machines is introduced as a general method for assessing the strength of stream ciphers. The statistical weakness can be used to reduce the uncertainty of unknown plaintext and also to reconstruct the unknown structure of a keystream generator, regardless of the initial state. The linear correlations in arbitrary keystream generators can be used for divide and conquer correlation attacks on the initial state based on known plaintext or ciphertext only. Linear cryptanalysis of irregularly clocked shift registers as well as of arbitrary shift register based binary keystream generators proves to be feasible. In particular, the direct stream cipher mode of block ciphers, the basic summation generator, the shrinking generator, the clock-controlled cascade generator, and the modified linear congruential generators are analyzed. It generally appears that simple shift register based keystream generators are potentially vulnerable to linear cryptanalysis. A proposal of a novel, simple and secure keystream generator is also presented.

1 Introduction

Keystream generators for additive stream cipher applications can generally be realized as autonomous finite-state machines whose initial state and possibly structure as well depend on a secret key. Their practical security can be defined as immunity to various types of divide and conquer attacks on secret key based on known plaintext or ciphertext only, for a survey see [29] and [12]. Apart from that, keystream pseudorandom sequences should have large period, high linear complexity [18], [28], and should satisfy the standard key-independent statistical tests, which should prevent the reconstruction of statistically redundant plaintext from known ciphertext.

---

* This research was supported in part by the Science Fund of Serbia, grant #0403, through the Institute of Mathematics, Serbian Academy of Arts and Sciences.

A general structure-dependent and initial-state-independent linear statistical weakness of arbitrary binary keystream generators is pointed out and analyzed in [13]. It is based on the local properties of the keystream sequence on blocks of consecutive bits whose size is larger than the memory size, and is measured in terms of an appropriate correlation coefficient. The linear weakness can be regarded as a generalization of the linear complexity which is different from the concept of the linear complexity stability introduced in [7]. An effective method for detecting the weakness based on the *linear sequential circuit approxima-tion* (LSCA) [11] of autonomous finite-state machines is presented in [13]. If the structure is key-independent, then the corresponding statistical test can be used for the reconstruction of a statistically redundant plaintext from known ci-phertext. If the structure is key-dependent, then the same test can also be used to determine the corresponding unknown key, which presents a specific divide and conquer attack. The linear statistical weakness and the corresponding linear models and correlation coefficients are described in Section 2, including some unpublished results from [14]. Section 3 contains the basic lines of the LSCA method for arbitrary binary keystream generators. Besides its potential to de-termine linear statistical weaknesses in the keystream sequence, it is shown that the LSCA method can also be used to find out linear correlations between the keystream sequence and appropriate sequences depending on individual initial state variables. The linear correlations can then be used for divide and con-quer attacks on the initial state of keystream generators, see [30], [21]. Linear cryptanalysis of additive stream ciphers thus essentially reduces to the LSCA method.

Section 4 is devoted to linear cryptanalysis of clock-controlled shift registers and arbitrary shift register based keystream generators. It turns out that clock-controlled shift registers possess a detectable linear statistical weakness [13], but are immune to linear correlation attacks resulting from individual linear models. Note that simultaneous use of many different linear models may open new possibilities for correlation attacks. Regularly clocked shift registers are potentially vulnerable to correlation attacks, especially if the feedback is linear and if they are autonomous. Keystream generators based on a small number of shift registers appear to be vulnerable to linear cryptanalysis.

Linear cryptanalysis is then applied to concrete additive stream ciphers in-cluding the direct stream cipher mode of block ciphers, the basic summation generator [19, 29], the shrinking generator [6, 25], the clock-controlled cascade generator [3, 17, 16, 4], and the modified linear congruential generators [4] and [2], see Section 5. In Section 6, a proposal of a novel, simple and secure keystream generator is presented, which incorporates the principles of linear congruential generators, clock-controlled generators, and combiners with memory.

2   Linear Models for Stream Ciphers

A binary autonomous finite-state machine or sequential circuit is defined by

$$S_{t+1} = \mathcal{F}(S_t), \quad t \geq 0 \tag{1}$$

$$y_t = f(S_t), \quad t \geq 0 \tag{2}$$

where $\mathcal{F} : \mathrm{GF}(2)^M \rightarrow \mathrm{GF}(2)^M$ is a next-state vector Boolean function, $f :$ $\mathrm{GF}(2)^M \rightarrow \mathrm{GF}(2)$ is an output Boolean function, $S_t = (s_{1t}, \ldots, s_{Mt})$ is the state vector at time t, M is the number of memory bits, y_t is the output bit at time t, and $S_0 = (s_{1,0}, \ldots, s_{M,0})$ is the initial state. A binary keystream generator can be defined as a binary autonomous finite-state machine whose initial state and the next-state and output functions are controlled by a secret key.

Given $\mathcal{F}$ and f, each output bit is a Boolean function of the initial state variables, that is, $y_t = f(\mathcal{F}^t(S_0))$ where $\mathcal{F}^t$ denotes the t-fold self-composition of $\mathcal{F}$ and $\mathcal{F}^0$ is the identity function, $t \geq 0$. If S_0 is assumed to be a uniformly distributed random variable, then the output bits become binary random variables. A basic design criterion for f and $\mathcal{F}$, related to the statistics of the output sequence, is that each bit y_t should be a balanced function of S_0. This is clearly satisfied if both $\mathcal{F}$ and f are balanced. However, the vector of $M+1$ consecutive output bits $(y_t, \ldots, y_{t-M})$ can not be a balanced function of S_0 for any $t \geq M$, since S_0 has dimension only M. Therefore, there must exist a linear function $L(y_t, \ldots, y_{t-M})$ that is a nonbalanced function of S_0 for each $t \geq M$. When the next-state function is balanced, it follows that the state vector S_t is a balanced random variable at any time $t \geq 0$, provided that S_0 is balanced. The probability distribution of the linear function $L(y_t, \ldots, y_{t-M})$, treated as a function of S_{t-M}, is then the same for each $t \geq M$ and there exists such a linear function that effectively depends on y_t. The probability distribution can be expressed in terms of the correlation coefficient to the constant zero function. This essentially means that an autonomous finite-state machine can equivalently be represented as a non-autonomous linear feedback shift register of length at most M with an additive input sequence of nonbalanced identically distributed binary random variables, that is, by a linear model

$$y_t = \sum_{i=1}^{M} a_i \, y_{t-i} + e_t, \quad t \geq M. \tag{3}$$

The variables are not independent. The linear function L specified by the feedback polynomial applied to the output sequence $\{y_t\}$ produces a nonbalanced sequence $\{e_t\}$. The standard chi-square frequency statistical test can then be applied to $\{e_t\}$. To distinguish this sequence from the purely random binary sequence with error probability less than about 10^{-3}, the length of the observed keystream sequence should not be larger than $10/c^2$, see [30], [20], for example. Note that $c = 1 - 2\Pr\{e_t = 1\}$. For each additional bit of uncertainty to be resolved, one needs to know an additional segment of the keystream sequence of the same length. Since the linear function L is not unique in general, the maximum effect will be achieved when the linear function with the correlation coefficient of maximum magnitude is used. If this value is smaller than $2^{-M/2}$, then the keystream generator is not vulnerable to this statistical test. However, for large M, which is often the case in practice, it appears very difficult, if not impossible, to determine the value of the maximum correlation coefficient.

Another fact is even more discouraging from the cryptographer's viewpoint. Namely, one could also consider linear functions of more than $M + 1$ consecutive output bits. In particular, since every linear function can be defined as a polynomial in the generating function domain, it follows that one should consider all the polynomial multiples of the polynomials corresponding to linear functions of at most $M + 1$ variables. It appears very difficult to control all the corresponding correlation coefficients. Apart from that, one may simultaneously use all the obtained linear models and thus significantly reduce the required length of the observed keystream sequence.

The following result [14] determines the total correlation between the output sequence and the all zero sequence for autonomous finite-state machines. Let the next-state function of a binary autonomous finite-state machine with M memory bits be balanced. Then for any $m \geq 1$, the sum $C(m)$ of the squares of the correlation coefficients between all nonzero linear functions of m successive output bits y_t^m and the constant zero function is the same for every $t \geq m - 1$ and satisfies $\underline{C}(m) \leq C(m) \leq \bar{C}(m)$, where $\bar{C}(m) = 2^m - 1, m \geq 1$, and

$$\underline{C}(m) = \begin{cases} 0, & 1 \leq m \leq M \\ 2^{m-M} - 1, & m \geq M + 1 \end{cases}. \tag{4}$$

The minimum value $\underline{C}(m)$ is achieved for all $m \geq 1$ if and only if any M consecutive output bits constitute a balanced function of the initial state variables. The maximum value $\bar{C}(m)$ is achieved for any $m \geq 1$ if and only if the output function is constant. For any m, the total correlation is distributed among $2^m - 1$ output linear functions. It then follows that for each $m > M$ the maximum absolute value of the correlation coefficients can not be smaller than approximately $2^{-M/2}$ which corresponds to the uniform distribution of correlation, provided that the minimum total correlation condition is satisfied. For the condition to hold it is necessary that the output function is balanced. Large memory size is therefore an important design criterion. It is of course clear that minimum total correlation does not guarantee the uniform distribution, as is demonstrated by a linear feedback shift register.

3   Linear Sequential Circuit Approximation

In order to find all the nonbalanced linear functions of at most $M + 1$ consecutive output bits whose existence is established in the previous section, one should determine the correlation coefficients for 2^M Boolean functions of M variables. Exhaustive search method has $O(2^{2M})$ computational complexity, which is not practically possible for large M.

Taking the linear sequential circuit approximation (LSCA) approach introduced in [11] for combiners with memory, we propose a LSCA method for autonomous finite-state machines which is a feasible procedure that with high probability yields nonbalanced linear functions of at most $M + 1$ consecutive output bits with comparatively large correlation coefficients. The LSCA method consists of two stages. First, find a linear approximation of the output function f

and each of the component functions of the next-state function $\mathcal{F}$. This enables one to express each of these $M + 1$ functions as the sum of a linear function and a nonbalanced function, whose correlation coefficient is different from zero. In practice, both the output function and the component next-state functions effectively depend on small subsets of the state variables or can be expressed as compositions of such functions. Therefore, the computational complexity of obtaining all the linear approximations along with the corresponding nonzero correlation coefficients is considerably smaller than $O((M + 1)M2^M)$, which is required by the direct application of the Walsh transform technique, see [29]. Finding good linear approximations of Boolean functions in real ciphers thus appears to be a feasible task.

Second, by virtue of the obtained linear approximations, put the basic equations (1) and (2) into the form

$$S_{t+1} = \mathbf{A}S_t + \Delta(S_t), \quad t \geq 0 \tag{5}$$

$$y_t = \mathbf{B}S_t + \varepsilon(S_t), \quad t \geq 0 \tag{6}$$

where the vectors are regarded as one-column matrices, $\mathbf{A}$ and $\mathbf{B}$ are binary matrices, and ε and all the components of $\Delta = (\delta_1, \ldots, \delta_M)$ are nonbalanced Boolean functions, called the noise functions. The main point now is to treat $\{\varepsilon(S_t)\}$ and $\{\delta_j(S_t)\}$, $1 \leq j \leq M$, as the input sequences so that (5) and (6) define a non-autonomous linear sequential circuit (LSC), see [11]. Then solve the LSC using the generating function technique and thus obtain

$$\mathbf{y} = \frac{1}{\varphi(z)} \sum_{j=1}^{M} g_j(z)\, s_{j0} + \frac{1}{\varphi(z)} \sum_{j=1}^{M} z\, g_j(z)\, \delta_j + \varepsilon \tag{7}$$

where $\mathbf{y}$, δ_j, and ε respectively denote the generating functions in variable z of the sequences $\{y_t\}$, $\{\delta_j(S_t)\}$, and $\{\varepsilon(S_t)\}$, and the polynomial $\varphi(z) = \sum_{i=0}^{M} \varphi_i z^i$, $\varphi_0 = 1$, is the reciprocal of the characteristic polynomial of the state-transition matrix $\mathbf{A}$. As a consequence of (7), we also get

$$\sum_{i=0}^{M} \varphi_i\, y_{t-i} = \sum_{i=0}^{M} \varphi_i\, \varepsilon(S_{t-i}) + \sum_{j=1}^{M} \sum_{i=0}^{M-1} g_{ji}\, \delta_j(S_{t-1-i}) \stackrel{\text{def}}{=} e_t(S_0), \quad t \geq M. \tag{8}$$

The computational complexity to obtain (7) and (8) is only $O(M^3)$. For each $t \geq M$, the noise function e_t is a sum of individual noise functions that are nonbalanced if $S_{t-i}, 0 \leq i \leq M$, are balanced. If one assumes that the next-state function $\mathcal{F}$ is balanced, then it follows that each of the individual noise functions in (8) is nonbalanced and identically distributed for every $t \geq M$, meaning that the corresponding correlation coefficients are nonzero and independent of t. In general, for random $\mathcal{F}$ and f, one should expect that the individual noise functions remain nonbalanced even if $S_{t-i}, 0 \leq i \leq M$, are not balanced functions of S_0 and that the resulting noise function e_t is also nonbalanced, for almost all $t \geq M$. This conclusion is justified by the following probabilistic result, see [13], which is also relevant for the linear cryptanalysis of block ciphers.

Lemma 1. Consider m Boolean functions of the same n variables with the correlation coefficients c_i to the constant zero function, $1 \le i \le n$. If the functions are chosen uniformly and independently at random, then for large 2^n the probability distribution of the correlation coefficient of their modulo two sum is asymptotically normal with the expected value $\prod_{i=1}^{m} c_i$ and the variance $O(\frac{m}{2^n})$.

$\diamond$

The described LSCA method is based on the linear approximations of the component next-state functions, which is a limitation. However, the method can be generalized to deal with the linear approximations of the linear combinations of the component next-state functions. To this end, let $\mathcal{L}$ denote an arbitrary balanced, that is, one-to-one linear function $\mathrm{GF}(2)^M \to \mathrm{GF}(2)^M$ and let $S_t' = \mathcal{L}(S_t)$ denote the transformed state vector at time t. Accordingly, one may put the equations (1) and (2) into an equivalent form $S_{t+1}' = \mathcal{L}\mathcal{F}\mathcal{L}^{-1}(S_t'), t \ge 0$, and $y_t = f\mathcal{L}^{-1}(S_t'), t \ge 0$, and then proceed with the basic LSCA method, as already defined. The linearization of the component functions of the transformed next-state function $\mathcal{L}\mathcal{F}\mathcal{L}^{-1}$ is essentially the same as the linearization of the linear combinations of the component functions of the original next-state function F.

Starting from (7), one may also develop divide and conquer correlation attacks on the individual bits of the initial state. The transfer function with respect to s_{j0} is given by $g_j'(z)/\varphi_j(z) = g_j(z)/\varphi(z)$ where $g_j'(z)$ and $\varphi_j(z)$ are relatively prime, $1 \le j \le M$. The denominator polynomials $\varphi_j(z)$ induce an equivalence relation among s_{j0} or just $j, 1 \le j \le M$. Let $\bar{\varphi}_j(z)$ denote the least common multiple of $\varphi_k(z)$ for all k not belonging to the equivalence class of j. Then for a single equation (7), the initial correlation attack on s_{j0} is possible if and only if the j-th component of the next-state function is linear ($\delta_j = 0$) and $\varphi_j(z) \nmid \bar{\varphi}_j(z)$. The degree of $\varphi_j(z)/(\varphi_j(z), \bar{\varphi}_j(z))$ determines the number of bits of uncertainty resolved by the attack on the equivalence class of s_{j0}. By subtracting the effect of the initially determined bits of the initial state from the right-hand side of (7), possibly including the noise functions as well, one then recomputes the output sequence and repeats the procedure iteratively. It follows that regularly clocked linear feedback shift registers are potentially vulnerable to correlation attacks. Note that for correlation attacks it may not be necessary to linearize the whole generator. Furthermore, if one simultaneously uses several linear sequential circuit approximations, then other possibilities for correlation attacks may exist as well.

It is desirable for the LSCA method to find a linear function/model with the maximum absolute value of the correlation coefficient. To this end, the number of noise terms in (8) should be small and their correlation coefficients should be large in magnitude. A reasonable approach is to repeat the procedure several times starting from the best linear approximations of the output and next-state functions. In fact, one should tend to find an optimum invertible set of the linear combinations of the component next-state functions that yields the maximum absolute value of the overall correlation coefficient. The power of the chi-square statistical test can be considerably improved by running the test on all the obtained linear models, rather than on a single one. In order to achieve the immunity to the LSCA attack, it follows that the memory size should be large and it appears recommendable that the output function and the linear combinations of the component next-state functions should have large distance from affine functions as well as that the component next-state functions should effectively depend on large subsets of the state variables.

4   Shift Register Based Keystream Generators

In this section, the LSCA method is applied to an arbitrary binary keystream generator consisting of regularly or irregularly clocked shift registers (SRs) combined by a function with or without memory. Clock-control sequences are produced either within the generator or by separate generators. One should first form the linear models for individual SRs: regularly clocked linear feedback SRs stay as they are, linear models for regularly clocked nonlinear feedback SRs are made by linearizing the feedback functions, and linear models for irregularly clocked SRs are formed as follows.

A clock-controlled shift register is a keystream generator consisting of a linear or nonlinear feedback shift register that is irregularly clocked according to an integer decimation sequence, which defines the number of clocks per output symbol and which is itself produced by a pseudorandom sequence generator, see [16] and [9]. More precisely, if $X = \{x_t\}_{t=0}^{\infty}$ denotes a regularly clocked shift register sequence and $D = \{d_t\}_{t=0}^{\infty}$ a decimation sequence, then the output sequence $Y = \{y_t\}_{t=0}^{\infty}$ is defined as a decimated sequence $y_t = x(\sum_{i=0}^{t} d_i)$, $t \geq 0$. First observe that a nonlinear feedback can in principle be treated in the same way as linear, except for the additive noise function. Second, assume a realistic probabilistic model for the decimation sequence, for example, assume that D is a sequence of identically distributed integer random variables with a probability distribution $\mathcal{P} = \{P(d)\}_{d \in \mathcal{D}}$ where $\mathcal{D}$ is the set of integers with positive probability. When $\mathcal{D}$ contains positive integers only, one can also define the deletion rate p_d as $1 - \frac{1}{\bar{d}}, \bar{d} = \sum_{d \in \mathcal{D}} dP(d)$.

We will distinguish between the two cases: the case with possible repetitions ($0 \in \mathcal{D}$) and the case without repetitions ($0 \notin \mathcal{D}$). In the first case, it is clear that regardless of the feedback $y_t + y_{t-1} = e_t, t \geq 1$, where the correlation coefficient of e_t is equal to $P(0)$. For the stop-and-go registers [3], for which $\mathcal{D} = \{0, 1\}$, $P(0) = 1/2$. In the second case, consider a clock-controlled linear feedback shift register of length r with the feedback polynomial $f(z) = 1 + \sum_{k=1}^{w} z^{i_k}, 1 \leq i_1 < \ldots < i_w = r$, where $W = w+1$ is the weight of $f(z)$. By using the LSCA method or directly, one can obtain [13] a linear model of the form

$$y_t + \sum_{k=1}^{w} y_{t-i_k} = e_t, \quad t \geq r' \tag{9}$$

where the polynomial $\hat{f}(z) = 1 + \sum_{i=1}^{r'} \hat{f}_i z^i = 1 + \sum_{k=1}^{w} z^{\hat{i}_k}, 1 \leq \hat{i}_1 < \ldots < \hat{i}_w = r'$, satisfies $\hat{i}_k - \hat{i}_{k-1} \leq i_k - i_{k-1}, 1 \leq k \leq w$, where $\hat{i}_0 = i_0 = 0$. Call $\hat{f}(z)$ a *shrunk polynomial* of $f(z)$. A shrunk polynomial of $f(z)$ is not unique but has the same weight as $f(z)$. It is possible to obtain an expression for the correlation coefficient of e_t for an arbitrary probability distribution $\mathcal{P} = \{P(d)\}_{d \in \mathcal{D}}$. For simplicity, we give only the expression for the geometric distribution $P(d) = p^{d-1}(1-p), d \geq 1$, which corresponds to the case of independent deletions with probability p. Note that an arbitrary $\mathcal{P}$ can be approximated by this distribution by setting $p = p_d$ where p_d is the deletion rate. It follows that the correlation coefficient in this case is given by

$$c = p^{r-r'}(1-p)^{r'+1} \prod_{k=1}^{w} \binom{\Delta_k}{\hat{\Delta}_k} \tag{10}$$

where $\Delta_k = i_k - i_{k-1} - 1$ and $\hat{\Delta}_k = \hat{i}_k - \hat{i}_{k-1} - 1, 1 \leq k \leq w$. Equation (10) has a clear combinatorial meaning in terms of the probability of decimation sequences. Namely, the correlation coefficient is equal to the probability of the event that the bits satisfying the feedback polynomial in the shift register sequence remain undeleted in such a way that they satisfy the shrunk feedback polynomial in the decimated sequence. It is assumed that the conditional correlation coefficient is equal to one when the event occurs and to zero otherwise. The·coefficient is maximized if $\hat{\Delta}_k = \lfloor (1-p)(\Delta_k + 1) \rfloor, 1 \leq k \leq w$. It follows that $c = 1$ if $p = 0$ and $c = 0$ if $p = 1$, which is natural. Suppose that r/w, Δ_k, $p\Delta_k$, and $(1-p)\Delta_k$ are all large. Then Stirling's approximation gives

$$c \sim (1-p) \left(\frac{2\pi p}{1-p} \right)^{-\frac{w}{2}} \left(\prod_{k=1}^{w} \Delta_k \right)^{-\frac{1}{2}} \tag{11}$$

which shows that the magnitude of c may be considerably larger than $2^{-r/2}$ let alone $2^{-M/2}$, M being the memory size of the whole generator. The smallest magnitude of c is obtained when the feedback taps are approximately equidistant. In this case the necessary length of the keystream sequence needed to detect the weakness is $(10/(1-p)^2)(2\pi p/(1-p))^w((r-w)/w)^w$. Given w, the larger the values of r and p the smaller the correlation coefficient, respectively. Given r and p, there exists an optimal value of w that minimizes the correlation coefficient. For a given feedback polynomial, one may use different shrunk polynomials or their polynomial multiples and thus obtain different linear models. This reduces the required length of the observed keystream sequence considerably. In addition to that, one may also use different polynomial multiples of the feedback polynomial, especially the ones with low weight, possibly much lower than for the original polynomial. So, the weakness is in general easily detected if the feedback polynomial is known.

By using the LSCA method from Section 3 one then develops a linear model for the function with memory, which is treated as a non-autonomous finite-state machine with purely random input sequences, see [11]. Consequently, one obtains a linear equation of the form (8) whose right-hand side contains the additional input sequences as well. The linear function on the left-hand side of (8) corresponds to the reciprocal $\varphi(z)$ of the characteristic polynomial of the state-transition matrix. Finally, one substitutes the outputs of the linearized SRs for the inputs to the linear sequential circuit and thus derives a linear model with the feedback polynomial being equal to the least common multiple of $\varphi(z)$ and the feedback polynomials of all the linearized SRs. For estimating the overall correlation coefficient, a reasonable assumption is that the noise sequences from the linearized SRs are mutually independent and independent from the noise sequences in the linear sequential circuit, unless the SRs are connected in a very special way. Various linear models are obtained by varying the linear models for irregularly clocked and nonlinear feedback SRs and the linear model for the function with memory. Polynomial multiples may also be used especially if they reduce the magnitude of the overall correlation coefficient. In fact, it is the polynomial multiples that make it very difficult to achieve the security against the LSCA attack.

5   Concrete Keystream Generators

We now apply the linear cryptanalysis to several types of the shift register based binary keystream generators.

Direct stream cipher mode for block ciphers

The Direct Stream Cipher (DSC) mode of operation of block ciphers can be defined as a particular case of the Output Feedback (OFB) mode in which only a single output bit is used to produce the keystream sequence and the initial state is key-controlled. The keystream bit can also be generated by a simple output function of several state bits, for example, by a modulo 2 addition. Since in this case a known plaintext does not provide pairs of block cipher inputs and outputs, the available cryptanalytic techniques for block ciphers [1] and [20] are not directly applicable. Note that some possibilities for differential and/or linear cryptanalysis of block ciphers in the CBC, CFB, or regular OFB mode have been explored in [26] and [27], assuming a partial knowledge of the ciphertext. The DSC mode of a block cipher is nothing but a stream cipher whose next-state function is defined by the block cipher, where the initial state is key-controlled. Therefore, the LSCA method for linear cryptanalysis of stream ciphers can be used. It may yield a linear statistical weakness of the keystream sequence and may be a basis for divide and conquer attacks on the secret key. The starting point of the method is to find an invertible set of the linear functions of the block cipher output with relatively large correlation coefficients to linear functions of the input. The next point is to solve the corresponding linear sequential circuit by the generating function technique. To minimize the resulting correlation co-efficient, the procedure is repeated several times using different invertible linear approximations of the block cipher. The method is computationally feasible for real ciphers. As a consequence, one can also obtain novel characteristics of block ciphers such as the characteristic polynomials and the corresponding correlation coefficients. Linear cryptanalysis of concrete block ciphers in the DSC mode is not an easy task and is out of the scope of this paper. It would be interesting to investigate whether the immunity of a block cipher to the linear cryptanalysis [20] in the ECB mode implies the immunity to the linear cryptanalysis in the DSC mode.

Basic summation generator [19, 29]

The basic summation generator is a combiner with one bit of memory and two regularly clocked linear feedback SRs. Its output function is already linear, whereas its next-state function can be linearized in several ways with large correlation coefficients. A good way is to take a linear function depending on a single input with the correlation coefficient $1/2$. The feedback polynomial of the corresponding linear model is just the least common multiple of the feedback polynomials of the two input SRs, or any of its polynomial multiples. The overall correlation coefficient is $(1/2)^W$ where W is the weight of this polynomial. In addition, as was already noted in [22], the output is correlated to the sum of one input and the linear transform $1 + z$ of the other with the correlation coefficient $1/2$ which is highly vulnerable to fast correlation attacks, see [21] and [5], for example. Linear cryptanalysis of a general summation generator consisting of more than two SRs, as is suggested in [19], remains an open problem.

Shrinking generator [6, 23], [25]

This is a single irregularly clocked linear feedback SR whose clock is controlled by another linear feedback SR in a manner that corresponds to independent deletions with probability $p = 1/2$. The principle has first appeared in [25], but is implicit in [10] as well. The two SRs may even be the same as was suggested in [23]. Section 4 then gives a linear model with the feedback polynomial equal to a shrunk polynomial of the feedback polynomial of the irregularly clocked SR or of any of its polynomial multiples. The correlation coefficient is given by (10) or (11) for $p = 1/2$. For a single shrunk polynomial of a polynomial of weight $W = w + 1$ and relatively large degree r, the required length of the keystream sequence to detect the weakness is thus about $40 (6.28 \, r/w)^w$, where the taps are assumed to be approximately equidistant. If for example $w = 4$, then the length is about $243 \, r^4$. When the feedback polynomial is key-dependent [6], the weakness may be used to determine the corresponding key from a known keystream sequence or even from ciphertext only. The results [6] of the statistical analysis of the shrinking generator are somewhat misleading for two reasons. First, the statistical properties are not nice on blocks whose length exceeds the length of the clock-controlled SR and, second, the key-dependent feedback polynomial is assumed to be selected uniformly at random.

Clock-controlled cascade generator [3, 17, 16, 4]

This is a cascade of K linear feedback SRs with the same feedback polynomial $f(z)$ of degree r. The first SR is clocked regularly and the others are clocked either k or m times per each output bit. For $k, m > 0$, by using a model for a single irregularly clocked shift register with deletion rate $p = 1 - 2/(k + m)$, one obtains a linear model with the feedback polynomial $\hat{f}(z) f(z)$ and the overall correlation coefficient $c^{(K-1)W}$ where W is the weight of $f(z)$, $\hat{f}(z)$ is a shrunk polynomial of $f(z)$, and c is given by (10) or (11). This is an approximation: the actual c is in fact different because the irregular clocking is constrained rather than independent. For a stop-and-go cascade ($k = 0, m > 0$), instead of $\hat{f}(z)$ one should use $1 + z$ and the correlation coefficient $c = 1/2$. Instead of $f(z)$ one may also take any of its polynomial multiples. For example, if $k = 1$ and $m = 2$, then $p = 1/3$ and $c \sim 0.66 \, (3.14 \, r/w)^{-w/2}$, so that the required length to detect the weakness is approximately $10 \, (2.25 \, (3.14 \, r/w)^w)^{(w+1)(K-1)}$. If one takes into account the constrained clocking, then the required length becomes $10 \, (2.25 \, (4.19 \, r/w)^w)^{(w+1)(K-1)}$. This length can be reduced considerably by using many different shrunk polynomials instead of a single one.

Apart from the described statistical weakness, the linear transform $\hat{f}(z)$ of the output of the cascade is correlated to the same linear transform of the output of the first shift register, with the correlation coefficient c^{K-1}. For $k = 1$ and $m = 2$, the required keystream sequence length for the successful correlation attack on the initial state of the first shift register is then $10 \, r \, (2.25 \, (4.19 \, r/w)^w)^{K-1}$. This of course implies an exhaustive search through all possible initial states. Fast correlation attacks might also be feasible, see [21] and [5], for example. This is in accordance with the recent statistical analysis of a stop-and-go cascade from [24]. Both the weaknesses diminish as K increases, but the efficiency of the generator remains the same. The choice of small SR length r does not seem to be appropriate, because it might be an open gate for algebraic cryptanalysis. On the other hand, if the SR length and K are both large, then the generator is not efficient.

Modified linear congruential generators [4], [2, 6]

This type of keystream generators is based on linear recursions modulo 2^m for a positive integer m, which may be chosen to be relatively large, such as 32, as was suggested in [2]. Since linear recurring truncated integer sequences are in principle predictable, for example, see [8], various modifications have been suggested. In [4], it is suggested to use a single truncated integer sequence generated by the bitwise sum modulo 2 (which is nonlinear modulo 2^m) of two feedbacks linear modulo 2^m. In [2, 6], it is proposed to use two simple linear recursions with fixed binary coefficients without truncation and the unconstrained clock-control principle [25, 6]. Linear cryptanalysis of modified linear congruential generators should start from a linear approximation of the feedback function, which is nonlinear over the binary field. A linearized scheme is then a set of non-autonomous binary linear feedback shift registers, one for each order of significance [4], with additive inputs whose correlation coefficients can be derived by using the results from [31]. Note that the shift register for the lowest order of significance is autonomous. The shift register lengths are upper-bounded by the order of the linear recursion which is relatively small compared to the memory size. Interestingly enough, it turns out that the correlation coefficient c_i remains biased when the order of significance i increases if the number w of integers to be added is odd. For example, $c_i \to -0.3333$ when $w = 3$ and $c_i \to 0.1333$ when $w = 5$, see

[31]. This is a potential trapdoor. If w is even, then c_i tends to zero like $2^{-iw/2}$. The results are slightly different for a modified recursion [4] with nonbinary coefficients. In any case, it follows that the cryptographic strength of the output binary sequences with respect to linear cryptanalysis strongly depends on the order of significance, which is not good. Possible use of low weight polynomial multiples modulo 2^m to reduce w might also be studied. Apart from that, simple modified linear congruences are potentially vulnerable to linear cryptanalysis modulo 2^m. For example, the modulo 2 sum of two integers, as suggested in [4], is correlated to their sum modulo 2^m, and the corresponding correlation coefficient c_i then behaves like 2^{-i}.

For the generator [2], $w = 2$ and there is no truncation, so that the linear statistical weakness of low order output binary sequences is easily detectable despite the irregular clocking. Note that the output feedforward function can be linearly approximated with large correlation coefficients and hence does not make much of a difference with respect to the linear cryptanalysis over the binary field.

6 Proposal

We now propose a novel scheme which is a self-clock-controlled modified linear congruential generator with a nonlinear feedforward function with memory. The first two parts are very simple to realize in software or hardware, whereas the third part is very simple to implement in hardware. The proposal is given a name GOAL.

First pick at random a primitive binary polynomial f of degree n not smaller than 100 and of weight $W = w + 1$ not smaller than 5. The polynomial should not have 'low' degree trinomial multiples, which is easily checked, and may be controlled by a secret key. The polynomial defines a linear congruence modulo 2^{32} with w nonzero binary coefficients. The initial conditions are controlled by the secret key. The 32-bit integer feedback is circularly shifted so that the least significant bit becomes the most significant one. The modified feedback is split into two 16-bit parts which are bitwise added modulo 2 to form the 16-bit output of the modified linear congruence.

Each of the 16 binary output sequences is transformed by a combiner with 15 bits of memory and a single input and output, respectively. All the 16 combiners have the same next-state function defined as a (16×15)-bit table, while the output function is the sum modulo 2 of the input bit and one of the state bits. The table is generated at random so that the maximum squared correlation coefficient between the input and output linear functions is 'close' to 2^{-16}, which is not difficult. This criterion along with relatively large memory size is in accordance with the results from [15]. The table can be stored on a single 1Mbit chip and may be controlled by the secret key. The 16 15-bit initial state vectors may also be controlled by the key. The individual combiners may be different, but that requires more space.

A constrained (1,2)-clock-control is defined by the sum modulo 2 of all the 16 bits from a previous output of the modified linear congruence that is not used in forming the current feedback. If the control bit is 1, then the output is discarded and the congruence is computed once more and transformed by the combiners with memory to form the current 16-bit integer output. Note that the constrained clocking is cryptographically weaker than unconstrained, but is faster and does not give rise to buffer-control problems. On the average, it takes 3 modified linear congruence computations to generate each 32 output bits in our scheme.

Preliminary analysis of the proposed generator suggests the following conclusions. Algebraic properties of the modified linear congruence, as the period and the linear complexity, and the distribution over a period of blocks of output integers whose size does not exceed n, may in principle be derived. For the generator as a whole, both the period and the linear complexity are almost certainly lower-bounded by 2^{n+5} and very likely by 2^{16n} as well. Furthermore, it may well be the case that they are close to 2^{32n}, see [9]. On the other hand, if one assumes that the modified linear congruence produces a purely random integer sequence, then the self-clock-control and the function with memory, defined as above, ensure that the output sequence is also purely random.

The generator is resistant against the linear cryptanalysis modulo 2, because of the circular shift feedback operation which results in balanced correlation coefficients in a linear model with the feedback polynomial $f(x^{32})$ of large degree and because of the clock-control and the function with memory applied before the clock-control. It is also immune to the linear cryptanalysis modulo 2^{32}, because of the circular shift operation which is nonlinear modulo 2^{32} and the feedforward bitwise addition modulo 2 which reduces 32-bit integers to 16-bit integers and because of the clock-control and the function with memory. Other divide and conquer attacks are very unlikely since the internal state variables of the proposed keystream generator as a binary autonomous finite-state machine are well mixed both in the modified linear congruence and in the clock-control. Finally, by changing the parameters it is easy to increase or decrease the security of the proposed generator. For example, instead of selecting a (16×15)-bit table at random, one may choose a table easy to realize in software.

7   Conclusion

By combining the recent results on a linear statistical weakness of arbitrary keystream generators [13] and on linear correlation properties of combiners with memory [11], a novel general method for assessing the strength of stream ciphers is proposed. The method is based on the linear sequential circuit approximation of finite-state machines and is called the linear cryptanalysis of stream ciphers. It results in a linear statistical weakness of the keystream sequence on blocks of consecutive output bits whose size is larger than the memory size as well as in correlations between feedforward linear transforms of the keystream sequence and linear transforms of the individual initial state variables. The statistical weakness can be used to reduce the uncertainty of unknown plaintext and also to reconstruct the unknown structure of a keystream generator, regardless of the initial state. Linear correlations can be used for divide and conquer attacks on the initial state of keystream generators based on known plaintext or ciphertext only, see [30], [21]. The effectiveness of linear cryptanalysis can be measured in terms of the corresponding correlation coefficients. Linear cryptanalysis of block ciphers [20] proves to be a special case of linear cryptanalysis of stream ciphers.

Linear cryptanalysis of irregularly clocked shift registers as well as of arbitrary binary keystream generators based on regularly or irregularly clocked shift registers, with linear or nonlinear feedback, combined by a function with or without memory is shown to be feasible. It turns out that clock-controlled shift registers possess a detectable linear statistical weakness, but are immune to linear correlation attacks resulting from individual linear models. However, simultaneous use of many different linear models may open new possibilities for correlation attacks. Regularly clocked shift registers are potentially vulnerable to correlation attacks, especially if the feedback is linear and if they are autonomous. In particular, the direct stream cipher mode of block ciphers, the basic summation generator, the shrinking generator, the clock-controlled cascade generator, and the modified linear congruential generators are analyzed. One may generally conclude that simple shift register based keystream generators are potentially vulnerable to linear cryptanalysis, especially if the number of shift registers is relatively small. A proposal of a novel, simple and secure keystream generator based on a modified linear congruential scheme, a self-clock-control principle, and combiners with memory is also presented.

References

1. E. Biham and A. Shamir, "Differential cryptanalysis of DES-like cryptosystems," *Journal of Cryptology*, 4(1):3–72, 1991.
2. U. Blöcher and M. Dichtl, "Fish: a fast software stream cipher," Fast Software Encryption – Cambridge '93, *Lecture Notes of Computer Science*, vol. 809, R. Anderson ed., Springer-Verlag, pp. 41–44, 1994.
3. W. G. Chambers and D. Gollmann, "Lock-in effect in cascades of clock-controlled shift registers," Advances in Cryptology – EUROCRYPT '88, *Lecture Notes in Computer Science*, vol. 330, C. G. Günther ed., Springer-Verlag, pp. 331–342, 1988.
4. W. G. Chambers, "Two stream ciphers," Fast Software Encryption – Cambridge '93, *Lecture Notes of Computer Science*, vol. 809, R. Anderson ed., Springer-Verlag, pp. 51–55, 1994.
5. V. Chepyzhov and B. Smeets, "On a fast correlation attack on stream ciphers," Advances in Cryptology – EUROCRYPT '91, *Lecture Notes in Computer Science*, vol. 547, D. V. Davies ed., Springer-Verlag, pp. 176–185, 1991.
6. D. Coppersmith, H. Krawczyk, and Y. Mansour, "The shrinking generator," Advances in Cryptology – CRYPTO '93, *Lecture Notes in Computer Science*, vol. 773, D. R. Stinson ed., Springer-Verlag, pp. 22–39, 1994.
7. C. Ding, G. Xiao, and W. Shan, *The Stability Theory of Stream Ciphers. Lecture Notes in Computer Science*, vol. 561, Springer-Verlag, 1991.

8. A. M. Frieze, J. Hastad, R. Kannan, J. C. Lagarias, and A. Shamir, "Reconstructing truncated integer variables satisfying linear congruences," *SIAM J. Comput.*, 17:262–280, 1988.

9. J. Dj. Golić and M. V. Živković, "On the linear complexity of nonuniformly decimated PN-sequences," *IEEE Trans. Inform. Theory*, 34:1077–1079, Sep. 1988.

10. J. Dj. Golić and M. J. Mihaljević, "A generalized correlation attack on a class of stream ciphers based on the Levenshtein distance," *Journal of Cryptology*, 3(3):201–212, 1991.

11. J. Dj. Golić, "Correlation via linear sequential circuit approximation of combiners with memory," Advances in Cryptology – EUROCRYPT '92, *Lecture Notes in Computer Science*, vol. 658, R. A Rueppel ed., Springer-Verlag, pp. 113–123, 1993.

12. J. Dj. Golić, "On the security of shift register based keystream generators," Fast Software Encryption – Cambridge '93, *Lecture Notes of Computer Science*, vol. 809, R. J. Anderson ed., Springer-Verlag, pp. 90–100, 1994.

13. J. Dj. Golić, "Intrinsic statistical weakness of keystream generators," *Pre-proceedings of Asiacrypt '94*, pp. 72–83, Wollongong, Australia, 1994.

14. J. Dj. Golić, "Linear models for keystream generators," to appear in *IEEE Trans. Computers*.

15. J. Dj. Golić, "Correlation properties of a general binary combiner with memory," to appear in *Journal of Cryptology*.

16. D. Gollmann and W. G. Chambers, "Clock controlled shift registers: a review," *IEEE J. Sel. Ar. Commun.*, 7(4):525–533, 1989.

17. D. Gollmann and W. G. Chambers, "A cryptanalysis of step$_{k,m}$-cascades," Advances in Cryptology – EUROCRYPT '89, *Lecture Notes in Computer Science*, vol. 434, J.-J. Quisquater and J. Vandewalle eds., Springer-Verlag, pp. 680–687, 1990.

18. J. L. Massey, "Shift-register synthesis and BCH decoding," *IEEE Trans. Inform. Theory*, 15:122–127, Jan. 1969.

19. J. L. Massey and R. A. Rueppel, "Method of, and apparatus for, transforming a digital sequence into an encoded form" U. S. Patent No. 4,797,922, 1989.

20. M. Matsui, "Linear cryptanalysis method for DES cipher," Advances in Cryptology – EUROCRYPT '93, *Lecture Notes in Computer Science*, vol. 765, T. Helleseth ed., Springer-Verlag, pp. 386–387, 1994.

21. W. Meier and O. Staffelbach, "Fast correlation attacks on certain stream ciphers," *Journal of Cryptology*, 1(3):159–176, 1989.

22. W. Meier and O. Staffelbach, "Correlation properties of combiners with memory in stream ciphers," *Journal of Cryptology*, 5(1):67–86, 1992.

23. W. Meier and O. Staffelbach, "The self-shrinking generator," *Pre-proceedings of Eurocrypt '94*, Perugia, Italy, pp. 201–210, 1994.

24. R. Menicocci, "Short Gollmann cascade generators may be insecure," CODES AND CYPHERS, Cryptography and Coding IV, P. G. Farrell ed., The Institute of Mathematics and its Applications, pp. 281–297, 1995.

25. M. J. Mihaljević, "An approach to the initial state reconstruction of a clock-controlled shift register based on a novel distance measure," Advances in Cryptology – AUSCRYPT '92, *Lecture Notes in Computer Science*, vol. 718, J. Seberry and Y. Zheng eds., Springer-Verlag, pp. 349–356, 1993.

26. K. Ohta and M. Matsui, "Differential attack on message authentication codes," Advances in Cryptology – CRYPTO '93, *Lecture Notes in Computer Science*, vol. 773, D. R. Stinson ed., Springer-Verlag, pp. 200–211, 1994.

27. B. Preneel, M. Nuttin, V. Rijmen, and J. Buelens, "Cryptanalysis of the CFB mode of the DES with a reduced number of rounds," Advances in Cryptology – CRYPTO '93, *Lecture Notes in Computer Science*, vol. 773, D. R. Stinson ed., Springer-Verlag, pp. 212–223, 1994.

28. R. A. Rueppel, *Analysis and Design of Stream Ciphers*. Berlin: Springer-Verlag, 1986.

29. R. A. Rueppel, "Stream ciphers," in *Contemporary Cryptology: The Science of Information Integrity*, G. Simmons ed., pp. 65–134. New York: IEEE Press, 1991.

30. T. Siegenthaler, "Decrypting a class of stream ciphers using ciphertext only," *IEEE Trans. Comput.*, 34:81–85, Jan. 1985.

31. O. Staffelbach and W. Meier, "Cryptographic significance of the carry for ciphers based on integer addition," Advances in Cryptology – CRYPTO '90, *Lecture Notes in Computer Science*, vol. 537, A. J. Menezes and S. A. Vanstone eds., Springer-Verlag, pp. 601–614, 1991.

Feedback with Carry Shift Registers over Finite Fields (Extended Abstract)

Andrew Klapper*

Dept. of Computer Science
763H Anderson Hall
University of Kentucky, Lexington
KY 40506-0046 USA
klapper@cs.uky.edu.

1 Introduction

The ideal cryptosystem would be one that can be proved secure against all possible cryptanalytic attacks (at least computationally secure). In practice, of course, we settle for cryptosystems that are secure against all *known* attacks (and in some cases merely rely on intuition rather than proof to believe they are secure). This is particularly true of private key systems. Thus the history of research on stream ciphers repeats a cycle in which a new system is developed, and perhaps proved secure against many existing attacks, then a new method is developed for attacking the new system. Future systems must be secure against the new attack. If the new attack is highly specialized, this tends to be easy – random unrelated methods of encryption are likely to be secure against the attack, or the attack may simply make no sense outside the context of the cryptosystem it was designed for. In some cases, however, a general purpose attack is developed that can potentially be used against a large class of cryptosystems. Furthermore, in some cases such attacks come with numeric measures of resistence to the attack. Such is the case, for example with the Berlekamp-Massey algorithm and linear span. All sequences used in stream ciphers must have large linear spans in order to resist the Berlekamp-Massey attack. This algorithm is based on the idea of synthesizing a linear feedback shift register (LFSR) that generates a given sequence, given a small number of bits of the sequence. Essential to make the algorithm work is the existence of an algebraic framework for the analysis of LFSR sequences, based on power series and polynomials over $GF(2)$.

In the Cambridge Algorithms Workshop in 1993, I described joint work with Mark Goresky in which we developed a new type of feedback register, feedback

* Project sponsored in part by the National Security Agency under Grant Number MDA904-91-H-0012. The United States Government is authorized to reproduce and distribute reprints notwithstanding any copyright notation hereon.

with carry shift registers (or FCSRs) [5]. These registers are equipped with an algebraic framework for analysis, analogous to that of LFSRs, but based on the 2-adic number rather than power series over $GF(2)$.

In that paper, we described the basic algebraic properties of FCSRs. As in the case of FCSRs, one can ask whether there is an algorithm which, given part of a binary sequence **a**, synthesizes a (minimal length) FCSR that generates **a**. We showed that the existence of such an algorithm implies that it is possible to crack Massey and Ruepell's summation combiner. We further argued that the *2-adic span*, the length of the smallest FCSR that generates a given sequence, is thus an important measure of security. A sequence must have large 2-adic span in order to be secure (though this of course does not guarantee security). At the time of the Cambridge workshop we believed that a variant of the Berlekamp-Massey algorithm for approximating rational numbers (due to Mandelbaum [6]) could be adapted to the case of LFSRs. This has proved to not work. However, we have since developed a provably correct analogue of the Berlekamp-Massey algorithm, based on De Weger's lattice theoretic approach to rational approximation [8]. We describe the algorithm here, but details of the proof of correctness and analysis will appear elsewhere.

We have further developed generalizations of FCSRs by replacing the 2-adic numbers by ramified extensions [2]. That is, by adjoining π, a real dth root of 2. We showed that these registers have a similar algebraic structure to that of FCSRs, and we thus get a security measure for each positive d. (Although it seems that the larger d is, the computationally harder, and hence less threatening, is an attack based on these registers.) Furthermore, we have recently shown that our rational approximation algorithm for FCSRs works at least in the case $d = 2$.

From the point of view of wanting to build more secure systems, we are thus left with the question of how we can generate sequences which resist these attacks. That is, sequences with large 2-adic span, and even large π-adic span, where π is a dth root of 2 with d small. We would further like to do so without sacrificing other measures of security.

We do not yet have an answer to this question, but the purpose of this paper is to introduce new feedback register based tools that may allow us to build such secure sequences. One method that has been used to increase linear span is to take a LFSR over an extension $GF(p^n)$ of $GF(p)$, p prime, and apply a nonlinear "feedforward" function to its output to obtain a binary sequence. When $p \neq 2$, these sequences can have very large linear spans [1], although they are vulnerable for other reasons [4]. When $p = 2$, their linear spans are only moderately larger than LFSR sequences [3]. We describe here an FCSR analogue of LFSRs over nonprime finite fields. Hopefully such registers can be used to build sequences that have large 2-adic span.

In this abstract we describe the algebraic basis for these registers; the construction of the registers; various algebraic properties of the sequences they generate; and conditions under which our rational approximation algorithm can be generalized. For each class of generalized FCSR for which the rational approximation algorithm works, we obtain a new cryptographic security measure.

2 Definitions

The constructions of both LFSRs and FCSRs over a field K can be based on the following algebraic machinery: A ring R with a valuation such that the maximal ideal I of the valuation is principal, $I = (\pi)$ and such that $R/I = K$. There must be a set $S \subseteq R$ that maps bijectively to K under reduction modulo I, and we must be able to write every element of R as a difference $x - y$ where x and y are finite sums of powers of π with coefficients in S. We denote by P the set of finite sums of powers of π with coefficients in S. Infinite sequences over K can then be identified with infinite sequences over S, which can be identified with power series in π with coefficients in S, which can be identified with elements of the completion R_π of R at the valuation. Every element of R_π can be identified with such a power series. The feedback register is then constructed to carry out division in R, producing such an element of R_π.

In the case of LFSRs, $R = K[x]$, the power series ring in one variable over K, $I = (x)$, and $S = K$. In the case of FCSRs, $R = \mathbf{Z}$, the integers, $I = (2)$ (or, more generally, any prime ideal), and $S = \{0, 1\}$. In the case of FCSRs over a ramified extension of $\mathbf{Z}$, if π is a real dth root of 2, then $R = \mathbf{Z}[\pi]$, $I = (\pi)$, and $S = \{0, 1\}$. It should be noted that in the first two cases R is a Euclidean domain (that is, we can carry out "division with remainder"), but in the third case, R is only a Euclidean domain for some values of d, and the question of which values of d give Euclidean domains is a quite subtle one. This turns out to have an impact on what can be done with these registers. In particular, our rational approximation algorithm only works in a Euclidean domain.

Definition 1. Let R, $I = (\pi)$, and S be as above. A *feedback with carry shift register over* R, I, S, or simply an R-FCSR, of length r, is specified by r elements of S, $q_1, \cdots, q_r$ (which can be identified with elements of $K = R/I$). The register consists of r cells for storing elements of S, some additional memory for storing a "carry", and, if the contents of the register are $(a_{n-1}, \cdots, a_{n-r})$ and the memory is m_{n-1}, circuitry for implementing the following operations:

A1. Form the number $\sigma_n = \sum_{k=1}^{r} q_k a_{n-k} + m$.

A2. Shift the contents one step to the right, outputting the rightmost element a_{n-r}.

A3. Let $\sigma_n = a_n + \pi m_n$, with $a_n \in S$ (σ_n can always be written this way). Place a_n into the leftmost cell of the shift register. Replace the memory m_{n-1} by m_n.

If K is finite, it is straightforward to design circuitry to implement these operations, though only practical if K is small. The memory, an arbitrary element of R, can be represented as a finite set of elements of S by writing it as a difference $x - y$, with $x, y \in P$. We refer to $q = \sum_{i=1}^{r} q_i 2^i - 1$ as the *connection number* because it is the analog to the connection polynomial in the usual theory of

LFSRs. Any periodic sequence of elements of S may be generated by such a FCSR,

In order to define FCSRs over extensions of $GF(2)$, it is necessary to find suitable R, I, and S. In this abstract we consider only rings in which the ideal (2) remains a prime ideal (extensions that are unramified at 2), and take $I = (2)$. In order for R to reduce to $K = GF(2^n)$, R must contain an element β that reduces modulo 2 to a primitive element of K. Thus the minimal polynomial $f(x)$ of β must reduce modulo 2 to a primitive polynomial over K. We assume that $R = \mathbf{Z}[\beta]$. Let $\bar{\beta}$ be the reduction of β modulo 2. Then $\bar{\beta}$ is primitive, so $1, \bar{\beta}, \bar{\beta}^2, \cdots, \bar{\beta}^{n-1}$ is a basis for K over $GF(2)$. Thus the set of linear combinations of $1, \beta, \beta^2, \cdots, \beta^{n-1}$ with coefficients in $\{0, 1\}$ maps bijectively to K modulo 2. We take this set as S. For example, if $n = 2$ then we can take $f(x) = x^2 - x - 1$. In the next section we assume R, I, and S are defined in this manner. As it turns out, the amount of auxiliary memory needed for a register tends to be smallest when the b_i are small, so we would like to take $b_i \in \{0, \pm 1\}$. However, other properties of the registers (such as the convergence rate of the rational approximation algorithm) may be superior for other choices of the b_i.

3   Properties of R-FCSRs

Many of the algebraic properties of FCSRs also hold for R-FCSRs. There are five different ways to view an infinite, eventually periodic sequence over $GF(2^n)$:

1. As a sequence $\mathbf{a} = a_0, a_1, \cdots, a_i \in GF(2^n)$.
2. As a sequence $\mathbf{a} = a_0, a_1, \cdots, a_i \in S$.
3. As an element α of the completion R_2 of R at 2.
4. As an R-rational number $p/q \in K$.
5. As the output stream of an R-FCSR.

Representations (1) and (2) are identified by reducing S modulo 2. Representations (2) and (3) are identified by associating the binary sequence $\mathbf{a}$ with the coefficients in the formal power series expression for α. The translation between representations (3) and (4) is essentially the same as the identification between real numbers whose decimal expansions are eventually periodic and rational numbers. To translate between representations (4) and (5) we have the following.

Theorem 2. *The output, $\mathbf{a}$, of an R-FCSR with connection number q, initial memory value m_{r-1}, and initial loading $a_{r-1}, a_{r-2}, \ldots, a_1, a_0$, is the bit sequence of the 2-adic representation of an R-rational number*

$$\alpha = \frac{\sum_{i=0}^{r-1} \sum_{j=0}^{r-i-1} q_i 2^i a_j 2^j - m_{r-1} 2^r}{q}. \tag{1}$$

Thus the denominator of α is equal to the connection integer q of the shift register.

Corollary 3. *Adding b to the memory adds* $-b2^r/q$ *to the output.*

The converse of Theorem 2, that every R-rational number p/q can be realized as the output of an R-FCSR, is true as well. To see this, we show how to construct the initial loading of an R-FCSR for certain p, and then use Corollary 3 to obtain initial loadings for other FCSR. Let $p = \sum_{i=0}^{r-1} p_i 2^i$ with $p_i \in S \cup -S$. Every element of R differs from some such a p by a multiple of 2^r. Thus if we can construct initial loadings for p/q with p of this type, then we can construct initial loadings for all p/q.

Theorem 4. *Given a connection number* $q = \sum_{i=0}^{r} q_i 2^i$ *with* $q_0 = -1$ *and* $q_1, \cdots, q_r \in S$, *and* $p = \sum_{i=0}^{r-1} p_i 2^i$ *with* $p_i \in S \cup -S$, *define* $a_0, \cdots, a_{r-1}$ *and* m_{r-1} *by the following procedure:*

B1. *Set* $m_{-1} = 0$ *and* $\sigma_0 = 0$.
B2. *For each* $i = 0, 1, \ldots, r-1$ *compute the following numbers:*

$$\sigma_i = \sum_{k=0}^{i-1} q_{i-k} a_k + m_{i-1} - p_i \in R.$$

Write

$$\sigma_i = a_i + 2m_i,$$

where $a_i \in S$, $m_i \in R$, *and the empty sum in* σ_0 *is interpreted as zero.*

If $(a_{r-1}, a_{r-2}, \ldots, a_1, a_0)$ *is used as the initial loading,* m_{r-1} *is used as the initial memory in an* R-FCSR *with connection number* q, *then the output sequence will correspond to the* R-*rational number* p/q.

Note that the memory may not be in P (the set of elements in R which are finite sums of powers of 2 with coefficients in S). However, it can always be represented as a difference of such elements since this is true of every element of R. Moreover, if $p \in -P$, i.e., if all the p_i are in $-S$, and $f(x) = x^n - \sum_{i=0}^{n-1} b_i x^i$ with $b_i \geq 0$, then $m_{r-1} \in P$.

It is natural to ask how large an auxiliary memory is needed. In the case where $p \in -P$, we have the following.

Theorem 5. *Suppose that the coefficients* b_i *in the polynomial* $f(x) = x^n - \sum_{i=0}^{n-1} b_i x^i$ *are all either 0 or 1. Let* q *be expressed is a polynomial in* α, $q + 1 = \sum_{i=0}^{r-1} t_i \beta^i$, *with* $t_i \in \mathbf{Z}$. *With the initial value constructed as in Theorem 4, the number of bits* M *needed for the initial memory value is bounded by*

$$M \leq n \log(\sum_{i=0}^{n-1} 2^i \mathrm{wt}(t_i) \leq n^2 + n \log(\max(\mathrm{wt}(t_i))),$$

where $\mathrm{wt}(t_i)$ *is the Hamming weight of the binary expansion of* t_i. *This bound continues to hold for all later values of the memory.*

For other choices of the b_i, a similar but higher bound can be given.

4   Rational Approximation and Security Measures

The role to be played by R-FCSRs in cryptography is two fold. First, as discussed in the introduction, they are potential tools for building sequences that have large 2-adic span, and thus that are secure against 2-adic rational approximation (Berlekamp-Massey) algorithms. However, there is also a possibility that rational approximation can be carried out over R – we would call this R-adic rational approximation – and that we can thus carry out this sort of attack on sequences over $GF(2^n)$.

As it turns out, the ingredients needed to generalize our 2-adic rational approximation algorithm appear to be quite restrictive. R must be a Euclidean domain, with a little extra. Specifically, we need a *norm function* $N : K - \{0\} \to \mathbf{Q}^+$ ($\mathbf{Q}^+$ denotes the positive rational numbers), that satisfies the following:

a. For all $a \in R$, $N(a) \in \mathbf{N}$ ($\mathbf{N}$ denotes the natural numbers).
b. For all $a, b \in K$, we have $N(ab) = N(a)N(b)$.
c. For all $a, b \in R$, there exist $q, p \in R$ so that $a = qb + p$ and $N(p) < N(b)$.

In addition, in order to ensure the algorithm converges rapidly, we need

d. There is a function $f : \mathbf{N} \to \mathbf{N}$ so that if $a \equiv b \bmod \pi^{f(k)}$, $N(a) < 2^k$, and $N(b) < 2^k$, then $a = b$.

For any pair of elements p and q of R, define

$$\Phi(p, q) = \max(N(p), N(q)).$$

Assume we have consecutive terms $a_0, a_1, \cdots$ of a sequence $\mathbf{a}$ of elements of K (or equivalently S). We can think of $\mathbf{a}$ as the R_π-adic expansion of a number α. We wish to determine a pair of elements (p, q) of R such that $\alpha = p/q$ and $\Phi(p, q)$ is minimal among all such pairs. In the rational approximation algorithm, given in Figure 1, the symbols $f = (f_1, f_2)$ and $g = (g_1, g_2)$ denote pairs of elements of R. With these ingredients, the rational approximation algorithm is described in Figure 1. Unlike De Weger's approach to rational approximation, our algorithm is adaptive. That is, the number of terms of known key does not need to be predetermined. The algorithm can continue to revise the approximation as long as new key terms are found. Note that the minimization steps can be carried out with a pair of divisions in R. These divisions can be computed since R is a Euclidean domain. An example of the execution of the algorithm is given in Table 1. The input used is the 2-adic expansion of $-252/269$, the first 30 bits of which are 001010000100100010000110010000. The table shows the values of k, α, g, and f through 15 iterations. The algorithm thus uses 17 bits of the sequence since the first two bits are zero. Note that $17 < 2 \lceil \log(269) \rceil$. Thereafter the value of g remains unchanged.

Theorem 6. *Suppose such a norm function exists for R. There is a rational approximation algorithm which, when given $f(2 \max(N(p), N(q)))$ terms of the expansion over S of an R-rational number p/q as input, will output an R-FCSR that outputs p/q.*

```
Rational_Approximation()
    begin
    Input aᵢs until the first a_{k-1} ≠ 0
    α = a_{k-1}π^{k-1}
    f = (0, π)
    g = (π^{k-1}, 1)
    while more input do
        input a_k
        α = α + a_kπ^k
        if α · g₂ − g₁ ≡ 0 ( mod π^{k+1}) then
            f = πf
        else if Φ(g) < Φ(f) then
            Let d minimize Φ(f + dg) with g + df ≡ 0 mod π^{k+1}
            ⟨g, f⟩ = ⟨f + dg, πg⟩
        else
            Let d minimize Φ(g + df) with g + df ≡ 0 mod π^{k+1}
            ⟨g, f⟩ = ⟨g + df, πf⟩
        fi fi
        k = k + 1
    od
    return g
    end
```

Fig. 1. Rational Approximation Algorithm.

It turns out that algebraic number fields that possess such norm functions are rare. For the fields that arise in the construction of FCSRs, as well as those that arise by taking extensions of $\mathbf{Z}$ that are ramified at 2, we have shown this in only two cases.

Corollary 7. *Such a norm function, and hence such a rational approximation algorithm exists for:*

1. *Ordinary FCSRs;*
2. *FCSRs based on $R = \mathbf{Z}[\pi]$, with $\pi^2 + 2 = 0$, $S = \{0, 1\}$, and $K = GF(2)$;*
3. *FCSRs based on $R = \mathbf{Z}[\beta]$, with $\beta^2 + \beta + 1 = 0$, $S = \{0, 1, \beta, 1 + \beta\}$, and $K = GF(4)$.*

In the unramified case (the case described in Sections 2 and 3), if $n = 2$ we have a Euclidean domain whenever b_0 is square free and divides b_1. However, if $b_0 = 1$, condition (d) does not hold. In other cases we do not yet know whether condition (d) holds. Thus the algorithm finds rational approximations, but we do not know how fast, or even whether, it converges. In the ramified case (that is, when R is formed by adjoining a real nth root of 2 to $\mathbf{Z}$), condition (d) always holds, but we do not know whether R is a Euclidean domain except when $n = 2$.

k	α	g	f
2	4	$(4, 1)$	$(0, 2)$
3	4	$(4, 1)$	$(0, 4)$
4	20	$(-4, 3)$	$(0, 8)$
5	20	$(-4, 3)$	$(0, 16)$
6	20	$(12, 7)$	$(-8, 6)$
7	20	$(4, 13)$	$(-16, 12)$
8	20	$(20, 1)$	$(8, 26)$
9	532	$(-12, 25)$	$(40, 2)$
10	532	$(28, 27)$	$(-24, 50)$
11	532	$(-52, 23)$	$(56, 54)$
12	4628	$(-52, 23)$	$(112, 108)$
13	4628	$(60, 131)$	$(-104, 46)$
14	4628	$(164, 85)$	$(-208, 92)$
15	4628	$(164, 85)$	$(-416, 184)$
16	70164	$(-252, 269)$	$(328, 170)$

Table 1. Execution of the Rational Approximation Algorithm for $-252/269$.

As a consequence of these considerations, we have new measures of cryptologic security.

Definition 8. If R, I, S is a ring over which FCSRs can be constructed, based on maximal ideal $I = (\pi)$ and lift S of R/I, then the *R-adic span* of a sequence **a** is the size of the smallest R-FCSR that outputs **a**.

Notice that we have been somewhat vague as to what is meant by the size of a FCSR. One reasonable definition might be the integer r, the number of terms in the expansion $q + 1 = \sum_{i=1}^{r} q_i 2^i$. This makes sense if p has fewer than r terms in its expansion. More generally, we might take the maximum number of bits (or S cells) required to store the contents of the register, including the extra memory, over the course of its execution. From the cryptographic point of view, what we want is that $\max(N(p), N(q))$ is bounded by the size of the register (or perhaps a multiple of the size). In this case, it follows that a sequence must have large R-adic span to be secure. This indeed holds in the two cases covered by Corollary 7.

5   Conclusions

We have described a general method for constructing feedback with carry shift registers over certain rings of algebraic integers, R. These registers are analogous to linear feedback shift registers. They can be thought of as generating sequences by carrying out division in the completion R_2 of the ring at the prime ideal (2). They carry similar algebraic structures to those of LFSRs.

The cryptographic importance of these registers is twofold. First, they are a potential source of cryptographically secure sequences for stream ciphers. As with LFSR sequences, there are many possible (as yet unexplored) ways to attempt to modify these sequences to make them secure. Second, these registers can be used for cryptanalysis if a rational approximation algorithm can be devised for R_2. Such an algorithm exists if R is a Euclidean domain with an extra condition on its norm. For only a few Rs we have shown that these conditions occur. It remains to be seen whether other such Rs have these properties and, if not, whether there is a different rational approximation algorithm that works.

Finally, it should be mentioned that further generalizations are possible. We have only considered the cases where R is totally ramified or purely unramified at (2), but more general extensions can be considered. We have also only considered here sequences over finite fields of characteristic two, but the same constructions can be carried out for primes other than two. The advantage might be in obtaining rings for which rational approximation works.

References

1. A. Chan and R. Games, On the linear span of binary sequences from finite geometries, q odd, *IEEE Trans. Info. Theory*, vol. IT-36 (1990) pp. 548-552.
2. M. Goresky and A. Klapper, Feedback registers based on ramified extensions of the 2-adic numbers – extended abstract, *Proceedings of Eurocrypt '94*, Perugia, Italy, May 1994.
3. E. L. Key, An Analysis of the structure and complexity of nonlinear binary sequence generators, *IEEE Trans. Info. Theory*, vol. IT-22 no. 6 (1976) pp. 732-736.
4. A. Klapper, The Vulnerability of Geometric Sequences Based on Fields of Odd Characteristic, *Journal of Cryptology*, **7** (1994) pp. 33-51.
5. A. Klapper and M. Goresky, Feedback Shift Registers, Combiners with Memory, and Arithmetic Codes, *University of Kentucky, Deptartment of Computer Science Technical Report No. 239-93*. Presented at *1993 Cambridge Workshop on Algorithms*.
6. D. Mandelbaum, An approach to an arithmetic analog of Berlekamp's algorithm. *IEEE Trans. Info. Theory*, vol. IT-30 (1984) pp. 758-762.
7. R. Rueppel, *Analysis and Design of Stream Ciphers*. Springer Verlag, New York, 1986.
8. B. M. M. de Weger, *Approximation lattices of p-adic numbers. J. Num. Th.* vol. 24 (1986) pp. 70-88.

A Free Energy Minimization Framework for Inference Problems in Modulo 2 Arithmetic

David J.C. MacKay*

Cavendish Laboratory
Cambridge CB3 0HE
United Kindom

mackay@mrao.cam.ac.uk

Abstract. This paper studies the task of inferring a binary vector s given noisy observations of the binary vector t = As modulo 2, where A is an $M \times N$ binary matrix. This task arises in correlation attack on a class of stream ciphers and in the decoding of error correcting codes.

The unknown binary vector is replaced by a real vector of probabilities that are optimized by variational free energy minimization. The derived algorithms converge in computational time of order between w_A and $N w_A$, where w_A is the number of 1s in the matrix A, but convergence to the correct solution is not guaranteed.

Applied to error correcting codes based on sparse matrices A, these algorithms give a system with empirical performance comparable to that of BCH and Reed-Muller codes.

Applied to the inference of the state of a linear feedback shift register given the noisy output sequence, the algorithms offer a principled version of Meier and Staffelbach's (1989) algorithm B, thereby resolving the open problem posed at the end of their paper. The algorithms presented here appear to give superior performance.

1 The problem addressed in this paper

Consider three binary vectors: s with components s_n, $n = 1 \ldots N$, and r and n with components r_m, n_m, $m = 1 \ldots M$, related by:

$$(\mathbf{As} + \mathbf{n}) \bmod 2 = \mathbf{r} \tag{1}$$

where A is a binary matrix. Our task is to infer s given r and A, and given assumptions about the statistical properties of s and n.

This problem arises, for example, in the decoding of a noisy signal transmitted using a linear code A. As a simple illustration, the (7,4) Hamming code takes

---

* Supported by the Royal Society Smithson Research Fellowship

$N = 4$ signal bits, s, and transmits them followed by three parity check bits. The $M = 7$ transmitted symbols are given by $t = \mathbf{A}s \bmod 2$, where

$$\mathbf{A} = \begin{bmatrix} 1 & 0 & 0 & 0 \\ 0 & 1 & 0 & 0 \\ 0 & 0 & 1 & 0 \\ 0 & 0 & 0 & 1 \\ 1 & 0 & 1 & 1 \\ 1 & 1 & 0 & 1 \\ 1 & 1 & 1 & 0 \end{bmatrix}$$

The noise vector n describes the corruption of these bits by the communication channel. The received message is $\mathbf{r} = \mathbf{t} + \mathbf{n} \bmod 2$. The receiver's task is to infer s given r and the assumed noise properties of the channel. For the Hamming code above, this is not a difficult task, but as N and M become large and as the number of 1s in the matrix $\mathbf{A}$ increases, the inference of s in a time less than exponential in N becomes more challenging, for general $\mathbf{A}$. Indeed, the general decoding problem for linear codes is NP-complete (Berlekamp, McEliece and van Tilborg 1978).

The problem defined in equation (1) also arises in the inference of the sequence of a linear feedback shift register (LFSR) from noisy observations (Meier and Staffelbach 1989, Mihaljević and Golić 1992, Mihaljević and Golić 1993, Anderson 1995).

This paper presents a fast algorithm for attempting to solve these tasks. The algorithm is similar in spirit to Gallager's (1963) soft decoding algorithm for low-density parity check codes.

Assumptions

I assume that the prior probability distribution of s and n is separable, *i.e.*, $P(\mathbf{s}, \mathbf{n}) = P(\mathbf{s})P(\mathbf{n}) = \prod_n P(s_n) \prod_m P(n_m)$. Defining the transmission $\mathbf{t}(\mathbf{s}) = \mathbf{A}s \bmod 2$, the probability of the observed data r as a function of s (the 'likelihood function') can be written:

$$P(\mathbf{r}|\mathbf{s}, \mathbf{A}) = \prod_m e_m^{t_m(\mathbf{s})} (1 - e_m)^{(1 - t_m(\mathbf{s}))},$$

where e is a vector of probabilities indexed by m given by $e_m = P(n_m = 1)$ if $r_m = 0$ and $e_m = P(n_m = 0)$ if $r_m = 1$. This likelihood function is fundamental to any probabilistic approach. The log likelihood can be written:

$$\log P(\mathbf{r}|\mathbf{s}, \mathbf{A}) = \sum_m t_m(\mathbf{s}) \log \frac{e_m}{1 - e_m} + \text{const.} \tag{2}$$

$$\equiv \sum_m t_m(\mathbf{s}) \, g_m(e_m) + \text{const.} \tag{3}$$

where $g_m(e_m) \equiv \log[e_m/(1 - e_m)]$. Thus the natural norm for measuring the distance between t and e is $\sum_m t_m g_m(e_m)$.

The task is to infer **s** given **A** and the data **r**. The posterior distribution of **s** is, by Bayes's theorem:

$$P(\mathbf{s}|\mathbf{r}, \mathbf{A}) = \frac{P(\mathbf{r}|\mathbf{s}, \mathbf{A})P(\mathbf{s})}{P(\mathbf{r}|\mathbf{A})}. \tag{4}$$

I assume our aim is to find the most probable **s**, *i.e.*, the`s that maximizes the expression (4) over **s**. I assume however that an exhaustive search over the 2^N possible sequences **s** is not permitted. One way to attack a discrete combinatorial problem is to create a related problem in which the discrete variables **s** are replaced by real variables, over which a continuous optimization can then be performed [see for example (Hopfield and Tank 1985, Aiyer, Niranjan and Fallside 1990, Durbin and Willshaw 1987, Peterson and Soderberg 1989, Gee and Prager 1994, Blake and Zisserman 1987)]. In the present context, the question then is 'how should one generalize the posterior probability (4) to the case where **s** is replaced by a vector with real components?' An appealing way to answer this question is to derive the continuous representation in terms of an approximation to probabilistic inference.

2 Free Energy Minimization

The variational free energy minimization method (Feynman 1972) takes an 'awkward' probability distribution such as the one in (4), and attempts to approximate it by a simpler distribution $Q(\mathbf{s}; \theta)$, parameterized by a vector of parameters θ. For brevity in the following general description I will denote the complex probability distribution $P(\mathbf{s}|\mathbf{A}, \mathbf{r})$ by $P(\mathbf{s})$. The measure of quality of the approximating distribution is the variational free energy,

$$F(\theta) = \sum_{\mathbf{s}} Q(\mathbf{s}; \theta) \log \frac{Q(\mathbf{s}; \theta)}{P(\mathbf{s})}.$$

The function $F(\theta)$ has a lower bound of zero which is attained only by a θ such that $Q(\mathbf{s}; \theta) = P(\mathbf{s})$ for all **s**. Alternatively, if $P(\mathbf{s})$ is not normalized, and we define $Z = \sum_{\mathbf{s}} P(\mathbf{s})$, then F has a lower bound of $-\log Z$, attained only by θ such that $Q = P/Z$. The variational method used in this paper is traditionally used in statistical Physics to estimate $\log Z$, but here, $\log Z$ is just an additive constant which we ignore.

When Q has a sufficiently simple form, the optimization of F over θ may be a tractable problem, even though F is defined by a sum over all values of **s**. We find a θ^* that minimizes $F(\theta)$ in the hope that the approximating distribution $Q(\mathbf{s}; \theta^*)$ will give useful information about $P(\mathbf{s})$. Specifically, we might hope that the **s** that maximizes $Q(\mathbf{s}; \theta^*)$ is a good guess for the **s** that maximizes $P(\mathbf{s})$. Generically, free energy minimization produces approximating distributions $Q(\mathbf{s}; \theta^*)$ that are more compact than the true distribution $P(\mathbf{s})$.

3   Free Energy Minimization for Equation (4)

I take Q to be a separable distribution,

$$Q(\mathbf{s}; \theta) \equiv \prod_n q_n(s_n; \theta_n)$$

with θ_n defining the probabilities q_n thus:

$$q_n(s_n = 1; \theta_n) = \frac{1}{1 + e^{-\theta_n}} \equiv q_n^1; \quad q_n(s_n = 0; \theta_n) = \frac{1}{1 + e^{\theta_n}} \equiv q_n^0.$$

This is a nice parameterization because the log probability ratio is $\theta_n = \log(q_n^1/q_n^0)$. The variational free energy is defined to be:

$$F(\theta) = \sum_{\mathbf{s}} Q(\mathbf{s}; \theta) \log \frac{Q(\mathbf{s}; \theta)}{P(\mathbf{r}|\mathbf{s}, \mathbf{A})P(\mathbf{s})}.$$

I now derive an algorithm for computing F and its gradient with respect to θ in a time that is proportional to the weight of $\mathbf{A}$, w_A (i.e., the number of ones in $\mathbf{A}$). The free energy separates into three terms, $F(\theta) = E_L(\theta) + E_P(\theta) - S(\theta)$, where the 'entropy' is:

$$S(\theta) \equiv -\sum_{\mathbf{s}} Q(\mathbf{s}; \theta) \log Q(\mathbf{s}; \theta) = -\sum_n \left[q_n^0 \log q_n^0 + q_n^1 \log q_n^1 \right],$$

with derivative: $\frac{\partial}{\partial \theta_n} S(\theta) = -q_n^0 q_n^1 \theta_n$; the 'prior energy' is:

$$E_P(\theta) \equiv -\sum_{\mathbf{s}} Q(\mathbf{s}; \theta) \log P(\mathbf{s}) = -\sum_n b_n q_n^1$$

where $b_n = \log[P(s_n = 1)/P(s_n = 0)]$, and has derivative $\frac{\partial}{\partial \theta_n} E_P(\theta) = -q_n^0 q_n^1 b_n$; and the 'likelihood energy' is:

$$E_L(\theta) \equiv -\sum_{\mathbf{s}} Q(\mathbf{s}; \theta) \log P(\mathbf{r}|\mathbf{s}, \mathbf{A}) = -\sum_m g_m \sum_{\mathbf{s}} Q(\mathbf{s}; \theta) t_m(\mathbf{s}) + \text{const.}$$

The additive constant is independent of θ and will now be omitted. To evaluate E_L, we need the average value of $t_m(\mathbf{s})$ under the separable distribution $Q(\mathbf{s}; \theta)$, that is, the probability that $\sum_n A_{mn} s_n \bmod 2 = 1$ under that distribution.

Forward algorithm. We can compute this probability for each m by a recursion involving a sequence of probabilities $p_{m,\nu}^1$ and $p_{m,\nu}^0$ for $\nu = 1 \ldots N$. These are defined to be the probabilities that the partial sum $t_m^{1\nu} = \sum_{n=1}^{\nu} A_{mn} s_n \bmod 2$ is equal to 1 and 0 respectively. These probabilities satisfy:

$$\left. \begin{array}{l} p_{m,\nu}^1 = q_\nu^0 p_{m,\nu-1}^1 + q_\nu^1 p_{m,\nu-1}^0 \\ p_{m,\nu}^0 = q_\nu^0 p_{m,\nu-1}^0 + q_\nu^1 p_{m,\nu-1}^1 \end{array} \right\} \text{ if } A_{m\nu} = 1;$$

$$\left. \begin{array}{l} p_{m,\nu}^1 = p_{m,\nu-1}^1 \\ p_{m,\nu}^0 = p_{m,\nu-1}^0 \end{array} \right\} \text{ if } A_{m\nu} = 0.$$

$$(5)$$

The initial condition is $p_{m,0}^1 = 0, p_{m,0}^0 = 1$. The desired quantity is obtained in a time that is linear in the weight of row m of $\mathbf{A}$:

$$\sum_{\mathbf{s}} Q(\mathbf{s}; \theta)\, t_m(\mathbf{s}) = p_{m,N}^1.$$

The quantity $p_{m,N}^1$ is a generalization to a continuous space of the product $t_m = \mathbf{A}_m \mathbf{s} \bmod 2$, with the satisfying property that if any one of the contributing terms q_ν^1 is equal to 0.5, then the resulting value of $p_{m,N}^1$ is 0.5.

The energy E_L is then given by:

$$E_L(\theta) = -\sum_m g_m p_{m,N}^1.$$

Reverse algorithm. To obtain the derivative of the energy E_L with respect to θ_n it is necessary to perform the same number of computations again. We introduce another 'reverse' sequence of probabilities $r_{m,\nu}^1$ and $r_{m,\nu}^0$ for $\nu = N \dots 1$, defined to be the probabilities that the partial sum $t_m^{\nu N} = \sum_{n=\nu}^N A_{mn} s_n \bmod 2$ is equal to 1 and 0 respectively. These can be computed by a recursive procedure equivalent to that in equation (5). Now note that $p_{m,N}^1$ can be written thus, for any n:

$$p_{m,N}^1 = q_n^0 \left(p_{m,n-1}^1 r_{m,n+1}^0 + p_{m,n-1}^0 r_{m,n+1}^1\right) + q_n^1 \left(p_{m,n-1}^1 r_{m,n+1}^1 + p_{m,n-1}^0 r_{m,n+1}^0\right).$$

Having pulled out the θ_n dependence˙ (in the two factors q_n^0 and q_n^1), it is easy to obtain the derivative:

$$\frac{\partial}{\partial \theta_n} E_L(\theta) = -q_n^0 q_n^1 \sum_m g_m d_{mn}$$

where $d_{mn} = \left(p_{m,n-1}^1 r_{m,n+1}^1 + p_{m,n-1}^0 r_{m,n+1}^0\right) - \left(p_{m,n-1}^1 r_{m,n+1}^0 + p_{m,n-1}^0 r_{m,n+1}^1\right).$

Total derivative

The derivative of the free energy is:

$$\frac{\partial F}{\partial \theta_n} = q_n^0 q_n^1 \left[\theta_n - b_n - \sum_m g_m d_{mn}\right]. \tag{6}$$

Optimizers

This derivative can be inserted into a variety of continuous optimizers. I have implemented both conjugate gradient optimizers and 'reestimation' optimizers and found the latter, which I now describe, to be superior. The reestimation method is motivated by the form of the derivative (6); setting it to zero, we obtain the iterative prescription:

$$\theta_n := b_n + \sum_m g_m d_{mn},$$

which I call the reestimation optimizer. It can be implemented with either *synchronous* or *sequential* dynamics, that is, we can update all θ_n simultaneously, or one at a time. The sequential reestimation optimizer is guaranteed to reduce the free energy on every step, because everything to the right of θ_n in equation (6) is independent of θ_n. The sequential optimizer can be efficiently interleaved with the reverse recursion; the reverse probability r_ν^1 is evaluated from $r_{\nu+1}^1$ for each m after θ_ν has been updated. The synchronous optimizer does not have an associated Lyapunov function, so it is not guaranteed to be a stable optimizer, but empirically it sometimes performs better.

Optimizers of the free energy can be modified by introducing 'annealing' or 'graduated non-convexity' techniques (Blake and Zisserman 1987, Van den Bout and Miller 1989), in which the non-convexity of the objective function F is switched on gradually by varying an inverse temperature parameter β from 0 to 1. This annealing procedure is intended to prevent the algorithm running headlong into the minimum that the initial gradient points towards. We define:

$$F(\theta, \beta) = \beta E_L(\theta) + \beta_P E_P(\theta) - S(\theta), \tag{7}$$

and perform a sequence of minimizations of this function with successively larger values of β, each starting where the previous one stopped. If we choose $\beta_P \equiv \beta$ then β influences both the likelihood energy and the prior energy, and if $\beta_P \equiv 1$ then β influences just the likelihood energy. The gradient of $F(\theta, \beta)$ with respect to θ is identical to the gradient of F in equation (6) except that the energy terms are multiplied by β_P and β. This annealing procedure is deterministic and does not involve any simulated noise.

Comments

None of these algorithms is expected to work in all cases, because a) there may be multiple free energy optima; b) the globally optimal s might not be associated with any of these free energy optima; c) the task is a probabilistic task, so even an exhaustive search is not guaranteed always to identify the correct vector s.

Any particular problem can be reparameterized into other representations $s' = sU$ with new matrices $A' = U^{-1}A$. The success of the algorithms is expected to depend crucially on the choice of representation. The algorithms are expected to work best if A is sparse and the true posterior distribution over s is close to separable.

Computational complexity

The gradient of the free energy can be calculated in time linear in w_A. The algorithms are expected to take of order 1, or at most N, gradient evaluations to converge, so that the total time taken is of order between w_A and $w_A N$.

The space requirements of the sequential reestimation optimizer are the most demanding (but not severe): memory proportional to w_A is required.

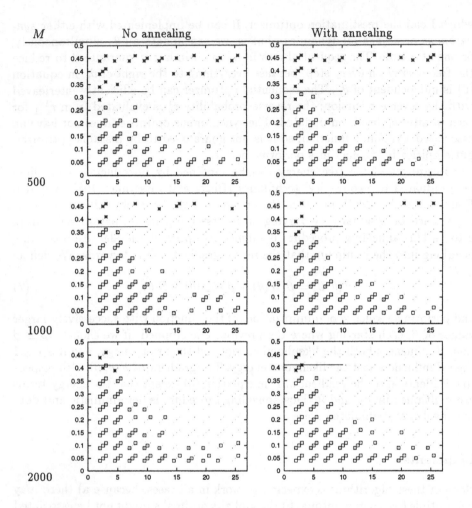

Fig. 1. Performance of synchronous reestimation algorithm without and with annealing: $N=50$

Horizontal axis: average number of 1s per row of $\mathbf{A}$. Vertical axis: noise level f_n. Three experiments were conducted at a grid of values. Outcome is represented by: box = 'correct'; star = 'ok'; dot = 'wrong'. Two points in each triplet have been displaced so that they do not overlap. The horizontal line on each graph indicates an information theoretic bound on the noise level beyond which the task is not expected to be soluble.

4   Demonstration on Random Sparse Codes A

Mock data were created as follows. The first N lines of $\mathbf{A}$ were set to the identity matrix, and of the remaining bits, a fraction f_A were randomly set to 1. This matrix can be viewed as defining a systematic error correcting code in which the signal $\mathbf{s}$ is transmitted, followed by $(M-N)$ sparse parity checks. Each component of $\mathbf{s}$ was set to 1 with probability 0.5. The vector $\mathbf{t} = \mathbf{A}\mathbf{s}\,\mathrm{mod}\,2$

was calculated and each of its bits was flipped with noise probability f_n. Four parameters were varied: the vector length N, the number of measurements M, the noise level f_n of the measurements, and the density f_A of the matrix $\mathbf{A}$.

When optimizing θ the following procedure was adopted. The initial condition was $\theta_n = 0, \forall n$. If annealing was used, a sequence of 20 optimizations was performed, with values of β increasing linearly from 0.1 to 1.0. Without annealing, just a single optimization was performed with $\beta = 1$. The optimizers were iterated until the gradient had magnitude smaller than a predefined threshold, or the change in F was smaller than a threshold, or until a maximum number of iterations was completed. At the end, s was guessed using the sign of each element of θ.

We can calculate a bound on the noise level f_n beyond which the task is definitely not expected to be soluble by equating the capacity of the binary symmetric channel, $(1 - H_2(f_n))$, to the rate of the code, N/M. Here $H_2(f_n)$ is the binary entropy, $H_2(f_n) = -f_n \log_2 f_n - (1 - f_n) \log_2(1 - f_n)$. This bound is indicated on the graphs that follow by a solid horizontal line. For noise levels significantly below this bound we expect the correct solution typically to have significantly greater likelihood than any other s.

Results

For the results reported here, I set N to 50 and M to 500, 1000 and 2000, and ran the synchronous reestimation algorithm multiple times with different seeds, with density f_A varying from 0.05 to 0.50, and error probability f_n varying from 0.05 to 0.45. In each run there are three possible outcomes: 'correct', where the answer is equal to the true vector s; 'wrong', where the answer is not equal to the true vector, and has a smaller likelihood than it; and 'ok', where the algorithm has found an answer with greater likelihood than the true vector (in which case, one cannot complain).

Annealing helps significantly when conjugate gradient optimizers are used, but does not seem to make much difference to the performance of the reestimation algorithms. As was already mentioned, the reestimators work much better than the conjugate gradient minimizers (even with annealing).

Figure 1 shows the outcomes as a function of the weight of $\mathbf{A}$ (x-axis) and error probability (y-axis). There seems to be a sharp transition from solvability to unsolvability. It is not clear whether this boundary constitutes a fundamental bound on what free energy minimization can achieve; performance might possibly be improved by smarter optimizers. Another idea would be to make a hybrid of discrete search methods with a free energy minimizer.

Experiments with larger values of N have also been conducted. The success region looks the same if plotted as a function of of the average number of 1s per row of $\mathbf{A}$ and the noise level.

Table 1. Error rates of error correcting codes using free energy minimization.

a) Sparse random code matrix (section 4). These results use $N = 100$ and $f_A = 0.05$, and various noise levels f_n and $M = 400, 1000, 2000$. The synchronous reestimation optimizer was used without annealing, and with a maximum of 50 gradient evaluations. For every run a new random matrix, signal and noise vector were created. One thousand runs were made for each set of parameters. The capacity of a binary symmetric channel with $f_n = 0.1, 0.15, 0.2$ is $0.53, 0.39, 0.28$ respectively.

b) LFSR code (section 5). The number of taps was 5, selected at random. The synchronous reestimation optimizer was used with the annealing procedure described in section 5.

a)

f_n	N	M	Number of errors	Information rate
0.1	100	400	14/1000	0.25
0.1	100	1000	0/1000	0.1
0.15	100	1000	3/1000	0.1
0.2	100	1000	54/1000	0.1
0.2	100	2000	11/1000	0.05

b)

f_n	k	N	Number of errors	Information rate
0.1	100	400	69/1000	0.25
0.15	100	1000	1/1000	0.1
0.2	100	2000	26/1000	0.05

Table 2. BCH codes and Reed-Muller codes.

For each noise level $f_n = 0.1, 0.15, 0.2$, I give n, k, t for selected BCH codes and Reed-Muller codes with $\epsilon^{block} = P(\text{more than } t \text{ errors}|n) < 0.1$, and rate $> 0.05, 0.03, 0.01$ respectively, ranked by ϵ^{block}. Values of n, k, t from Peterson and Weldon (1972).

$f_n = 0.1$					$f_n = 0.15$					$f_n = 0.2$				
n	k	t	ϵ^{block}	rate	n	k	t	ϵ^{block}	rate	n	k	t	ϵ^{block}	rate
BCH CODES					BCH CODES					BCH CODES				
63	10	13	0.003	0.159	127	8	31	0.002	0.063	1023	26	239	0.004	0.025
127	29	21	0.008	0.228	511	40	95	0.011	0.078	255	9	63	0.028	0.035
1023	153	125	0.009	0.150	63	7	15	0.021	0.111	511	19	119	0.030	0.037
63	16	11	0.021	0.254	127	15	27	0.022	0.118	255	13	59	0.093	0.051
511	85	63	0.037	0.166	255	21	55	0.002	0.082	1023	46	219	0.123	0.045
REED-MULLER CODES					1023	91	181	0.008	0.089	511	28	111	0.152	0.055
128	8	31	8.8e-7	0.063	1023	101	175	0.028	0.099	REED-MULLER CODES				
64	7	15	4.4e-4	0.109	255	29	47	0.056	0.114	1024	11	255	5.7e-5	0.011
1024	56	127	0.0055	0.055	REED-MULLER CODES					512	10	127	0.0034	0.020
32	6	7	0.012	0.188	256	9	63	2e-5	0.035	128	8	31	0.098	0.063
512	46	63	0.038	0.0898	128	8	31	0.0021	0.063					
16	5	3	0.068	0.313	32	6	7	0.096	0.188					

Potential of this method for error correcting codes

Might an error correcting code using this free energy minimization algorithm be practically useful? I restrict attention to the ideal case of a binary symmetric channel and make comparisons with BCH codes, which are described by Peterson and Weldon (1972) as "the best known constructive codes" for memoryless noisy channels, and with Reed-Muller (RM) codes. These are multiple random error correcting codes that can be characterized by three parameters (n, k, t). The block length is n (this section's M), of which k bits are data bits (this section's N) and the remainder are parity bits. Up to t errors can be corrected in one block. How do the information rate and probability of error of a sparse random code using free energy minimization compare with those of BCH codes?

To estimate the probability of error of the present algorithm I made one thousand runs with $N = 100$ and $f_A = 0.05$ for a small set of values of M and f_n (table 1a). A theoretical analysis will be given, along with a number of other codes using free energy minimization, in (MacKay and Neal 1995).

For comparison, table 2 shows the best BCH and RM codes appropriate for the same noise levels, giving their probability of block error and their rate. I assumed that the probability of error for these codes was simply the probability of more than t errors in n bits. In principle, it may be possible in some cases to make a BCH decoder that corrects more than t errors, but according to Berlekamp (1968), "little is known about... how to go about finding the solutions" and "if there are more than $t + 1$ errors then the situation gets very complicated very quickly."

Comparing tables 1a and 2 it seems that the new code performs as well as or better than equivalent BCH and RM codes, in that no BCH or RM code has both a greater information rate and a smaller probability of block error.

The complexity of the BCH decoding algorithm is $n \log n$ (here, $n \Leftrightarrow M$), whereas that of the FE algorithm is believed to be $w_{A(r)}M$ where $w_{A(r)}$ is the number of 1s per row of $\mathbf{A}$, or at worst $w_{A(r)}MN$. There is therefore no obvious computational disadvantage.

5 Application to a cryptanalysis problem

The inference of the state of a shift register from probabilistic observations of the output sequence is a task arising in certain code breaking problems (Meier and Staffelbach 1989, Anderson 1995).

A cheap 'stream cipher' for binary communication is obtained by sending the bitwise sum modulo 2 of the plain text to be communicated and one bit of a linear feedback shift register (LFSR) of length k bits that is configured to produce an extremely long sequence of pseudo-random bits. This cipher can be broken by an adversary who obtains part of the plain text and the corresponding encrypted message. Given k bits of plain text, the state of the shift register can be deduced, and the entire encrypted message decoded. Even without a piece of plain text, an adversary may be able to break this code if the plain text has high redundancy (for example, if it is an ASCII file containing English words), by guessing part of the plain text.

To defeat this simple attack, the cipher may be modified as follows. Instead of using one bit of the shift register as the key for encryption, the key is defined to be a binary function of a subset of bits of the shift register. Let the number of bits in that subset be h. If this function is an appropriately chosen many-to-one function, it might be hoped that it would be hard to invert, so that even if an adversary obtained a piece of plain text and encrypted text, he would still not be able to deduce the underlying state of the shift register. However, such codes can still be broken (Anderson 1995). Consider a single bit moving along the shift register. This bit participates in the creation of h bits of the key string. It is possible that these h emitted bits together sometimes give away information about the hidden bit. To put it another way, consider the augmented function that maps from $2h - 1$ successive bits in the shift register to h successive key bits. Think of the single bit of the preceding discussion as being the middle bit of these $2h - 1$ bits; call this middle bit b. Write down all 2^h possible outputs of this mapping, and run through all 2^{2h-1} possible inputs, counting how often each output was produced by an input in which $b = 1$, and how often $b = 0$. If these two counts are unequal for any of the 2^h outputs, then the occurrence of such an output in the key sequence gives information about the bit b.

In principle, sufficient amounts of such information can be used to break the code. But if computations that scale exponentially with k are assumed not feasible, it may be difficult to use this information. Two algorithms are given by Meier and Staffelbach (1989) for the case where every bit in the shift register sequence has been observed with uniform probability of error; I study the same case here. The methods derived from free energy minimization lead to an algorithm similar to their algorithm B, and thus constitute a solution to the 'open problem' posed at the end of their paper.

Derivation

There are two ways in which the cryptanalysis problem above can be mapped onto this paper's equation $\mathbf{As+n} = \mathbf{r}$. One might define $\mathbf{s}$ to be the initial state of the shift register, and $\mathbf{r}$ to be the observed noisy sequence, with $\mathbf{A}$ representing the dependence of the mth emitted bit on the initial state, and $\mathbf{n}$ being the noise vector. This representation, however, defines a matrix $\mathbf{A}$ that becomes increasingly dense with 1s as one descends from the top row to the bottom row. It seems likely that a second representation, inspired by the methods of Meier and Staffelbach (1989), will make better use of large quantities of data.

In the second approach, we define $\mathbf{s}$ to be the noise vector. Let the linear feedback shift register's true sequence be $\mathbf{a}_0$, and let the observed noisy sequence be $\mathbf{a}_1 = (\mathbf{a}_0 + \mathbf{s}) \bmod 2$. Following the notation of Meier and Staffelbach (1989), let the number of bits of state in the shift register be k and let the number of observed bits (*i.e.*, the length of $\mathbf{a}_1$ and $\mathbf{s}$) be N. The algorithms of Meier and Staffelbach (1989) centre on the examination of low-weight parity checks that are satisfied by $\mathbf{a}_0$. If the shift register employs t taps, then for $N \gg k$, a large number $M \simeq N \log(N/2k)$ of relations involving $t + 1$ bits of $\mathbf{a}_0$ can be written down. [An even larger number of relations involving $t + 2$, $t + 3$, etc. bits are also available. An attack based on this method is expected to do best if such relations are also included.] Putting these relations in an $M \times N$ parity matrix $\mathbf{A}$, we have:

$$\mathbf{A}\mathbf{a}_0 \bmod 2 = 0$$

so that

$$\mathbf{A}\mathbf{s} \bmod 2 = \mathbf{r} \tag{8}$$

where $\mathbf{r} = \mathbf{A}\mathbf{a}_1 \bmod 2$ is the 'syndrome' vector listing the values of the parity checks. Equation (8) defines our problem. The vector $\mathbf{s}$ is to be inferred, and, unlike equation (1), there is no noise added to $\mathbf{A}\mathbf{s}$. However, we can define a sequence of problems of the form $\mathbf{A}\mathbf{s} + \mathbf{n} \bmod 2 = \mathbf{r}$ such that the real task is the limiting case in which the noise level goes to zero. Then we can apply the free energy method of this paper to these problems. There are two important differences from the demonstration in section 4. First, there is a non-uniform prior probability over $\mathbf{s}$, favouring low weight $\mathbf{s}$. Second, $\mathbf{A}$ is not a full rank matrix; there are many (2^k) values of $\mathbf{s}$ satisfying equation (8), one for each of the possible initial shift register states. Our task is to find the solution that has maximum prior probability, *i.e.* (in the case of uniform noise level) the $\mathbf{s}$ with the smallest number of 1s.

Demonstration

This demonstration uses random polynomials rather than the special polynomials that are used to make encryption keys, but this is not expected to matter. Test data were created for specified k and N using random taps in the LFSR and random observation noise with fixed uniform probability. The parameter β was initially set to 0.25. For each value of β, a sequential reestimation optimization was run until the decrease in free energy was below a specified tolerance (0.001). β was increased by factors of 1.4 until either the most probable vector under $Q(\mathbf{s}; \theta)$ satisfied all the relations (8), or until a maximum value of $\beta = 4$ was passed. The parameter β_P was set to 1 so that the prior probability was not influenced by the annealing.

In the sequential reestimation procedure the bits $1 \ldots N$ of $\mathbf{s}$ may be updated in any order. I have experimented with various orders: a) the order $1 \ldots N$; b) a random order; c) an order ranked by the 'suspicion' associated with a bit, defined for each bit, following Meier and Staffelbach (1989), to be the fraction of the parity checks in which it is involved that are violated. It seemed plausible that if the most suspect bits are modified first, the algorithm would be less likely to modify good bits erroneously. Experimentally however, I found the difference in performance using these different orderings to be negligible.

Results are shown in figure 2 for $(k, N) = (50, 500)$, $(100, 1000)$ and $(50, 5000)$, using each of the three orderings of bits. A dot represents an experiment. A box represents a success, where the algorithm returned the correct error vector. On each of these graphs I also show three lines. A horizontal line, as in section 4,

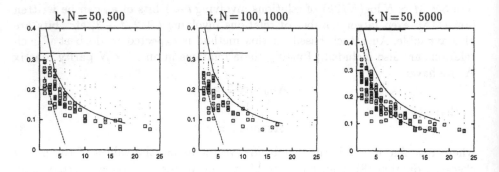

Fig. 2. Results for cryptanalysis problem as a function of number of taps (horizontal axis) and noise level (vertical).

shows the information theoretic bound above which one does not expect to be able to infer the shift register state because the data are too noisy. The two curved lines are from tables 3 and 5 of Meier and Staffelbach (1989), and show (lower line) the limit at which their 'algorithm B appeared to be very successful in most experiments' and (upper line) the theoretical bound beyond which their approach is definitely not feasible. Apparently the algorithm of the present paper not only works fine well beyond the lower line of Meier and Staffelbach, but it frequently finds the correct answer at parameter values right up to the upper line.

Discussion

The forward and backward calculations of the free energy algorithm are similar to calculations in algorithm B of Meier and Staffelbach (1989), but in detail the mapping from $[0,1]^N \rightarrow [0,1]^N$ has a different form. Also, Meier and Staffelbach employ multiple 'rounds' in which the data vector a_1 is changed. This procedure has no analogue in the present algorithm. This algorithm has a well-defined objective function which is guaranteed to decrease during the sequential reestimation algorithm, or which may be minimized by other methods.

A 'bitwise Bayesian' approach to this problem has also been given by Gallager (1963) and Mihaljević and Golić (1993). Their iterative procedure is similar to Meier and Staffelbach's, having two phases in each iteration. In the first phase an inference is made considering each bit individually, assuming a simplified distribution for the other bits. In the second phase a bit-by-bit decision (error-correction) is made. In the present paper, in contrast, the joint posterior distribution of all unknowns given all available information is written down (equation 4), and the iterative procedure optimizes an approximation to this true posterior distribution. No decisions are made until the end of the optimization. These algorithms are similar in that a simplified separable distribution over s is employed, but the details of this distribution's evolution are different.

The McEliece (1978) public-key cryptosystem depends for its security on the intractability of the general decoding problem $\mathbf{As+n} = \mathbf{r}$. The present algorithm is not expected to have any application as regards that cryptosystem, as the free energy approximation only appears to be effective when the matrix $\mathbf{A}$ is sparse.

The LFSR system as an error correcting code

The inference of the state of the linear feedback shift register was motivated as a cryptanalytic application, but the LFSR can also be viewed as a linear error correcting code known as a 'shortened cyclic code', for which the present work offers a decoding algorithm.

Encoding: A rate k/N and a feedback polynomial with t taps are chosen. A signal of length k bits is used as the initial state of an LFSR. The shift register is iterated for N cycles, the resulting sequence of N bits $\mathbf{a_0}$ being transmitted. This procedure defines a linear code with generator matrix $\mathbf{G}$ in which the first k rows are the identity matrix, and successive rows become increasingly dense.

Decoding: The linear code has an $(N-k) \times N$ parity matrix $\mathbf{H}$ such that for any $\mathbf{a_0}$ generated by the code, $\mathbf{Ha_0} = 0$. In this special case where the code is generated by an LFSR, $\mathbf{H}$ can be written as a sparse matrix with just weight $t+1$ per row, each row describing the parity check on one cycle of transmission. Furthermore, we can write many more equally sparse parity checks by adding rows of $\mathbf{H}$. The matrix $\mathbf{A}$ of all parity checks of weight $t+1$ is created. This has of order $N \log(N/2k)$ rows. We evaluate the extended syndrome vector $\mathbf{r} = \mathbf{Aa_1} \bmod 2$. Denoting the noise vector by $\mathbf{s}$, we solve the problem $\mathbf{As} \bmod 2 = \mathbf{r}$ using the free energy minimization method as described earlier. The complexity of this decoding is believed to be $(t+1)N \log(N/2k)$.

As in section 4, I estimated the probability of error of this system by making one thousand runs with $k = 100$ for a small set of values of N and f_n. The results (table 1b) seem comparable with the theoretical performance of BCH codes (table 2), though not as good as the results for sparse random code matrices (table 1a) for $f_n = 0.1$.

6   Conclusion

This paper has derived a promising algorithm for inference problems in modulo 2 arithmetic. Applied to decoding a random sparse error correcting code, this algorithm gives a error/rate trade-off comparable and possibly superior to that of BCH and Reed-Muller codes. For the cryptanalysis problem, the algorithm supersedes Meier and Staffelbach's algorithm B, working close to the theoretical limits given in their paper. The linear feedback shift register system also shows potential for implementation as an error correcting code.

Acknowledgements

I thank Ross Anderson, Radford Neal, Roger Sewell, Robert McEliece and Malcolm MacLeod for helpful discussions, and Mike Cates, Andreas Herz and a referee for comments on the manuscript.

Appendix: Pseudo–code

Here follows pseudo-code in C style for the sequential reestimation algorithm. The vector θ of the text is called $\mathbf{x}$ in this appendix. For a more efficient implementation, the matrix $\mathbf{A}$ should be represented in the form of two lists of indices m and n such that $A_{mn} = 1$. I do not include the calculations required for the termination conditions given in the text.

This routine requires as arguments an initial condition $\mathbf{x}$, a matrix $\mathbf{A}$, an observation vector $\mathbf{g}$ as defined in section 1, and a value for β, a prior bias $\mathbf{b}$ as defined in section 3. The final answer is returned in $\mathbf{x}$; a reconstructed binary vector can be obtained by taking the sign of $\mathbf{x}$.

The routine makes use of arrays q0[n] and q1[n] which contain the probabilities q, and arrays p0[m][n] and p1[m][n] which contain both the forward and reverse probabilities. Again, for efficient implementation, these arrays should be represented as lists.

This code is free software as defined by the Free Software Foundation.

```
sequential_optimizer        /* arguments :                              */
                            /*                the arrays have indices in */
                            /*                the following ranges:      */
( double *x ,               /*      x[n] :    x[1] ... x[N]              */
  double **A ,              /*      A[m][n] : A[1][1] ... A[M][N]        */
  double *g ,               /*      g[m] :    g[1] ... g[M]              */
  double beta ,
  double *bias ,            /*      bias[n] : bias[1] ... bias[N]        */
  int N ,
  int M
)
{
  double *q0 , *q1 ;        /*      q0[n]    : q0[1] ... q0[N]           */
  double **p0 , **p1 ;      /*      p0[m][n] : p0[1][0] ... p0[M][N+1]   */

  for ( m = 1 ; m <= M ; m ++ ) {          /* set up boundary conditions */
    p0[m][0]   = 1.0 ; p1[m][0]   = 0.0 ; /*      for forward pass       */
    p0[m][N+1] = 1.0 ; p1[m][N+1] = 0.0 ; /*      and reverse pass       */
  }

  do {
    for ( n = 1 ; n <= N ; n ++ ) {               /* forward pass        */
      q1[n] = 1.0 / ( 1.0 + exp ( - x[n] ) ) ;
      q0[n] = 1.0 / ( 1.0 + exp (   x[n] ) ) ;
      for ( m = 1 ; m <= M ; m ++ ) {
```

```
      if ( A[m][n] == 0  ) {
        p0[m][n] = p0[m][n-1] ;
        p1[m][n] = p1[m][n-1] ;
      }
      else {
        p0[m][n] = q0[n] * p0[m][n-1] + q1[n] * p1[m][n-1] ;
        p1[m][n] = q0[n] * p1[m][n-1] + q1[n] * p0[m][n-1] ;
      }
    }
  }
  for ( n = N ; n >= 1 ; n -- ) {               /* backward pass       */
    gradient_n = 0.0 ;
    for ( m = 1 ; m <= M ; m ++ ) {
      if ( A[m][n] ) {                          /* accumulate gradient */
        gradient_n -= g[m] *
          ( p1[m][n-1] * p1[m][n+1] + p0[m][n-1] * p0[m][n+1] -
              p0[m][n-1] * p1[m][n+1] - p1[m][n-1] * p0[m][n+1]    ) ;
      }
    }
    gradient_n *= beta ;
    gradient_n -= bias[n] ;
    x[n] = - gradient_n ;
    q1[n] = 1.0 / ( 1.0 + exp ( - x[n] ) ) ;
    q0[n] = 1.0 / ( 1.0 + exp (   x[n] ) ) ;
    for ( m = 1 ; m <= M ; m ++ ) {
      if ( A[m][n] == 0 ) {
        p0[m][n] = p0[m][n+1] ;
        p1[m][n] = p1[m][n+1] ;
      }
      else {
        p0[m][n] = q0[n] * p0[m][n+1] + q1[n] * p1[m][n+1] ;
        p1[m][n] = q0[n] * p1[m][n+1] + q1[n] * p0[m][n+1] ;
      }
    }
  }
} until ( converged ) ;
}
```

References

Aiyer, S. V. B., Niranjan, M. and Fallside, F. (1990). A theoretical investigation into the performance of the Hopfield model, *IEEE Trans. on Neural Networks* 1(2): 204–215.

Anderson, R. J. (1995). Searching for the optimum correlation attack, *in* B. Preneel (ed.), *Fast Software Encryption* Lecture Notes in Computer Science, Springer-Verlag, pp. 137–143 (these proceedings).

Berlekamp, E. R. (1968). *Algebraic Coding Theory*, McGraw-Hill, New York.

Berlekamp, E. R., McEliece, R. J. and van Tilborg, H. C. A. (1978). On the intractability of certain coding problems, *IEEE Transactions on Information Theory* 24(3): 384–386.

Blake, A. and Zisserman, A. (1987). *Visual Reconstruction*, MIT Press, Cambridge Mass.

Durbin, R. and Willshaw, D. (1987). An analogue approach to the travelling salesman problem using an elastic net method, *Nature* **326**: 689–91.

Feynman, R. P. (1972). *Statistical Mechanics*, W. A. Benjamin, Inc.

Gallager, R. G. (1963). *Low density parity check codes*, number 21 in *Research monograph series*, MIT Press, Cambridge, Mass.

Gee, A. H. and Prager, R. W. (1994). Polyhedral combinatorics and neural networks, *Neural Computation* **6**: 161–180.

Hopfield, J. J. and Tank, D. W. (1985). Neural computation of decisions in optimization problems, *Biological Cybernetics* **52**: 1–25.

MacKay, D. J. C. and Neal, R. M. (1995). Error correcting codes using free energy minimization, in preparation.

McEliece, R. J. (1978). A public-key cryptosystem based on algebraic coding theory, *Technical Report DSN 42-44*, JPL, Pasadena.

Meier, W. and Staffelbach, O. (1989). Fast correlation attacks on certain stream ciphers, *J. Cryptology* **1**: 159–176.

Mihaljević, M. J. and Golić, J. D. (1992). A fast iterative algorithm for a shift register initial state reconstruction given the noisy output sequence, *Advances in Cryptology - AUSCRYPT'90*, Vol. 453, Springer-Verlag, pp. 165–175.

Mihaljević, M. J. and Golić, J. D. (1993). Convergence of a Bayesian iterative error-correction procedure on a noisy shift register sequence, *Advances in Cryptology - EUROCRYPT 92*, Vol. 658, Springer-Verlag, pp. 124–137.

Peterson, C. and Soderberg, B. (1989). A new method for mapping optimization problems onto neural networks, *Int. Journal Neural Systems*.

Peterson, W. W. and Weldon, Jr., E. J. (1972). *Error-Correcting Codes*, 2nd edn, MIT Press, Cambridge, Massachusetts.

Van den Bout, D. E. and Miller, III, T. K. (1989). Improving the performance of the Hopfield-Tank neural network through normalization and annealing, *Biological Cybernetics* **62**: 129–139.

Truncated and Higher Order Differentials

Lars R. Knudsen

Aarhus University, Denmark
email:ramkilde@daimi.aau.dk

Abstract. In [6] higher order derivatives of discrete functions were considered and the concept of higher order differentials was introduced. We introduce the concept of truncated differentials and present attacks on ciphers presumably secure against differential attacks, but vulnerable to attacks using higher order and truncated differentials. Also we give a differential attack using truncated differentials on DES reduced to 6 rounds using only 46 chosen plaintexts with an expected running time of about the time of 3,500 encryptions. Finally it is shown how to find a minimum nonlinear order of a block cipher using higher order differentials.

1 Introduction

Differential cryptanalysis [1] was introduced by Biham and Shamir. Lai considered higher order derivatives of discrete functions [6] and the concept of higher order differentials was introduced. As a special case binary functions were considered, which is relevant for cryptanalysis of block ciphers. The cryptographic significance of higher order differentials was discussed, but no applications given. Knudsen and Nyberg [8] showed that block ciphers exist secure against a differential attack using first order differentials, as proposed by Biham and Shamir.

In this paper we introduce the concept of **truncated** differentials, i.e. differentials where only a part of the difference in the ciphertexts (after a number of rounds) can be predicted. We show examples of Feistel block ciphers secure against a differential attack using first order differentials, but vulnerable to a differential attack using truncated differentials and higher order differentials, thus illustrating that one should be careful when claiming for resistance against differential attacks. Finally, we give a method of how to find a minimum nonlinear order of a block cipher using higher order differentials.

2 Differential Attacks

In this paper we consider Feistel ciphers. A **Feistel cipher** with block size $2n$ and with r rounds is defined as follows. The round function g is

$$g : GF(2)^n \times GF(2)^n \times GF(2)^m \rightarrow GF(2)^n \times GF(2)^n$$
$$g(X, Y, Z) = (Y, f(Y, Z) + X)$$

where f can be any function taking two arguments of n bits and m bits respectively and producing n bits. '+' is a commutative group operation on the set of n bit blocks.

Given a plaintext $P = (P^L, P^R)$ and r round keys $K_1, K_2, ..., K_r$ the ciphertext $C = (C^L, C^R)$ is computed in r rounds. Set $C_0^L = P^L$ and $C_0^R = P^R$ and compute for $i = 1, 2, ..., r$

$$(C_i^L, C_i^R) = (C_{i-1}^R, f(C_{i-1}^R, K_i) + C_{i-1}^L)$$

Set $C_i = (C_i^L, C_i^R)$ and $C^L = C_r^R$ and $C^R = C_r^L$.

Traditionally, the round keys $(K_1, K_2, ..., K_r)$, where $K_i \in GF(2)^m$, are computed by a key schedule algorithm on input a master key K.

The differential attacks exploit that pairs of plaintexts with certain differences yield other certain differences in the corresponding ciphertexts with a non-uniform probability distribution. For a pair of plaintexts, which are not discarded by a filtering process, see [1, 2], one tries for all values of the round key in the last round, if the expected difference in the ciphertexts occur. This is repeated several times and the most suggested value is taken to be the value of the secret key of the last round. Now all ciphertexts can be decrypted one round and a weaker cipher attacked in the same way but with a smaller complexity.

The signal to noise ratio, S/N [1, 2], is the number of times the right key is counted over the number of times a random key is counted.

$$S/N = \frac{|K| \times p}{\gamma \times \lambda}$$

where p is the probability of the differential used in the attack, $|K|$ is the number of possible values of the key, we are looking for, γ is the number of keys suggested by each pair of plaintexts and λ is the ratio of non-discarded pairs to all pairs, see [1, 2] for further details. For our attacks in this paper $\lambda = 1$. If $S/N \leq 1$ then a differential attack will not succeed.

Sometimes one also calls the function f, the round function. We adopt this convention for convenience, since it should cause no confusion.

For the remainder of this paper we will assume that the round keys are independent and uniformly random and of size n, i.e. half the block size. The difference of two quantities is always taken to be the operation for which the difference is independent on the value of the inserted key. Therefore when considering differences for the round function f we will write $f(x)$ instead of $f(x, k)$. We will assume that the difference of two quantities chosen in an attack is the exclusive-or operation, if not stated explicitly otherwise. The complexity of the attacks is measured as the number of encryptions of the full cipher that an attacker has to perform for success.

3   Truncated Differentials

In a conventional differential attack on a $2n$ bit Feistel cipher, a differential is a tool to predict an n bit value of the ciphertext after a certain number of rounds.

One defines a difference of two bit strings of equal length. Then (a, b) is called an i round differential, if a difference a in two plaintext blocks yields a difference b in the two ciphertext blocks after i rounds of encryption. But as we will show now it is not always necessary to predict the full n bit value. Even a 1 bit value suffices in some cases. A differential that predicts only parts of an n bit value is called a *truncated differential*. More formally, let (a, b) be an i-round differential. If a' is a subsequence of a and b' is a subsequence of b, then (a', b') is called an i round truncated differential.

In [7] it is shown that the functions $f(x) = x^{-1}$ in $GF(2^n)$, where $f(x) = 0$ for $x = 0$, are differentially 2-uniform for odd n and differentially 4-uniform for even n, i.e. the highest probability of a non-trivial one round differential is $2/2^n$ and $4/2^n$ respectively. In both cases the nonlinear order of the outputs is $n - 1$ [7]. As an example consider a 5 round cipher using as round function

$$f(x, k) = (x \oplus k)^{-1}$$

in $GF(2^n)$ for n odd. From the results of [8] this cipher is highly resistant against differential attacks using full differentials, since any 3 round differential has a probability of at most 2^{3-2n} according to Th. 2 of [8], that is, using differentials, where full n bit differences are used. In an attack counting on the round key of the last round the signal to noise ratio is

$$S/N < \frac{2^n \times 2^{3-2n}}{1 \times 1} < 1$$

for $n > 3$ and the attack will not succeed. In an attack counting on the round keys of the last two rounds only a 2 round differential is needed. And since the concepts of characteristics and differentials coincide for 2 rounds in a Feistel cipher it is easy to see that there exists a differential with a probability of $2/2^n$ and that this differential obtains a maximum probability. The signal to noise ratio is

$$S/N = \frac{2^{2n} \times 2^{1-n}}{1 \times 1} = 2^{n+1}$$

and the attack will succeed with complexity 2^n chosen plaintexts and running time of about 2^{3n}.

However, for every non-trivial input difference to one round there are only 2^{n-1} possible differences in the outputs, each one with a probability of $2/2^n$, since the round function is differentially 2-uniform and the exclusive-or operation is commutative. That is, for a non-trivial input difference we get one bit of information about the output differences. From this fact we can construct a 2 round differential of probability one, where only one bit of the differences after 2 rounds of encryption is predicted. In a differential attack counting on the round keys of the last two rounds for every pair of plaintexts only half the possible values of the keys will be suggested. We obtain

$$S/N = \frac{2^{2n} \times 1}{2^{(2n-1)} \times 1} = 2$$

and the attack will succeed with sufficiently many pairs of chosen plaintexts. We implemented the attack on a 5 round 18 bit cipher with a key of 45 bits using as round function $f(x) = x^{-1}$ in $GF(2^9)$. Using 18 pairs of chosen plaintexts in 100 tests only one pair of keys was found, the right keys in the fourth and fifth rounds.

The attack can be generalised and the following result holds.

Theorem 1. *Let $f(x, k) : GF(2^n) \times GF(2^n) \to GF(2^n)$ be the round function in a 5 round Feistel cipher with block size $2n$ bits using 5 round keys, each of size n bits. Let α ($\neq 0$) be an input difference for which only a fraction W of all output differences are possible. Then a differential attack using truncated differentials has a complexity of $2L$ chosen plaintexts and a running time of about $L \times 2^{2n}$, where L is the smallest integer s.t. $(W)^L < 2^{-2n}$. The value of L is at most $2n + 1$.*

Proof: Consider the following attack.

1. Let α be the non-trivial difference of two inputs to f, for which only a fraction W of the output differences can occur.
2. Compute a table T (initialised to zero in all entries), s.t. for
   $i = 0, .., 2^n - 1$, $T[f(i) \oplus f(i \oplus \alpha))] = 1$.
3. Choose plaintext P_1 at random and set $P_2 = P_1 \oplus (\alpha \parallel 0)$.
4. Get the encryptions C_1 and C_2 of P_1 and P_2.
5. For every value k_5 of the round key RK_5 do
   (a) Decrypt the ciphertexts C_1, C_2 one round using k_5. Denote these ciphertexts D_1, D_2.
   (b) For every value k_4 of the round key RK_4 do
       i. Calculate $t_i = f(D_i^R \oplus k_4)$ for $i = 1, 2$.
       ii. If $T[t_1 \oplus t_2 \oplus D_1^L \oplus D_2^L] > 0$ then output k_5 and k_4.

Since the nonlinear order of $f(x)$ can be as high as $n - 1$, the information about the output differences we get from a given input difference is not necessarily easily determined. Therefore we may have to compute a table T, s.t. for a given input difference α, if $T[\beta] > 0$ then an output difference β is possible. The inputs to the first round are equal and the inputs to the second round has difference α. That is, we can compute a fraction W of all possible values of the output difference of the fourth round from the right halves of the ciphertexts and from the values in table T. Upon termination about $W \times 2^{2n}$ of the possible values of (RK_4, RK_5) have been suggested, one of which is the right pair of keys. By repeating the attack sufficiently many times only one unique pair of keys, the right pair of keys, will be left suggested. Any other keys will be suggested with probability W for each run of the above attack. Therefore after trying L pairs of plaintexts any key but the right key, is suggested L times with a probability of $(W)^L$ and if $(W)^L < 2^{-2n}$ with a high probability the right keys are uniquely determined. Finally, note that since $W \le 1/2$, $\min_L : (1/2)^L < 2^{-2n} = 2n + 1$.□

The attack can be extended to work on ciphers with any number of rounds by counting on all but the first three round keys.

4   Higher Order Differentials

In [6] the definition of derivatives of cryptographic functions was given.

Definition 2 (Lai [6]). Let $(S, +)$ and $(T, +)$ be Abelian groups. For a function $f : S \mapsto T$, the derivative of f at the point $a \in S$ is defined as

$$\Delta_a f(x) = f(x + a) - f(x).$$

The i'th derivative of f at the point $a_1, ..., a_i$ is defined as

$$\Delta_{a_1,...,a_i}^{(i)} f(x) = \Delta_{a_i}(\Delta_{a_1,...,a_{i-1}}^{(i-1)} f(x)).$$

Note that the characteristics and differentials used by Biham and Shamir in their attacks correspond to the first order derivative described by Lai. Therefore it seems natural to extend the notion of differentials into **higher order differentials**.

Definition 3. A one round differential of order i is an $(i + 1)$-tuple $(\alpha_1, ..., \alpha_i, \beta)$, s.t. $\Delta_{\alpha_1,...,\alpha_i}^{(i)} f(x) = \beta$.

When considering functions over $GF(2)$ the points $a_1, ..., a_i$ must be linearly independent for the i'th derivative not to be trivial zero.

Proposition 4 (Lai [6]). Let $L[a_1, a_2, ..., a_i]$ be the list of all 2^i possible linear combinations of $a_1, a_2, ..., a_i$. Then

$$\Delta_{a_1,...,a_i}^{(i)} f(x) = \sum_{\gamma \in L(\alpha_1,...,\alpha_i)} f(P \oplus \gamma).$$

If a_i is linearly dependent of $a_1, ..., a_{i-1}$, then

$$\Delta_{a_1,...,a_i}^{(i)} f(x) = 0.$$

Proposition 5 (Lai [6]). Let $ord(f)$ denote the nonlinear order[1] of a multivariable polynomial function $f(x)$. Then

$$ord(\Delta_a f(x)) \leq ord(f(x)) - 1.$$

This leads to the following Corollary.

Corollary 6. If $\Delta_{a_1,...,a_i} f(x)$ is not a constant, then the nonlinear order of f is greater than i.

Proof: From Prop. 5 it follows that

$$ord(f) \geq ord(\Delta_{a_1} f(x)) + 1 \geq \cdots\cdots\cdots \geq ord(\Delta_{a_1,...,a_i} f(x)) + i.$$

$\square$

---

[1] In [6] called the nonlinear degree.

4.1 Attacks using higher order differentials

In the previous section we showed how to exploit partial information of differentials. One may ask the following question: does round functions exist, which does not leak any partial information for any non-trivial difference? The answer is positive and in the following we give an example of a 5 round Feistel cipher, for which the round function is differentially 1-uniform i.e. for every non trivial input difference all output differences occur exactly once. We show that differential attacks on this cipher using higher order differentials are much more efficient than conventional differential attacks. We generalise the result to any 5 round Feistel cipher.

Theorem 7. *Let $f(x, k) = (x + k)^2 \bmod p$, p prime, be the round function in a Feistel cipher of block size $log_2 p^2$, where '+' is addition modulo p and the difference of two quantities, x and y, is $x - y \bmod p$. f is differentially 1-uniform, a non-trivial one round differential has a probability of $1/p$. Secondly, the second order derivative of f is constant.*

Proof: To prove the first statement, consider a fixed $a \neq 0 \bmod p$. Then

$$f(x) - f(x + a) =_p f(y) - f(y + a) \Leftrightarrow$$
$$x^2 - (x^2 + a^2 + 2ax) =_p y^2 - (y^2 + a^2 + 2ay) \Leftrightarrow$$
$$2ax =_p 2ay \Leftrightarrow 2a(x - y) =_p 0 \Leftrightarrow x =_p y$$

since p is prime. To prove the second statement, let a_1, a_2 be constants, then

$$\begin{aligned}
\Delta_{a_1, a_2} f(x) &= f(x + a_1 + a_2) - f(x + a_1) - f(x + a_2) + f(x) \\
&= x^2 + (a_1 + a_2)^2 + 2(a_1 + a_2)x - (x^2 + a_1^2 + 2a_1 x) \\
&\quad - (x^2 + a_2^2 + 2a_2 x) + x^2 \\
&= (a_1 + a_2)^2 - a_1^2 - a_2^2 \\
&= 2a_1 a_2.
\end{aligned}$$

□

Theorem 8. *Let $f(x, k) = (x + k)^2 \bmod p$, p prime, be the round function in a 5 round Feistel cipher of block size $log_2 p^2$ with independent round keys, i.e. a key size of $5 \times log_2 p$. A differential attack using first order differentials needs about $2p$ chosen plaintexts and has a running time of about p^3.*

Proof: When doing a differential attack counting on the round key in the fifth round of the above cipher we need a 3 (or 4) round differential. It is easy to see that there exists a 3 round differential with a probability of $1/p$ and that this differential obtains a maximum probability. We obtain

$$S/N = \frac{p \times 1/p}{1 \times 1} = 1$$

This attack is not possible, since the right key cannot be distinguished from other random keys. When doing a differential attack counting on the round keys in both the fourth and fifth rounds we need only a 2 round differential. There exists a 2 round differential with a probability of $1/p$, which is a maximum probability for the above cipher. In this case we obtain

$$S/N = \frac{p^2 \times 1/p}{1 \times 1} = p$$

This attack is possible. We need about $2p$ chosen plaintexts and for every pair of plaintexts we do two rounds of encryption for every p^2 possible keys of the fourth and fifth rounds. Therefore we obtain a complexity of about p^3.   □

Theorem 9. *Let $f(x,k) = (x+k)^2$ mod p, p prime, be the round function in a 5 round Feistel cipher of block size $\log_2 p^2$ with independent round keys, i.e. a key size of $5 \times \log_2 p$. A differential attack using second order differentials needs about 8 chosen plaintexts with a running time of about p^2.*

Proof: Consider $\Delta_{\alpha,\beta} f(x)$ where $\alpha = a \parallel 0$ and $\beta = b \parallel 0$ for some fixed a, b, i.e the left halves of α and β are zero. See Fig. 1, where $(0,0)$ denotes the trivial second order derivative of f and where in the second round the second order derivative is $(a, b, 2 \times a \times b)$. Consider the following attack

1. Choose plaintext P_1 at random.
2. Set $P_2 = P_1 + \alpha$, $P_3 = P_1 + \beta$ and $P_4 = P_1 + \alpha + \beta$.
3. Get the encryptions $C_1, ..., C_4$ of $P_1, ..., P_4$
4. For every value k_5 of the round key RK_5 do
   (a) Decrypt all ciphertexts $C_1, ..., C_4$ one round using k_5. Denote these 4 ciphertexts $D_1, ..., D_4$.
   (b) For every value k_4 of the round key RK_4 do
      i. Calculate $t_i = f(D_i^R + k_4)$ for $i = 1, .., 4$.
      ii. If $(t_1 + t_4 - (t_2 + t_3)) - (D_1^L + D_4^L - (D_2^L + D_3^L)) = 2 \times a \times b$ then output k_5 and k_4.

Here X^L and X^R denote the left and right halves of X respectively. In the first round all inputs to the f-function are equal. In the second round the inputs form a second order differential with $(a, b, 2 \times a \times b)$. Since this differential has probability 1 according to Th. 7, the difference in the four inputs to the third round is $\Gamma = 2 \times a \times b$. Therefore the difference in the outputs of the fourth round can be computed as the exclusive-or sum of Γ and of the right halves of the ciphertexts. Upon termination a few keys will have been suggested, among which the right keys appear, since the two round second order differential has probability 1. Therefore by repeating this attack a few times only one value of (RK_4, RK_5) is suggested every time. This value is guaranteed to be the secret fourth and fifth round key. The signal to noise ratio of the attack is

$$S/N = \frac{p^2 \times 1}{1 \times 1} = p^2$$

where we have assumed that one key in average is suggested by each pair of plaintexts. Now it is trivial to find the remaining three round keys by similar

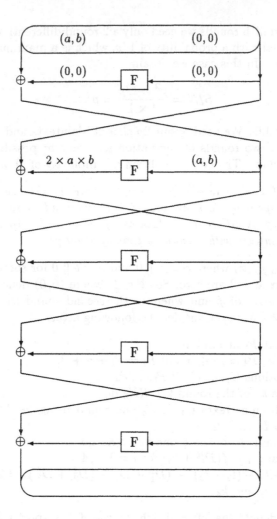

Fig. 1. A second order differential of a five round Feistel cipher

attacks on cryptosystems with less than five rounds. As in [1, 2] we can pack the chosen plaintexts in economical structures, thus as an example obtain four second order differentials from 8 chosen plaintexts.□

If the prime p above is of cardinality, say about 2^{25}, according to Th. 8 a differential attack using first order differential has a complexity of about 2^{75} using about 2^{26} chosen plaintexts, i.e. not at all a practical attack. According to Th. 9 a differential attack using second order differentials has a complexity of about 2^{50} using only about 8 chosen plaintexts, a practical attack or at least not far from being one.

The attack in the proof of Th. 9 can be applied to any 5 round Feistel cipher, where the round function contains no expansion and where the output coordinates are quadratic, i.e. the nonlinear order of f is 2. Furthermore the attack can be converted into an attack on any 5 round Feistel cipher. For convenience let us now consider functions over $GF(2)$. We state explicitly the definition of higher order differentials for this important case.

Definition 10. A one round differential of order i is an $(i+1)$-tuple $(\alpha_1, ..., \alpha_i, \beta)$, s.t. all α_j's are linearly independent and

$$\sum_{\gamma \in L(\alpha_1, ..., \alpha_i)} F(P \oplus \gamma) = \beta.$$

It is seen there are 2^i plaintexts in an i-order differential.

Theorem 11. *Let $f(x, k)$ be the round function in a 5 round Feistel cipher of block size $2n$ with independent round keys, i.e. a key size of $5 \times n$ bits. Assume that the nonlinear order of f is r. Then a differential attack using r-order differentials needs about 2^{r+1} chosen plaintexts with a running time of about 2^{2n+r}.*

Proof: According to Prop. 6 the r-order derivative of a function of nonlinear order r is a constant. Therefore we can obtain a 2 round r-order differential with probability 1 and do a similar attack as in the proof of Th. 9.   □

To illustrate the above attack, we consider now the differentially uniform mappings $f(x) = x^{2^k+1}$ in $GF(2^n)$ described in [8].

Lemma 12. *Consider the permutation $f(x) = x^{2^k+1}$ in $GF(2^n)$ for n odd and $gcd(k, n) = 1$. f is differentially 2-uniform and the second order derivative of f, $\Delta_{\alpha,\beta}f(x)$ is a constant with the value $\Gamma = \alpha \times \beta \times (\alpha^{2^k-1} \oplus \beta^{2^k-1})$, where $'\times'$ is multiplication in $GF(2^n)$.*

Proof: The first statement is proved in [8] and that the second derivative is a constant follows from Prop. 5. The actual constant can be computed in a straightforward way and is omitted here (see [5]).   □

We implemented the attack of Th. 11 counting on both the fourth and fifth round key using second order differentials in a five round Feistel cipher with $f(x)$ of Lemma 12 as round function and with $n = 9$ and $k = 1$, i.e. a 18 bit cipher with a 45 bit key. In 100 tests using 12 chosen plaintexts only one pair of keys was suggested and every time this pair was the right values of the fourth and fifth secret round keys. By using quartets as defined in [1, 2] the number of chosen plaintexts can be reduced to about 8. Note that for this cipher the probability of any 3 round differential of first order is at most 2^{3-2n} [8], where $2n$ is the block size. Also note that the example cipher of [8] has 6 rounds, and is therefore not vulnerable to the above attacks.

The outputs of S-box	Does not affect S-boxes
1	1, 7
2	2, 6
3	3, 1
4	4, 2
5	5, 8
6	6, 4
7	7, 5
8	8, 3

Table 1. Flow of the S-box output bits.

5 Truncated Differentials of the DES

For the DES [9] there are truncated differentials with probability one. When two inputs to the F-function are equal in the inputs to an S-box, the outputs from that S-box are always equal, independent of the values of the inputs to other S-boxes. These truncated differentials are used to a wide extent in Biham and Shamirs attacks on the DES [1, 2].

The output of an S-box affects the inputs of at most six S-boxes in the following round, because of the P-permutation, see Table 1. This fact can be used to construct a four round truncated differential for the DES with probability one, which gives knowledge about the difference of eight bits in the ciphertext after four rounds. Consider a pair of plaintexts where the right halves are equal and the left halves differ, such that the inputs to only one S-box, say S-box 1, are different after the E-expansion. The first round in the differential holds always, and in the second round the outputs of all S-boxes except S-box 1 are equal. In the inputs to the third round the inputs of two S-boxes, S-boxes 1 and 7, are always equal, since S-box 1 does not affect these S-boxes according to Table 1. Therefore the outputs of these S-boxes are equal, and the xor of eight bits in the right halves of the ciphertexts after three rounds are known, since the xor in the inputs in the second round is known. The right halves after three rounds equal the left halves after four rounds, therefore the xor of eight bits after four rounds of encryption are known with probability one. This differential can be used to attack the DES with 6 rounds in a differential attack using only a few chosen plaintexts as we will show in the next section.

5.1 Attack on 6 round DES.

In this section we consider the DES [9] reduced to 6 rounds. We take the first 6 rounds of the standard and omit the initial and final permutation, since they are of no importance for our attack.

Theorem 13. *There exists a differential attack on DES with 6 rounds, which finds the secret key using 46 chosen plaintexts in expected time the time of about 3,500 encryptions, which can be done in a few seconds on a PC.*

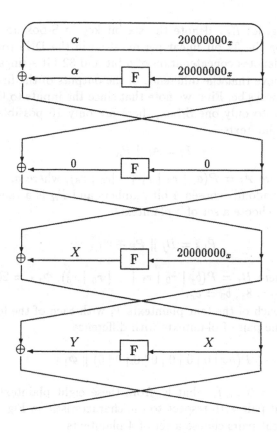

Fig. 2. A 4 round differential of DES.

Proof: We consider a differential chosen plaintext attack using the differential in Fig. 2 and a similar differential where all the quantities 20000000_x are replaced by 40000000_x. Assume first that the outputs of the first round have difference α. The inputs to the third round differ in only two bits both affecting only S-box 1. According to the above discussion, the inputs with difference X to the fourth round are equal in the inputs to the S-boxes 1 and 7. Therefore eight bits of the difference Y are zero. Since the difference of the inputs to the third round is known, the attacker knows eight bits of the difference of the outputs of the F-function in the sixth round, since he knows the difference in the ciphertexts. These eight bits are the output bits of S-boxes 1 and 7. The attacker now tries for all 64 possible values of the key whether the inputs to S-box 1 yield the computed expected output difference, and does the same for S-box 7. For every pair of ciphertexts used in the analysis for both S-boxes the attacker will get an average of 4 suggested key values, among which the right key values appear, since the used differential has probability one. By trying a few pairs, e.g. four pairs with a high probability only one key value, the right key value, will be left suggested by all pairs.

In the following, let $K_{i,j}$ denote the six bit key in S-box no. j in the i'th round and let P be the 32 bit linear permutation in the DES round function, see [9]. '|' and '||' denotes concatenation of 4 bit and 32 bit strings respectively.

We assumed above that the difference of the outputs of the first round is α, which it will not always be. First we note that since the inputs to the first round differ in the inputs to only one S-box, there are only 16 possible values of α. Choose a set of 4 plaintexts

$$P_i = A_i \parallel P_R$$

for $i = 0, ..., 3$, where $A_i = P(a_i \mid r_0 \mid r_1 \mid ...|r_5 \mid r_6)$, where $a_i = i$, each of 4 bits, the r_k's are randomly chosen 4 bit numbers and P_R is a randomly chosen 32 bit string. Next choose a set of 4 plaintexts

$$P_{1,j} = B_j \parallel P_R \oplus \Phi_{1,1}$$

for $j = 0, ..., 3$, where $B_j = P(b_j \mid r_0 \mid r_1 \mid ... \mid r_5 \mid r_6)$, $\Phi_{1,1} = 20000000_x$ and $b_0 = 0_x$, $b_1 = 4_x$, $b_2 = 8_x$, $b_3 = c_x$.

By combining each of the four plaintexts P_i with each of the four plaintexts $P_{1,j}$ one obtains one pair of plaintexts with difference

$$P(h_x \mid 0 \mid 0 \mid 0 \mid 0 \mid 0 \mid 0 \mid 0) \parallel \Phi_{1,1} \qquad (1)$$

for all values of $h = 0, ..., f_x$, that is, from these eight plaintexts one pair of plaintexts is a right pair with respect to the characteristic in Fig. 2.

To get more right pairs choose a set of 4 plaintexts

$$P_{2,j} = B_j \parallel P_R \oplus \Phi_{1,2}$$

for $j = 0, ..., 3$, where $\Phi_{1,2} = 40000000_x$, and a set of 4 plaintexts

$$P_{3,j} = A_i \parallel P_R \oplus \Phi_{1,1} \oplus \Phi_{1,2}$$

for $i = 0, ..., 3$.

By combining the set $P_{2,j}$ with the set $P_{3,j}$ one obtains another pair of plaintexts with difference (1) for all values of $h = 0, ..., f_x$.

By combining the set $P_{1,j}$ with the set $P_{2,j}$ and combining the set P_i with the set $P_{3,j}$ one obtains 2 pairs of plaintexts with difference

$$P(h_x \mid 0 \mid 0 \mid 0 \mid 0 \mid 0 \mid 0 \mid 0) \parallel \Phi_{1,2}$$

for all values of $h = 0, ..., f_x$. Note that the characteristics just defined both affect the same S-box in the first round. Get the encryptions of the 16 plaintexts $P_i, P_{1,j}, P_{2,j}$ and $P_{3,j}$.

The attack proceeds as follows.

1. For every value $k_{1,1}$ of the key $K_{1,1}$ in S-box 1 in the first round do

(a) Let $k_{1,*}$ be the 48 bit key obtained from the concatenation of the value of $k_{1,1}$ and 42 randomly chosen bits.
Compute $c_0 = F(k_{1,*}, P_R)$ and $c_1 = F(k_{1,*}, P_R \oplus \Phi_{1,1})$. Now $c_0 \oplus c_1 = P(y \mid 0 \mid 0 \mid ... \mid 0)$ for some hex value y. Find the plaintext P_i and $P_{1,j}$, such that $c_0 \oplus c_1 = A_i \oplus B_j$. The pair of plaintexts P_i and $P_{1,j}$ is a right pair with respect to the characteristic in Fig. 2. Next compute $c_2 = F(k_{1,*}, P_R \oplus \Phi_{1,2})$ and $c_3 = F(k_{1,*}, P_R \oplus \Phi_{1,1} \oplus \Phi_{1,2})$. Find the plaintext $P_{2,j}$ and $P_{3,j}$, such that $c_2 \oplus c_3 = B_j \oplus A_i$. The pair of plaintexts $P_{2,j}$ and $P_{3,j}$ is a right pair with respect to the characteristic in Fig. 2.

Repeat this procedure finding 2 right pairs P_i and $P_{2,j}$, $P_{1,j}$ and $P_{3,j}$ for the second characteristic.

(b) Use the four right pairs in the differential attack described above. First do the attack on S-box 1 in the last round. If one key value $k_{6,1}$ of $K_{6,1}$ is suggested by all four pairs, perform the differential attack on S-box 7 in the last round. If one key value $k_{6,7}$ of $K_{6,7}$ is suggested by all four pairs, take $k_{6,1}$ and $k_{6,7}$ as the key values of $K_{6,1}$ and $K_{6,7}$ and take $k_{1,1}$ as the value of $K_{1,1}$.

The above attack finds 18 key bits with a high probability. In step 1(a) above we need not do a complete evaluation of the F-function, only the computation of the one S-box involved is needed. For every value of $K_{1,1}$ we do 4 S-box evaluations. Then for every value of $K_{6,1}$ we do 8 S-box evaluations, one for each of the 8 ciphertexts in the 4 pairs. The search for $K_{6,7}$ is done only when one key value of $K_{6,1}$ is suggested all four times. Totally the time used is about the time of 2^{15} S-box evaluations, about the time of 500 encryptions of six round DES. Note that the differential used in the attack has probability one. More key bits can be found in similar attacks by plaintexts yielding other characteristics.

With an additional 2 sets of each 16 plaintexts involving other S-boxes in the first round one finds 54 key bits. By a careful choice of each of the 2 sets one of the plaintext P_i in the above described attack can be reused. Since the DES has dependent round keys some of the key bits tried in the first and in the sixth round are identical. Using the S-boxes 1, 2 and 5 in the first round is an optimal choice and the attack finds 45 bits of the 56 bit secret key. The remaining 11 bits can be found by exhaustive search. The attack needs a total of 46 plaintexts and runs in time about 3,500 encryptions of six round DES, which can be done in a few seconds on a PC.   □

There are possible variations of the above attack, which are listed in Table 2. It should be noted that the linear attack combined with differential 'techniques' by Hellman and Langford [4] exploits the same phenomenon as in our attack, but the two attacks are different. Finally we note that in [10] Preneel et al. considered, what they call *reduced exors*, in differential attacks on the DES in CFB mode. The reduced exors have some resemblance with truncated differentials.

No. of chosen plaintexts	No. of key bits found
7	8
16	18
31	33
46	45

Table 2. Complexities of our attacks on DES with 6 rounds.

6 Computing the Nonlinear Order

In [11] it was considered to cryptanalyse the DES by the method of formal coding. The conclusion was that this is hardly possible. It was shown also that the nonlinear order of any of the 8 S-boxes in the DES is 5. An open question is, what is the order of the outputs for the full 16 round DES. In general, a cipher will be vulnerable to attacks like the method of formal coding if the nonlinear order of the outputs is too low. Higher order differentials can be used to determine a lower bound of the nonlinear order of a block cipher.

Test for nonlinear order

Input: $E_K(\cdot)$, a block cipher, a key K, plaintexts $x_1 \neq x_2$ and r, an integer.

Output: $i \leq r$, a minimum nonlinear order of E_K.

Let $a_1, a_2, ..., a_i$ be linearly independent.

1. Set $i = 1$
2. Compute $y_1 = \Delta_{a_1,...,a_i} E_K(x_1)$ and $y_2 = \Delta_{a_1,...,a_i} E_K(x_2)$
3. If $y_1 = y_2$ output i and stop
4. If $i \geq r$ output i and stop
5. Set $i = i + 1$ and go to step (2)

If in step (3), $y_1 \neq y_2$ then the nonlinear order is greater than i according to Prop. 6. If $y_1 = y_2$ then the nonlinear order may be greater than i, because it is possible for other values of x'_1 and x'_2 that $y'_1 \neq y'_2$. However the above test must stop, since if the i'th derivative of f is constant, then the $i + r$'th derivative of f is zero for all $r > 0$. Also, note that computing an i'th order derivative of f, is equivalent to computing two times an $i - 1$'st order derivative of f. Therefore the values of y_1, y_2 can be stored and re-used in following steps.

To test a block cipher E, pick a random key K and two random plaintexts and run the test for nonlinear order. If the output of the test is d then the nonlinear order of E_K is at least d. Repeat this procedure for as many keys and plaintexts as desired. The input r and the test in step (4) is necessary for block ciphers like the DES and r should be chosen not much greater than 32, since it takes about 2^r encryptions to check a nonlinear order of r.

7 Concluding Remarks and Open Problems

We have shown applications for truncated and higher order differentials. We presented ciphers secure against conventional differential attacks, but vulnerable to attacks using either truncated or higher order differentials. We presented a differential attack on DES with 6 rounds using truncated differentials with complexity of about 46 chosen plaintexts and a running time of about the time of 3,500 encryptions. Finally we presented a method to test the nonlinear order of a block cipher using higher order differentials.

In the above attacks we have exploited the small number of rounds in the Feistel ciphers we have analysed. It is an open problem, whether differential attacks based on higher order differentials are applicable to ciphers with more than 5 rounds. This seems to require a method of iterating higher order differentials to more than two rounds in the same way as with first order differentials. Truncated differentials can be combined with conventional differentials to refine attacks using the latter. It is an open problem whether truncated differentials can improve the attacks on DES [1, 2] for more than 6 rounds.

8   Acknowledgements

The author wish to thank Luke O'Connor, Bart Preneel and an anonymous referee for helpful comments which improved this paper.

References

1. E. Biham and A. Shamir. Differential cryptanalysis of DES-like cryptosystems. *Journal of Cryptology*, 4(1):3–72, 1991.
2. E. Biham and A. Shamir. *Differential Cryptanalysis of the Data Encryption Standard*. Springer Verlag, 1993.
3. M.E. Hellman, R. Merkle, R. Schroeppel, L. Washington, W. Diffie, S. Pohlig, and P. Schweitzer. Results of an initial attempt to cryptanalyze the NBS Data Encryption Standard. Technical report, Stanford University, U.S.A., September 1976.
4. M. E. Hellman and S. K. Langford. Differential–linear cryptanalysis. In Y. G. Desmedt, editor, *Advances in Cryptology - Proc. Crypto'94, LNCS 839*, pages 26–39. Springer Verlag, 1994.
5. L.R. Knudsen. *Block Ciphers - Analysis, Design and Applications*. PhD thesis, Aarhus University, Denmark, 1994, DAIMI PB – 485.
6. X. Lai. Higher order derivatives and differential cryptanalysis. In *Proc. "Symposium on Communication, Coding and Cryptography", in honor of James L. Massey on the occasion of his 60'th birthday, Feb. 10-13, 1994, Monte-Verita, Ascona, Switzerland*, 1994. To appear.
7. K. Nyberg. Differentially uniform mappings for cryptography. In T. Helleseth, editor, *Advances in Cryptology - Proc. Eurocrypt'93, LNCS 765*, pages 55–64. Springer Verlag, 1993.
8. K. Nyberg and L.R. Knudsen. Provable security against differential cryptanalysis. In E.F. Brickell, editor, *Advances in Cryptology - Proc. Crypto'92, LNCS 740*, pages 566–574. Springer Verlag, 1993.
9. National Bureau of Standards. Data encryption standard. Federal Information Processing Standard (FIPS), Publication 46, National Bureau of Standards, U.S. Department of Commerce, Washington D.C., January 1977.

10. B. Preneel, M. Nuttin, V. Rijmen, and J. Buelens. Differential cryptanalysis of the CFB mode. In D.R. Stinson, editor, *Advances in Cryptology - Proc. Crypto'93, LNCS 773*, pages 212–223. Springer Verlag, 1993.
11. I. Schaumüller-Bichl. The method of formal coding. In *Cryptography - Proc., Burg Feuerstein, 1992, LNCS 149*, pages 235–255. Springer Verlag, 1982.

SAFER K–64: One Year Later

James L. Massey

Signal & Information Processing Laboratory
Swiss Federal Institute of Technology
ETH Zentrum
CH-8092 Zurich, Switzerland

1 Introduction

Since we introduced the cipher SAFER K–64 (an acronym for Secure and Fast Encryption Routine with a user–selected Key of 64 bits) one year ago at the predecessor to this workshop [MAS94], we have been pleasantly surprised by the rapidity of its acceptance within the cryptographic users' community. Undoubtedly the foremost reason for this is the non-proprietary character of SAFER K–64, which makes it unusally attractive to users. Although our design of SAFER K–64 was sponsored by Cylink Corporation (Sunnyvale, CA, USA), Cylink has explicitly relinquished any proprietary rights to this algorithm. This largesse on the part of Cylink was motivated by the reasoning that the company would gain more from new business than it would lose from competition should many new users adopt this publicly available cipher. SAFER K–64 has not been patented and, to the best of our knowledge, is free for use by anyone without fees of any kind and with no violation of any rights of ownership, intellectual or otherwise. Indeed, one way in which we have become aware of applications of SAFER K–64 is via the requests that we have received from users for written assurance of the non–proprietary character of SAFER K–64 (and of SAFER K–128 that is described in the next section).

Almost immediately upon the announcement of SAFER K–64, we began to receive requests for a version of this cipher with a 128–bit user–selected key. In many ways, 128 is a natural key length because the cipher uses 128 bits from the key schedule within each round. The Special Projects Team of the Ministry of Home Affairs, Singapore, took the initiative to design a key schedule to be used with the basic SAFER algorithm for a 128–bit user–selected key. We found their key schedule to be very attractive because, when the two halves of the 128–bit key are the same 64–bit string, it produces the same round keys as does the key schedule for SAFER K–64 when its user–selected key is this same 64–bit string. The designers have renounced all proprietary rights to this 128–bit key schedule and have authorized us both to announce their key schedule and to standardize its use. We do this in Section 2 of this paper where we refer to the resultant cipher as SAFER K–128. Hereafter, we will say simply 'SAFER' when our remarks apply to both SAFER K–64 and SAFER K–128.

A second factor in the quick popularity of SAFER is its byte orientation. Within the enciphering and deciphering processes, all operations are on bytes, which makes SAFER especially attractive for implementation on smart cards with 8-bit internal processors. This fact played an important role in the tentative selection by Singaporean planners of SAFER K-128 as the standard cipher within the island–wide information system being planned for the turn of the century. A prototype smart-card implementation of SAFER was found there to run about 2.5 times as fast as a fully optimized smart-card implementation of the Data Encryption Standard (DES).

We have received several enquiries about our reasons for choosing the 'logarithm' and 'exponential' functions to provide the 'nonlinearities' in SAFER that are required for good 'confusion'. To answer these questions, we give in Section 3 an analysis to show that these functions well resemble 'randomly chosen' functions. Further justification for the choice of these nonlinearities is given in the paper [VAU95] in this volume, which shows that other choices would have given a much weaker cipher.

One of the novel features of SAFER was the use of a new linear transform to provide the "diffusion" that a good cipher requires, i.e., to ensure that small changes in each round input result in large changes in the round output. We called this transform the Pseudo-Hadamard Transform (PHT) as it differs from the conventional Hadamard (or Walsh-Hadamard) transform only enough to make it invertible over the ring of integers modulo 256. Again we have been questionned, sometimes skeptically, as to how good this diffusion is. In Section 4, we give a detailed discussion of the diffusing capability of the PHT, not only to answer these questions but also because the results are essential to the cryptanalysis in Section 6. We were remiss in [MAS94] in not mentionning two earlier applications in cryptography of transform techniques similar to the Hadamard transform and we are pleased to remedy this omission here. Huber [HUB90] also used the "butterfly with decimation" structure of the Hadamard transform within an encryption round to provide diffusion, but replaced the linear "butterflys" with two–input two–output nonlinear functions to obtain the required invertibility of the transform. Schnorr, in a paper presented in the rump session at CRYPTO '91, cf. [SCH92], used the "butterfly with decimation" structure of the fast Walsh-Hadamard transform to obtain diffusion within a hashing function.

For a cipher to gain popularity, there must be a general belief that it is 'secure'. The resistance of a cipher to differential cryptanalysis, introduced by Biham and Shamir [BIH90], is perhaps the best measure available today of its security.We are aware of several privately conducted and proprietary differential cryptanalyses of SAFER, all of which have reached the conclusion that SAFER is secure against differential cryptanalysis, but there has been some disagreement about how many rounds of SAFER are required for this security. In [MAS94], we recommended the use of six rounds in SAFER K-64 but allowed optionally up to ten rounds. In Section 6 we give our own detailed differential cryptanalysis, which shows that six rounds of SAFER K-64 suffices for protection against differential cryptanalysis. The next best measure today of a cipher's security is its resistance to linear cryptanalysis, introduced by Matsui [MAT93, MAT94]. We have had no reports from others on the strength of SAFER against linear cryptanalysis, but together with our students [PER94, HAR95b, HAR95a] we have undertaken the linear cryptanalysis of SAFER. Because of the lengthy treatment that is required to do justice to the differential cryptanalysis of SAFER, we will not discuss this work further here, except to mention that it indicates that SAFER is even more secure against linear cryptanalysis than against differential cryptanalysis, which is the reverse of the situation for DES.

Very recently, Knudsen [KNU95] has pointed out a 'weakness' in SAFER when this cipher is used within a public hashing scheme. We discuss this 'weakness' in Section 7 where we also give a specification for its avoidance. We close in Section 8 with some remarks.

2   SAFER K–128

SAFER K–64 with r rounds uses $2r+1$ 64–bit subkeys that are derived from the 64–bit user–selected key according to the key schedule shown in Fig. 1. We now define SAFER K–128 as the cipher whose encryption round structure, output transformation and key biases are identical to those of SAFER K–64 but whose $2r + 1$ 64–bit subkeys are derived from the 128–bit user–selected key according to the key schedule shown in Fig. 2. As mentioned above, this latter key schedule was designed by the Special Projects Team of the Ministry of Home Affairs, Singapore. We recommend that $r = 10$ rounds of encryption be used with SAFER K–128 and specify that not more than 12 rounds be used.

The left and right halves of the 128–bit user–selected key are denoted as K_a and K_b, respectively, in Fig. 2 where, as in [MAS94], we abide by the convention that more significant bits and bytes are to the left. Upon comparing Figs. 1 and 2, one sees immediately that if the righthalf key K_b in Fig. 2 coincides with the 64–bit user–selected key K_1 in Fig. 1, then the same subkeys K_1, K_3, K_5, ... are generated by both key algorithms. Similarly, if the lefthalf key K_a in Fig. 2 coincides with the 64–bit user–selected key K_1 in Fig. 1, then the same subkeys K_2, K_4, K_6, ... are generated by both key algorithms. Thus, when *both* K_a and K_b in Fig. 2 coincide with the 64–bit user–selected key K_1 in Fig. 1, then *all* subkeys produced by both key schedules are the same. This is a very desirable feature as it permits a user with an implementation of SAFER K–128 to encipher and decipher for SAFER K–64 whenever desired.

Appendix B contains a TURBO PASCAL program that implements encryption for the full r-round SAFER K–128 cipher. This program should be taken as the official definition of the SAFER K–128 encryption algorithm. Appendix C gives two examples of 12–round encryption (i.e., $r = 12$) that the reader may find useful in checking his or her own implementation of this cipher.

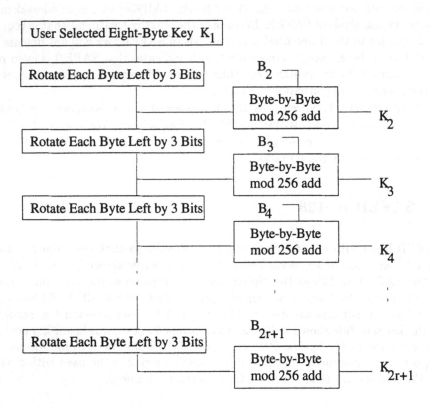

Fig. 1. Key Schedule for SAFER K-64.

3   The Nonlinearities of SAFER

We begin by recalling the encryption round structure of SAFER shown in Fig. 3. The first step within the i-th round is the Mixed XOR/Byte-Addition of the round input with the subkey K_{2i-1}. The eight resulting bytes are then individually subjected to one of two different transformations, namely: (1) the operation labelled "$45^{(\cdot)}$" in Fig. 3 to denote that if the input byte is the integer j then the output byte is 45^j modulo 257 (except that this output is taken to be 0 if if the modular result is 256, which occurs for $j = 128$) and (2) the operation labelled "$\log_{45}$" in Fig. 3 to denote that if the byte is the integer j then the output byte is $\log_{45}(j)$ (except that this output is taken to be 128 if the input is $j = 0$), i.e., the power to which one must raise 45 to obtain j modulo 257. Because 257 is a prime, arithmetic modulo 257 is the arithmetic of the finite field GF(257). The element 45 is a primitive element of this field, i.e., its first 256 powers generate all 256 non-zero field elements. Thus the mapping $45^{(\cdot)}$ is an invertible linear 2-to-one let 1-to-a-logic type. The inverse map log(·) is just the inverse of the mapping 9².

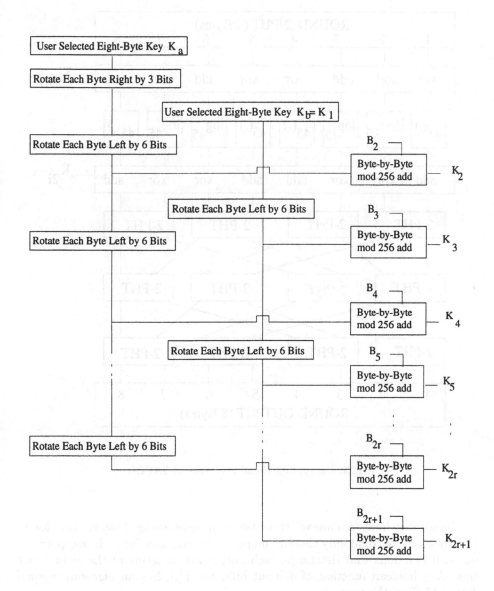

Fig. 2. Key Schedule for SAFER K–128.

an invertible mapping from one byte to one byte. The mapping $\log_{45}(.)$ is just the inverse of the mapping $45^{(.)}$.

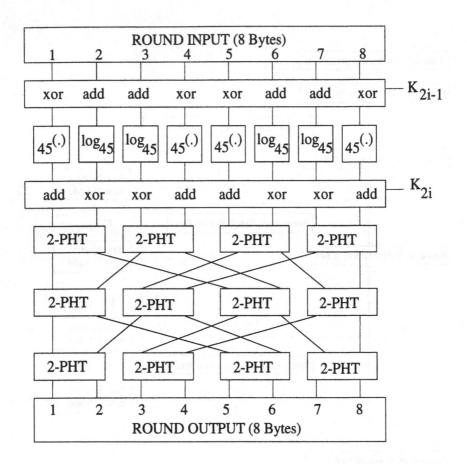

Fig. 3. Encryption round structure of SAFER.

To see just how "nonlinear" these two mappings are or, better, how closely they resemble a "randomly chosen" mapping, we consider for each mapping the boolean functions that determine each output bit in terms of the eight input bits. Any boolean function of 8 input bits, say $f(.)$, has an *algebraic normal form* (ANF) of the type

$$f(x_1, x_2, \ldots x_8) = a_0 + a_1 x_1 + a_2 x_2 + \cdots + a_8 x_8$$
$$+ a_{1,2} x_1 x_2 + a_{1,3} x_1 x_3 + \cdots + a_{7,8} x_7 x_8$$
$$+ \cdots + a_{1,2,3,4,5,6,7,8} x_1 x_2 x_3 x_4 x_5 x_6 x_7 x_8 \ . \tag{1}$$

The coefficients on the right are elements of the finite field GF(2) and the addition is addition in this field, which is just the XOR operation. The *nonlinear order* of a product of variables is the number of variables in that product; the nonlinear order of the function itself is the maximum nonlinear order of a product of variables appearing with a non–zero coefficient in its ANF. Each boolean function of eight bits uniquely determines the coefficients of its ANF and, conversely, any choice of these coefficients determines such a function. Choosing such a function f uniformly at random from the set of all 2^{256} such functions is thus equivalent to choosing the coefficients on the right in (1) by coin–tossing. It follows that, in a randomly chosen function, the number of terms of nonlinear order i that appear is binomially distributed from 0 to $\binom{8}{i}$ with mean $\binom{8}{i}/2$. In a randomly chosen function, the number of terms of nonlinear order i that appear should be rather close to this mean.

Table 1. The number of terms of nonlinear order i, $0 \leq i \leq 8$, in the boolean functions corresponding to the eight output bits of the exponential mapping $45^{(\cdot)}$.

order i	$\binom{8}{i}$	bit1	bit2	bit3	bit4	bit5	bit6	bit7	bit8
0	1	0	0	0	0	0	0	0	1
1	8	3	4	5	3	4	2	2	6
2	28	17	22	16	17	16	14	11	9
3	56	36	27	29	27	33	30	14	13
4	70	52	40	38	39	28	32	10	15
5	56	35	25	22	24	24	18	8	8
6	28	15	16	8	15	12	11	1	4
7	8	2	4	3	5	2	1	0	0
8	1	0	0	0	0	0	0	0	0

Table 1 shows the number of terms of each nonlinear order i that appear in the boolean function for the j-th output bit in the function $45^{(\cdot)}$ where $j = 1$ and $j = 8$ denote the most significant and least significant bits of the output, respectively, for $i = 1, 2, \ldots, 8$. The maximum possible number of terms $\binom{8}{i}$ is also indicated. One sees immediately that, in each output bit position j, the number of terms appearing is remarkably close to the mean number $\binom{8}{i}/2$ for a randomly chosen function. Table 2 is a similar table for the function $\log_{45}(\cdot)$ and again the agreement is remarkably close.

It is interesting to observe that the number of terms of nonlinear order i in the least significant bit (bit 8) function for the mapping $\log_\omega(\cdot)$ is invariant to the choice of primitive element ω in $GF(257)$. The reason is that, independently of the choice of ω, ω^k is a quadratic residue (or "square") just when k is even and hence its logarithm will have least significant bit 0 just in this case. But if

Table 2. The number of terms of nonlinear order i, $0 \leq i \leq 8$, in the boolean functions corresponding to the eight output bits of the logarithmic mapping $\log_{45}(.)$.

order i	$\binom{8}{i}$	bit1	bit2	bit3	bit4	bit5	bit6	bit7	bit8
0	1	1	0	0	0	0	0	0	0
1	8	5	4	4	4	0	0	0	0
2	28	12	12	13	14	22	9	11	18
3	56	26	25	27	28	24	24	38	19
4	70	39	38	34	33	43	34	26	38
5	56	25	27	18	28	29	30	19	22
6	28	13	16	15	14	17	11	9	11
7	8	1	4	2	2	3	2	1	2
8	1	0	0	0	0	0	0	0	0

ω is primitive in $GF(257)$, $\xi = \omega^i$ is also primitive if and only if i is odd. Hence any non–zero γ in $GF(257)$ is an even power of ω if and only if it is an even power of ξ and thus the least signficant bit functions in the mappings $\log_\omega(.)$ and $\log_\xi(.)$ coincide. In general, however, all the other output bit functions of the mapping $\log_\omega(.)$ and all the output bit functions of the mapping $\omega^{(\cdot)}$ will depend on the choice of ω. However, the variation with ω is not substantial— our conclusions about Tables 1 and 2 would still apply had we chosen any other primitive element, say $\omega = 3$, of $GF(257)$ to define the exponential and logarithmic mappings and SAFER so modified would be essentially as secure as for our choice of $\omega = 45$. This choice was rather arbitrary and was motivated primarily by the apparent "randomness" in the sequence of key biases that it produces, cf. [MAS94].

4 Pseuo-Hadamard Transform

The purpose of the Pseudo-Hadamard Transform (PHT) section in Fig. 3 is to provide SAFER with *diffusion*, i.e., to ensure that small changes in round inputs cause large changes in round outputs. Because the PHT is linear over the ring of integers modulo 256 and because "differences" can be taken conveniently as byte differences modulo 256 at the output of the eight nonlinear channels in Fig. 3, diffusion is well measured by how well the PHT converts low weight inputs into high weight inputs. Here and hereafter, *weight* means the number of non-zero bytes. We now treat this question in some detail as the results are essential to the differential cryptanalysis that will be carried out in Section 6.

If the input to the PHT is the eight-byte row $\mathbf{v} = [v_1, v_2, \ldots v_8]$, then the output is the eight-byte row

$$\mathbf{V} = [V_1, V_2, \ldots V_8] = \mathbf{vM} \, ,$$

where

$$M = \begin{bmatrix} 8 & 4 & 4 & 2 & 4 & 2 & 2 & 1 \\ 4 & 2 & 4 & 2 & 2 & 1 & 2 & 1 \\ 4 & 2 & 2 & 1 & 4 & 2 & 2 & 1 \\ 2 & 1 & 2 & 1 & 2 & 1 & 2 & 1 \\ 4 & 4 & 2 & 2 & 2 & 2 & 1 & 1 \\ 2 & 2 & 2 & 2 & 1 & 1 & 1 & 1 \\ 2 & 2 & 1 & 1 & 2 & 2 & 1 & 1 \\ 1 & 1 & 1 & 1 & 1 & 1 & 1 & 1 \end{bmatrix} \tag{2}$$

is the 8×8 matrix that we will refer to as the *PHT matrix*. The i-th row of M is just the PHT V of the input row v that is all-zero except in the i-th byte where it contains a 1. From (2), the action of the PHT matrix M on the inputs v of weight 1 is evident. These results are given in Table 4 of Appendix A for outputs with weight up to 4. The only weight–1 input v giving an output also of weight 1 is $[128, 0, 0, 0, 0, 0, 0, 0]$ as follows from the facts that only the first row of M contains a single 1 and that for non–zero a $2a = 0$ if and only if $a = 128$. Similarly, it is easy to check that there are 3 different weight–1 inputs that give weight–2 outputs, none whatsoever that give weight–3 outputs, and only 5 that give weight–4 outputs. One sees from Table 4 that the PHT diffuses weight–1 inputs exceedingly well.

The situation is not so much different for weight–2 inputs. In Table 6 we list all 33 weight–2 inputs that produce a PHT of weight between 1 and 3 inclusive. In particular, we note that only three weight–2 inputs produce an output of weight 1. There are nine weight–2 inputs that produce outputs also of weight 2, the most interesting of these being $[0, 128, 0, 128, 0, 0, 0, 0]$, $[0, 0, 128, 0, 0, 0, 128, 0]$ and $[0, 0, 0, 0, 128, 128, 0, 0]$, all of which reproduce themselves. Such replicating patterns might well represent a "weakness" that one could exploit in differential cryptanalysis were it not for the fact, which will be seen in Section 6, that byte differences of 128 cannot propagate unchanged through the nonlinear section of SAFER. From Table 4 one must conclude that the PHT also diffuses weight–2 inputs admirably well.

There are roughly 2^{13} weight–2 inputs, which is a fraction about 2^{-9} of the total number of weight–2 inputs, that produce PHT outputs of weight 4. There are 9 "isolated" weight–2 inputs, listed in Table 5, that produce weight–4 outputs, but these are of little use in differential cryptanalysis because of the plethora of 128's in the output—here "isolated" refers to the fact that the only non–zero multiples of these inputs that have weight 4 and produce weight–4 outputs are the trivial multiples by 1 and −1. The remaining weight–2 inputs play a rather important role in the differential cryptanalysis of SAFER in Section 6—we call them *one-dimensional* weight–2 inputs to emphasize that they appear in sets containing all the non–zero multiples of some weight–2 input, excluding possibly the non–zero multiples by 64, 128 and -64 when these have the effect of reducing either the weight of the input or the weight of the output, or both. This makes it possible to tabulate all these inputs in a compact way as we have done in Table 7. The last entry in this table indicates, for example, that all the non–zero multiples of $[0, 0, 0, 0, 0, 0, -1, 2]$, whose PHT is $[0, 0, 1, 1, 0, 0, 1, 1]$, are weight–2 inputs, except the multiple by 128, and produce weight–4 outputs.

There are no weight–3 inputs that give a PHT of weight 1. The lists of weight–3 inputs that produce PHT outputs with weights 2 and 3 are given in Tables 9 and 10, respectively. It is evident that the PHT diffuses even weight–3 inputs very well.

We will also have use in the differential cryptanalysis of SAFER for the list of weight–4 inputs that give a PHT of weight 1. There are only five of these and they are listed in Table 8.

5   Recognition of Certain Markov Ciphers

Differential cryptanalysis, originated by Biham and Shamir [BIH90], is a general attack on iterated ciphers, i.e., on ciphers that consist of many applications in cascade of the same round function. Our discussion of differential cryptanalysis will follow the treatment in [LAI91], which introduced and exploited the notion of a Markov cipher.

Differential cryptanalysis requires that one specify a notion of difference for round inputs and round outputs. In an iterated cipher, the round input and round output must take values in the same set G. In general, one can specify the difference ΔX between two round inputs (or two round outputs) X and $X^\star$ in the manner

$$\Delta X = X \otimes (X^\star)^{-1} \tag{3}$$

where $\otimes$ is a group operation on G and where $(X^\star)^{-1}$ denotes the group inverse of $X^\star$. The cipher is then said to be a *Markov cipher* if, when the round key is chosen uniformly at random and applied to two distinct round inputs X and $X^\star$, the conditional probability $P(\Delta Y = \beta \mid \Delta X = \alpha, X = \gamma)$ for the difference of the corresponding distinct round outputs Y and $Y^\star$ is independent of γ. In other words, the conditional probability of an output difference depends only on the input difference and not on the particular value of either input. It was shown in [LAI91] that, for a Markov cipher in which the round keys are chosen independently and uniformly at random [which is the universal assumption in differential cryptanalysis], the sequence of round differences is a Markov chain for which the uniform probability distribution is a stationary distribution. It follows that if this Markov chain has a steady-state probability distribution, then this must also be the uniform distribution.

We now prove a proposition that is very useful in identifying many commonly used block ciphers as Markov ciphers.

Proposition 1. *An iterated cipher in which the round input X and round output Y take values in a set G and for which the round function has the form*

$$Y = f(S, Z_b) \quad where \quad S = X \otimes Z_a \ ,$$

where $\otimes$ is a group operation on G and where $Z = (Z_a, Z_b)$ is the round key, is a Markov cipher for differences defined by $\Delta X = X \otimes (X^\star)^{-1}$ and $\Delta Y =$

$Y \otimes (Y^\star)^{-1}$. *Moreover, if $\oslash$ is any other group operation on G and the output difference $\tilde{\Delta}Y$ is defined as $\tilde{\Delta}Y = Y \oslash (Y^\star)^{-I}$, where $(Y^\star)^{-I}$ is the group inverse of $Y^\star$ with respect to the group operation $\oslash$, then the conditional probability $P(\tilde{\Delta}Y = \beta \mid \Delta X = \alpha, X = \gamma)$ is also independent of γ.*

Remark 1: Because a cipher must be decryptable, it follows that the function $f(S, Z_b)$ in this proposition, for every value of the partial key Z_b, must be an invertible function of S. No other assumption on this function is needed.

Remark 2: The latter part of the proposition, which seems unmotivated at this point, will be seen to be useful in the differential cryptanalysis of SAFER.

Proof: It suffices to prove that $P(\tilde{\Delta}Y = \beta \mid \Delta X = \alpha, X = \gamma)$ is independent of γ, since choosing $\oslash = \otimes$ implies the first claim of the proposition. To do this, we begin by noting that

$$P(\tilde{\Delta}Y = \beta \mid \Delta X = \alpha, X = \gamma) =$$
$$\sum_{\delta \in G} P(\tilde{\Delta}Y = \beta, Z_a = \delta \mid \Delta X = \alpha, X = \gamma) =$$
$$\sum_{\delta \in G} P(Z_a = \delta \mid \Delta X = \alpha, X = \gamma) P(\tilde{\Delta}Y = \beta \mid \Delta X = \alpha, X = \gamma, Z_a = \delta) \ .$$

$$(4)$$

But Z_a is uniformly random over G and jointly independent of X and $X^\star$ so that

$$P(Z_a = \delta \mid \Delta X = \alpha, X = \gamma) = 1/N \qquad (5)$$

where N is the cardinality of G. Moreover, because $S = X \otimes Z$, it follows that

$$\Delta S = (X \otimes Z) \otimes (X^\star \otimes Z)^{-1} = X \otimes Z \otimes Z^{-1} \otimes (X^\star)^{-1} = X \otimes (X^\star)^{-1} = \Delta X$$

where we used the fact that the inverse of a group product is the product of the inverses in reverse order. Thus,

$$P(\tilde{\Delta}Y = \beta \mid \Delta X = \alpha, X = \gamma, Z_a = \delta) =$$
$$P(\tilde{\Delta}Y = \beta \mid \Delta X = \alpha, X = \gamma, \Delta S = \alpha, S = \gamma \otimes \delta, Z_a = \delta) =$$
$$P(\tilde{\Delta}Y = \beta \mid \Delta S = \alpha, S = \gamma \otimes \delta) \qquad (6)$$

because, given ΔS and S, $\tilde{\Delta}Y$ has no further dependence on X and ΔX. Substituting (5) and (6) into (4) gives

$$P(\tilde{\Delta}Y = \beta \mid \Delta X = \alpha, X = \gamma) = (1/N) \sum_{\delta \in G} P(\tilde{\Delta}Y = \beta \mid \Delta S = \alpha, S = \gamma \otimes \delta) \ ,$$

which, because $\gamma \otimes \delta$ ranges over all the elements of G in this sum, is equivalent to

$$P(\tilde{\Delta}Y = \beta \mid \Delta X = \alpha, X = \gamma) = (1/N) \sum_{g \in G} P(\tilde{\Delta}Y = \beta \mid \Delta S = \alpha, S = g)$$

and hence is independent of γ, as was to be shown. $\square$

6   Differential Cryptanalysis of SAFER

As can be seen from Fig. 3, at the beginning of a round, SAFER combines the 8-byte round input $\mathbf{X} = [X_1, X_2, \ldots X_8]$ bytewise with the 8-byte first half $\mathbf{Z}_a = [Z_{a1}, Z_{a2}, \ldots Z_{a8}]$ of the round key to produce the 8- byte input $\mathbf{S} = [S_1, S_2, \ldots S_8]$ to the nonlinear operations in the manner that $\mathbf{S} = \mathbf{X} \otimes \mathbf{Z}_a$ where

$$\otimes = [\oplus, +, +, \oplus, \oplus, +, +, \oplus] \; ;$$

here $\oplus$ denotes the bitwise XOR operation on bytes and $+$ denotes usual byte addition, i.e., addition modulo 256. It follows that $\otimes$ is a group operation on the set G of 8-byte words. We then obtain as an immediate consequence of Proposition 1:

Corollary 2. *SAFER is a Markov cipher when the difference ΔV between 8-byte words $\mathbf{V}$ and $\mathbf{V}^*$ is defined in the manner $\Delta V = [V_1 \oplus V_1^*, V_2 - V_2^*, V_3 - V_3^*, V_4 \oplus V_4^*, V_5 \oplus V_5^*, V_6 - V_6^*, V_7 - V_7^*, V_8 \oplus V_8^*]$*

We now draw upon the latter part of Proposition 1 to establish a fact that will be especially useful in the differential cryptanalysis of SAFER.

Corollary 3. *When all output differences in SAFER are defined as byte differences modulo 256, i.e., $\tilde{\Delta} V = [V_1 - V_1^*, V_2 - V_2^*, V_3 - V_3^*, V_4 - V_4^*, V_5 - V_5^*, V_6 - V_6^*, V_7 - V_7^*, V_8 - V_8^*]$, then $P(\tilde{\Delta} Y = \beta \mid \Delta X = \alpha, X = \gamma)$ is independent of γ.*

6.1   Byte Differentials and Quasi-differentials

The detailed differential cryptanalysis of SAFER is facilitated by consideration of the input $\mathbf{S} = [S_1, S_2, \ldots S_8]$ to the PHT section in 3. Note that S_j is given by

$$S_j = 45^{(X_j \oplus Z_{aj})} + Z_{bj} \; , \qquad j \in \{1, 4, 5, 8\} \tag{7}$$

where $\mathbf{Z}_a$ and $\mathbf{Z}_b$ are the left and right halves of the round key, respectively. We thus refer to bytes 1, 4, 5, and 8 as the *exponential bytes*. Similarly, one notes that

$$S_j = \log_{45}(X_j + Z_{aj}) \oplus Z_{bj} \; , \qquad j \in \{2, 3, 6, 7\} \tag{8}$$

and we thus refer to bytes 2, 3, 6 and 7 as the *logarithmic bytes*. We will call a pair (α, τ), considered as the value of $(\Delta X_j, \Delta S_j)$, an exponential *byte differential* for $j \in \{1, 4, 5, 8\}$ and a logarithmic byte differential for $j \in \{2, 3, 6, 7\}$. Of interest greater than that of the exponential byte differentials are the exponential byte *quasi–differentials* where the output difference is taken as the modulo 256 difference $\tilde{\Delta} S_j$ rather than as the XOR difference ΔS_j.

The principal properties of the byte differentials and quasi–differentials are summarized in Table 3. When a difference ΔV or $\tilde{\Delta} V$ is a modulo 256 difference, then interchanging the inputs X and X^* negates this difference but has no effect on differences ΔV that are XOR differences. It follows that for logarithmic byte differentials, where both input and output differences are modulo 256 differences,

$$P(\Delta S = \tau \mid \Delta X = \alpha) = P(\Delta S = -\tau \mid \Delta X = -\alpha) \; .$$

Similarly for exponential byte quasi–differentials, where only the output difference is modulo 256,

$$P(\Delta S = \tau \mid \Delta X = \alpha) = P(\Delta S = -\tau \mid \Delta X = \alpha) \ .$$

These two facts are stated in the first section of Table 3. The other entries in this table were determined by direct computation of the transition probabilities $P(\Delta S = \tau \mid \Delta X = \alpha)$ and $P(\tilde{\Delta} S = \tau \mid \Delta X = \alpha)$ with the help of (7) and (8) when the bytes Z_{aj} and Z_{bj} are chosen uniformly at random over the 256 possible byte values.

Table 3. Properties of byte differentials for SAFER.

logarithmic	exponential conventional	exponential quasi-
input difference: mod 256 output difference: mod 256	input difference: XOR output difference: XOR	input difference: XOR output difference: mod 256
$P(\tau \mid \alpha) = P(-\tau \mid -\alpha)$		$P(\tau \mid \alpha) = P(-\tau \mid \alpha)$
$P(128 \mid 128) = 0$	$P(128 \mid 128) = 0$	$P(128 \mid 128) = 0$
$P(128 \mid \alpha) = 2^{-7}$ for α odd	$P(\tau \mid 128) = 2^{-7}$ for τ odd	$P(\tau \mid 128) = 2^{-7}$ for τ odd
$P(128 \mid \alpha) = 0$ for α even		$\mathrm{avg}[P(128 \mid \alpha)] = 2^{-8.2}$ for α odd
$\max P(\tau \mid \alpha) = 2^{-6.4}$ occurs for $(\alpha, \tau) \in$ $\{(128, 48), (128, -48)\}$	$\max P(\tau \mid \alpha) = 2^{-5}$ occurs for $(\alpha, \tau) \in$ $\{(-16, 32), (103, 64),$ $(18, 128), (-108, 128),$ $(48, 128), (-78, 128),$ $(54, 128), (-115, 128),$ $(-23, 128), (102, 128),$ $(-2, 128), (103, -64)\}$	$\max P(\tau \mid \alpha) = 2^{-4.7}$ occurs for $(\alpha, \tau) \in$ $\{(79, 68), (79, -68)\}$
$\max P(\tau \mid 128) = 2^{-6.2}$ occurs for $\tau \in \{48, -48\}$	$\max P(128 \mid \alpha) = 2^{-5}$ occurs for $\alpha \in$ $\{18, 48, 54, 102, -115,$ $-108, -78, -23, -2\}$	$\max P(128 \mid \alpha) = 2^{-5}$ occurs for $\alpha \in$ $\{18, 48, 54, 102, -115,$ $-108, -78, -23, -2\}$

It will be convenient in the differential cryptanalysis of SAFER to have available the relations between byte differentials and byte quasi-differentials that are given in the following proposition.

Proposition 4. *For byte differences $\Delta V = V \oplus V^{\star}$ and $\tilde{\Delta} V = V - V^{\star}$,*
a) $\tilde{\Delta} V = 0$ if and only if $\Delta V = 0$;
b) $\tilde{\Delta} V = 128$ if and only if $\Delta V = 128$; and
c) $\tilde{\Delta} V$ is odd if and only if ΔV is odd.

Proof: Relation a) is trivial. Relation b) follows from the fact that $\Delta V = 128$ if and only if V and V^* differ in the most significant bit only, which is also the necessary and sufficient condition for $\tilde{\Delta}V = 128$. Finally, ΔV is odd if and only if V and V^* differ in the least significant bit only, which is also the necessary and sufficient condition for $\tilde{\Delta}V$ to be odd. $\square$

6.2   The PHT and Byte Differentials

We have already defined $\mathbf{S} = [S_1, S_2, \ldots S_8]$ as the input to the PHT section in 3. Thus, the round output $\mathbf{Y} = [Y_1, Y_2, \ldots Y_8]$ is given by $\mathbf{Y} = \mathbf{SM}$ where $\mathbf{M}$ is the PHT matrix of (2) and where all the arithmetic is modulo 256. It follows that when each component of $\Delta\mathbf{S}$ is a modulo 256 difference, i.e., when $\Delta S_j = S_j - S_j^*$ as is the case in the logarithmic bytes and as is also the case in the exponential bytes when quasi-differences are used, then

$$\tilde{\Delta}\mathbf{Y} = \mathbf{SM} - \mathbf{S}^*\mathbf{M} = (\tilde{\Delta}\mathbf{S})\mathbf{M} \ . \tag{9}$$

The simple relation (9) is the primary reason that it is more natural to use quasi-differentials rather than ordinary differentials in the differential cryptanalysis of SAFER.

6.3   One–round and Two–round Quasi-differentials

We now get to the heart of the differential cryptanalysis of SAFER, i.e., to the finding of the most probable $(r - 1)$-round quasi-differentials for $r = 2, 3, \ldots$. It was shown in [LAI91] that an r-round cipher is immune from differential cryptanalysis just when all its $(r - 1)$- round differentials (or quasi-differentials) are essentially equally likely. Thus, SAFER is immune from differential cryptanalysis when $(\Delta\mathbf{X}, \tilde{\Delta}\mathbf{Y}(r - 1))$ takes on every possible value (α, β) with probability about $1/(2^{64} - 1) \approx 2^{-64}$ when $\mathbf{X} = \mathbf{Y}(0) = \alpha$ is the plaintext and $\mathbf{Y}(i)$ is the output of the i-th round. It is convenient for a one-round quasi-differential $(\Delta\mathbf{X}(i), \tilde{\Delta}\mathbf{Y}(i))$ to consider also the PHT input $\mathbf{S}(i)$ at mid–round. To emphasize the role of $\mathbf{S}(i)$, we will write one–round quasi–differentials in *expanded view* as $(\Delta\mathbf{X}(i), \tilde{\Delta}\mathbf{S}(i), \tilde{\Delta}\mathbf{Y}(i))$. It follows from (9) that

$$\tilde{\Delta}\mathbf{Y}(i) = (\tilde{\Delta}\mathbf{S}(i))\mathbf{M}$$

where $\mathbf{M}$ is the PHT matrix of (2). The probability of the transition from $\Delta\mathbf{X}(i)$ to $\tilde{\Delta}\mathbf{Y}(i)$ is just the probability of the transition from $\Delta\mathbf{X}(i)$ to $\tilde{\Delta}\mathbf{S}(i)$ because the transition from $\tilde{\Delta}\mathbf{S}(i)$ to $\tilde{\Delta}\mathbf{Y}(i)$ is deterministic. Note that the probability of a transition from $\Delta\mathbf{X}(i)$ to $\tilde{\Delta}\mathbf{S}(i)$ is the product of the probabilities of the byte differentials (in the logarithmic bytes) and the byte quasi–differentials (in the exponential bytes) for the corresponding bytes of $\Delta\mathbf{X}(i)$ and $\tilde{\Delta}\mathbf{S}(i)$. It follows then from consideration of Table 3 that the probability of such a transition decreases as the number of bytes specified in $\tilde{\Delta}\mathbf{S}(i)$ increases, which number will generally be the same as the weight of $\Delta\mathbf{X}(i)$. Finding high probability quasi–differentials for several rounds is thus mostly a matter of finding quasi–differentials whose evolution has input differences of weight as small as possible in every round. To a good first approximation, the probability of an i–round quasi–differential decreases as the total weight of the round inputs increases.

Table 3, which directly gives the probability of one–round byte differentials and quasi–differentials, immediately provides the justification of the following two claims in which, for brevity here and later, we have written 0^j to denote j successive zero bytes.

Claim 1 *The 1–round quasi–differential with the expanded view*

$$([79, 0^7], [68, 0^7], [32, 16, 16, -120, 16, -120, -120, 68])$$

has probability $2^{-4.7}$ and is a most likely 1–round quasi–differential for SAFER.

It follows from Table 3 that there are 8 such most probable quasi–differentials since any of the four exponential bytes could be chosen as the single non–zero byte and since a value of -68 in this byte of $\tilde{\Delta}S(1)$ would do just as well as the value 68.

Claim 2 *The 1–round differential with the expanded view*

$$([18, 0^7], [128, 0^7], [0^7, 128])$$

has probability 2^{-5} and is a most likely 1–round differential for SAFER.

It follows from Table 3 that there are 48 such most probable differentials since again any of the four exponential bytes could be chosen as the single non-zero byte and since there are 12 pairs of values for these non-zero bytes of $\Delta X(1)$ and $\tilde{\Delta}S(1)$ that have this same maximum probability.

Claims 1 and 2 illustrate interestingly that the most likely one–round quasi–differential is slightly more probable than the most likely one–round differential, which is another argument in favor of considering the former type of 'differential' rather than the latter.

Finding the most probable two-round quasi–differential is not much more difficult.

Claim 3 *The 2–round quasi–differential* $([18, 0^7], [1, 1, 1, 1, 1, 1, 1, 1])$ *with the expanded view*
(round 1) $([18, 0^7], [128, 0^7], [0^7, 128])$
(round 2) $([0^7, 128], [0^7, 1], [1, 1, 1, 1, 1, 1, 1, 1])$
has probability 2^{-12} and is a most likely 2–round quasi–differential for SAFER.

This claim requires more justification. Recall from the discussion in Section 5 that differences at round inputs must be of the type ΔX rather than of the type $\tilde{\Delta}X$. Thus, one cannot immediately set $\tilde{\Delta}Y(1)$ equal to $\Delta X(2)$. However, when each component of $\tilde{\Delta}Y(1)$ is either 0 or 128, it follows from Proposition 4 that this equality does hold. From Table 4, we recall that there is a unique PHT input of weight 1, namely $[128, 0^7]$, that gives an output also of weight 1, namely

$[0^7, 128]$. Thus, the two-round quasi-differential in Claim 3 is the unique (up to the choice of an odd byte value for the bytes of $\Delta\mathbf{Y}(2)$, which we have arbitrarily taken as 1) such two-round quasi-differential that has weight-1 inputs to each round—thus it has maximum probability. This probability is the product of the transition probability 2^{-5} from the 18 in the first byte (which is an exponential byte) of $\Delta\mathbf{X}(1)$ to the 128 in the first byte of $\tilde{\Delta}\mathbf{S}(1)$ and the transition probability 2^{-7} from the 128 in the eighth byte (which is also an exponential byte) of $\Delta\mathbf{X}(2)$ to the 1 in the eighth byte of $\tilde{\Delta}\mathbf{S}(2)$. There are $9 \times 128 = 1152$ such most probable two-round quasi-differentials, corresponding to the 9 choices seen in Table 3 for the first byte of $\Delta\mathbf{X}(1)$ and to the 128 choices of an odd number for the eighth byte of $\tilde{\Delta}\mathbf{S}(2)$.

6.4   Three–round Quasi–differentials

Finding the most probable three–round quasi–differential is a much more intricate matter. We begin by stating the solution.

Claim 4 *The 3–round quasi–differential* $([0^3, 18, 0^4], [0^3, 128, 0^4])$ *with the expanded view*

(1) $([0^3, 18, 0^4], [0^3, 128, 0^4], [0, 128, 0, 128, 0, 128, 0, 128])$

(2) $([0, 128, 0, 128, 0, 128, 0, 128], [0, b, 0, -b, 0, -b, 0, b]$: *b odd,* $[b, 0, b, 0^5])$

(3) $([c, 0, b, 0^5]$: *c odd,* $[128, 0, 128, 0^5], [0^3, 128, 0^4])$,

has probability $2^{-41.6}$ *and is a most likely 3–round quasi–differential for SAFER.*

We first show that this three–round quasi–differential has the claimed probability $2^{-41.6}$. From Table 3 we see that the transition from 18 to 128 in an exponential byte has probability 2^{-5}, which is thus the probability of the first–round transition. Because each byte of $\tilde{\Delta}\mathbf{Y}(1)$ is either 0 or 128, it follows from Proposition 4 that $\Delta\mathbf{X}(2)$ coincides with $\tilde{\Delta}\mathbf{Y}(1)$. The second round requires transitions in logarithmic bytes 2 and 6 from 128 to b and $-b$, respectively, where b can be any odd number. All byte transitions are independent because the corresponding keys for each byte are independent. A direction computation gives

$$\sum_{b \text{ odd}} P_{\log}(b \mid 128) P_{\log}(-b \mid 128) = 2^{-7.4}$$

where $P_{\log}(b \mid a)$ is the probability of the byte quasi–differential $(\Delta X, \tilde{\Delta}S) = (a, b)$. Again from Table 3 we see that the transitions from 128 to b and $-b$ (which is also odd) in exponential bytes 4 and 8 each have probability 2^{-7}. Thus the transition in round two has probability $2^{-(7.4+7+7)} = 2^{-21.4}$. It follows further from Proposition 4 that an odd value b in exponential byte 1 of $\tilde{\Delta}\mathbf{Y}(2)$ will give an odd value c, not necessarily the same as b, in byte 1 of $\Delta\mathbf{X}(3)$. From Table 3, we see that the transition from b in (logarithmic) byte 3 of $\Delta\mathbf{X}(3)$ to 128 in byte 3 of $\Delta\mathbf{S}(3)$ has probability 2^{-7}. The probability of the transition from the odd c in exponential byte 1 of $\Delta\mathbf{X}(3)$ to 128 in byte 1 of $\tilde{\Delta}\mathbf{S}(3)$ can be well approximated by the average probability for such c, which from Table 3 is seen to be $2^{-8.2}$. Hence, the transition in round 3 has probability essentially equal to $2^{-15.2}$. The probability of the 3–round differential in the claim is thus $2^{-5} \times 2^{-21.4} \times 2^{-15.2} = 2^{-41.6}$, as was to be shown.

It is interesting to note that the above 3–round differential consists of 128 different "characteristics" [to use the language of Biham and Shamir [BIH90]], one for each odd byte value b that specifies the four non–zero bytes of $\tilde{\Delta}S(2)$. An i–round characteristic is a sequence consisting of the first–round input and the outputs of rounds 1, 2, ... i. The probability of a differential is the sum of the probabilities of all the characteristics of which it is composed. It is often the case that the probability of a differential is dominated, and thus well approximated, by the probability of its most likely characteristic. However, many of its 128 characteristics contribute substantially to the probalility of the differential in Claim 4.

We now begin the rather tedious, but essential, task of showing that the 3–round differential in Claim 4 does indeed have maximum probability. Note that the sum of the weights of the three round inputs is 7—thus our task is to show that there exists no 3–round differentials having round inputs whose weights sum to 6 or less and that any whose weights sum to 7 have probability no greater than that in Claim 4.

We begin by considering differentials for which the first–round input has weight 1. If the second–round input also has weight 1, then the second–round output must have weight 8—as follows from the proof of Claim 3—and hence the differential has very low probability. Suppose then that the second–round input has weight 2. From Table 4 we see that the two non–zero bytes must be bytes 4 and 8, or bytes 6 and 8, or bytes 7 and 8. But the third–round input cannot then have weight 1 since, by Table 6, the two non–zero bytes in the round–2 input would then have had to be bytes 1 and 2, or bytes 1 and 3, or bytes 1 and 5. Nor could the third–round input have weight 2, since Table 6 shows that the two non–zero bytes in the round–2 input would then have had to be bytes 2 and 3, or bytes 2 and 4, or bytes 2 and 5, or bytes 2 and 6, or bytes 3 and 4, or bytes 3 and 5, or bytes 3 and 7, or bytes 5 and 6, or bytes 5 and 7. Nor could the third–round input have weight 3, since Table 6 shows that the two non–zero bytes in the round–2 input would then have had to be bytes 1 and 2, or bytes 1 and 3, or bytes 1 and 4, or bytes 1 and 5, or bytes 1 and 6, or bytes 1 and 7. The third–round input can indeed have weight 4, which gives round–input weights that sum to 7, but to give larger probability than the differential in Claim 4 at least three of the non–zero bytes would have to be logarithmic bytes—Table 7 shows that all four bytes then must be logarithmic bytes (bytes 2, 3, 6 and 7) and that the two non–zero bytes in the round–2 input would have had to be bytes 2 and 5, or bytes 4 and 7, which is again a contradiction. That the second–round input cannot have weight 3 follows immediately from Table 4. Still considering a weight–1 first–round input, suppose that the second–round input has weight 4. From Table 4, these non–zero bytes must be bytes 4, 6, 7 and 8, or bytes 2, 4, 6 and 8, or bytes 3, 4, 7 and 8, or bytes 5, 6, 7 and 8. It follows then from Table 8 that the third–round input cannot have weight 1. The third–round input can indeed have weight 2, which gives round input weights that again sum to

7, but the probability of such a differential will not be larger than that of the differential in Claim 4 since only two of the four non–zero bytes in the round–2 input are logarithmic bytes. We conclude that no three–round differential with a weight–1 first–round input can have larger probability than the differential in Claim 4.

We now consider the case where the first–round input has weight 2. Suppose that the second–round input has weight 1. From Table 6 it follows that this non–zero byte must be byte 4, or byte 6, or byte 7. It then follows further from Table 4 that the input to round three must have weight at least 4—when this weight is 4, the differential is less probable than that in Claim 4 because there is no "one–dimensional" intermediate set of mid-round outputs. Suppose next that the second–round input has weight 2. It then follows from Table 6 that the two non–zero bytes in the second–round input must be bytes 2 and 4, or bytes 2 and 6, or bytes 3 and 4, or bytes 3 and 7, or bytes 4 and 6, or bytes 4 and 7, or bytes 5 and 6, or bytes 5 and 7, or bytes 6 and 7. None of these pairs can give a third–round input of weight 1 or weight 3 as follows from Table 6. Several of these pairs can be seen from Table 6 to admit third-round inputs of weight 2 but require byte transitions from 128 to 128 in the second round and hence, by Table 3, give probability 0 for the second–round transition. The second–round input can indeed have weight 4 and, in fact, the differential of Claim 4 is of this type and was chosen to give a round–3 input of weight 1 via a one–dimensional intermediate set of mid-round outputs so as to maximize its probability in this class.

We now must consider the case when the first–round input has weight 3. Table 9 shows that weight 1 is impossible for the second–round input and that weight 2 is possible only if the two non-zero bytes are bytes 2 and 8, or bytes 3 and 8, or bytes 4 and 8, or bytes 5 and 6, or bytes 6 and 8, or bytes 7 and 8. But, according to Table 6, none of these pairs can lead of a third–round input with weight less than 4. Hence, a three–round differential with first–round input of weight 3 will be much less probable than that in Claim 4.

That weight–4 first–round inputs cannot give a three–round differential with probability larger than that in Claim 4 will be evident from the treatment of 4–round differentials that follows. First–round inputs of weight 5 or more obviously need not be considered.

6.5   Four–round Quasi–differentials

In light of the lengthy argument required to establish Claim 4 for three–round differentials, the reader will be pleasantly surprised to see that the four–round case follows from the former with very little additional work. In fact, the most likely four–round differential begins with the previously determined most likely three–round differential.

Claim 5 *The 4-round quasi-differential* $([0^3, 18, 0^4], [2, 1, 2, 1, 2, 1, 2, 1])$ *with the expanded view*

(1) $([0^3, 18, 0^4], [0^3, 128, 0^4], [0, 128, 0, 128, 0, 128, 0, 128])$
(2) $([0, 128, 0, 128, 0, 128, 0, 128], [0, b, 0, -b, 0, -b, 0, b] : b \ odd, [b, 0, b, 0^5])$
(3) $([c, 0, b, 0^5] : c \ odd, [128, 0, 128, 0^5], [0^3, 128, 0^4]),$
(4) $([0^3, 128, 0^4], [0^3, 1, 0^4], [2, 1, 2, 1, 2, 1, 2, 1]$
has probability $2^{-48.6}$ *and is a most likely 4-round quasi-differential for SAFER.*

The probability of the fourth–round transition is the probability of the byte quasi–differential (128, 1) [where 1 could be replaced by any odd byte value], which from Table 3 is seen to be 2^{-7}. Thus, this four–round differential has probability $2^{-41.6} \times 2^{-7} = 2^{-48.6}$ as claimed. Because the additional fourth round has a weight–1 input, essentially the same arguments as were just used for the 3–round case establish that this four–round differential likewise has maximum probability.

Note that the last three rounds of the above four–round differential constitute a three–round differential whose first–round input has weight 4. This is the most probable three-round differential of this type, but its probability $2^{-43.6}$ is smaller by a factor of 4 than the differential in Claim 4.

6.6   Five–rounds and More Quasi–differentials

It is an unrewardingly tedious task to try to determine precisely the most probable differentials for SAFER for five or more rounds. The four–round differential of Claim 5 ends with a weight-8 output and hence cannot be extended with an additional round to obtain a highly probable five–round differential. Nor can an additional low–weight round be placed before these four rounds. The analysis that we have done suggests that one will need to specify at least two more byte transitions to create a good five–round differential than were necessary to specify in order to create the most likely four–round differential. One expects very conservatively that the probability of the most probable five–round differential differs by a factor of 2^{-8} [the average probability of a byte transition] or less from that of the most probably four–round differential. With virtually no doubt then, the most probable five–round differential for SAFER will have probability at most 2^{-57}. This is close enough to the average differential probability of 2^{-64} that the attack to find the key of six–round SAFER K–64 by differential cryptanalysis would require more computation than a brute–force exhaustive key search. For this reason, we abide by our original recommendation of six rounds (with a maximum of ten rounds) for SAFER K–64. For six–round SAFER K–128, however, exhaustive key search would be much more complex than the attack by differential cryptanalysis, which is why we have recommended at least ten rounds (with a maximum of twelve rounds) be used with this cipher. It could mislead users were we to allow a 128–bit key rather than a 64–bit key when the security against differential cryptanalysis would not be substantially enhanced by the longer key.

7   A Hashing 'Weakness' in SAFER

Having announced a freely available and non–proprietary cipher, we consider it our responsibility to inform present and prospective users of this cipher should any significant weaknesses be found in it. The first such 'weakness' of which we are aware was discovered by Knudsen [KNU95] two months after the oral presentation of this paper and concerns the use of SAFER for hashing.

It is not uncommon to use secret-key ciphers within a public hashing scheme, cf. [LAI93]. The strength of the cipher for such hashing depends on the difficulty of producing 'collisions', i.e., of finding two distinct plaintext/key pairs that yield the same ciphertext. When the plaintext and ciphertext are 64 bit strings, the median number of distinct plaintext/key pairs that must be chosen uniformly at random before such a collision is found is about 2^{32}. By some very clever cryptanalysis, Knudsen devised a method to produce such collisions for *six–round* SAFER K–64 after choosing only about 2^{24} distinct plaintext/key pairs, i.e., about 256 times as fast as by random guessing. (Because SAFER K–128 reduces to SAFER K–64 when the two halves of the 128–bit key coincide, Knudsen's attack also applies to SAFER K–128.)

Knudsen exploited the fact, which can be seen from Fig. 1 for SAFER K–64, that changing one byte of the secret key K_1 changes only the byte in this same position in all $2r + 1$ round keys. This fact appears to be irrelevant for encryption because of the diffusing effect of the PHT, cf. Section 4, but it has significant implications for hashing. Two round keys differing in only one byte will sometimes encrypt a round input to the the same round output. Knudsen was able to select two secret keys differing in only one byte in such a way that both keys encrypt between 2^{22} and 2^{28} plaintexts in the same way for six rounds. This is the phenomenon that he exploited to produce collisions about 256 times faster than by random guessing when *six–round* SAFER is used within standard hashing schemes. He also found pairs of secret keys that encrypt about 2^{15} plaintexts in the same way for *eight rounds*, but this is not enough to give an advantage over random guessing in producing collisions. H also determined that there are no pairs of secret keys that encrypt many plaintexts in the same way for *ten or more rounds*.

Knudsen [KNU95] suggested a new key schedule that could be used with "SAFER" and would completely remove the hashing 'weakness' that he exploited, but that is somewhat more complicated than the original key schedules, which are described in Section 2. Although adopting Knudsen's key schedule would certainly be a more elegant cure for the hashing 'weakness' in SAFER, it seems preferable to us (in deference to the many users who have already implemented SAFER in software or in silicon) to abide by the original and simpler key schedules and merely to *specify that at least ten rounds of SAFER be used whenever SAFER is embedded in a hashing scheme* so that the hashing 'weakness' vanishes.

8   Concluding Remarks

We have attempted in the above to give a fairly complete picture of present knowledge concerning the security of SAFER. We will continue our own analysis of SAFER and will disseminate as rapidly as possible any 'weaknesses' in SAFER that we ourselves find or that are brought to our attention.

It is a pleasure here to acknowledge the contributions of the following Armenian scientists to the differential cryptanalysis of SAFER that was reported here: G. H. Khachatrian, M. K. Kuregian, and S. S. Martirossian. Their earlier studies, to which we were privy, were very helpful to us, but the responsibility for any errors in the analysis given in this paper rests of course with us.

A   Tables of PHT Correspondences

Table 4. Weight–1 inputs giving a PHT of weight 1, 2, 3 or 4.

input byte	input value	PHT bytes	PHT values
1	64	4 6 7 8	128 128 128 64
1	128	8	128
1	-64	4 6 7 8	128 128 128 -64
2	128	6 8	128 128
3	128	4 8	128 128
4	128	2 4 6 8	128 128 128 128
5	128	7 8	128 128
6	128	5 6 7 8	128 128 128 128
7	128	3 4 7 8	128 128 128 128

Table 5. Isolated weight–2 inputs giving a PHT of weight 4.

input bytes		input values		PHT bytes	PHT values
1	2	64	64	2 5 6 8	128 128 -64 128
1	2	-64	-64	2 5 6 8	128 128 64 128
1	3	64	64	2 3 4 8	128 128 -64 128
1	3	-64	-64	2 3 4 8	128 128 64 128
1	5	64	64	3 5 7 8	128 128 -64 128
1	5	-64	-64	3 5 7 8	128 128 64 128
2	7	128	128	3 4 6 7	128 128 128 128
3	6	128	128	4 5 6 7	128 128 128 128
4	5	128	128	2 4 6 7	128 128 128 128

Table 6. Weight–2 inputs giving a PHT of weight 1, 2 or 3.

input bytes	input values	PHT bytes	PHT values
1	2 64 128	4 7 8	128 128 -64
1	2 64 -64	2 5 6	128 128 64
1	2 128 128	6	128
1	2 -64 64	2 5 6	128 128 -64
1	2 -64 128	4 7 8	128 128 64
1	3 64 128	6 7 8	128 128 -64
1	3 64 -64	2 3 4	128 128 64
1	3 128 128	4	128
1	3 -64 64	2 3 4	128 128 -64
1	3 -64 128	6 7 8	128 128 64
1	4 64 128	2 7 8	128 128 -64
1	4 128 128	2 4 6	128 128 128
1	4 -64 128	2 7 8	128 128 64
1	5 64 128	4 6 8	128 128 -64
1	5 64 -64	3 5 7	128 128 64
1	5 128 128	7	128
1	5 -64 64	3 5 7	128 128 -64
1	5 -64 128	4 6 8	128 128 64
1	6 64 128	4 5 8	128 128 -64
1	6 128 128	5 6 7	128 128 128
1	6 -64 128	4 5 8	128 128 64
1	7 64 128	3 6 8	128 128 -64
1	7 128 128	3 4 7	128 128 128
1	7 -64 128	3 6 8	128 128 64
2	3 128 128	4 6	128 128
2	4 128 128	2 4	128 128
2	5 128 128	6 7	128 128
2	6 128 128	5 7	128 128
3	4 128 128	2 6	128 128
3	5 128 128	4 7	128 128
3	7 128 128	3 7	128 128
5	6 128 128	5 6	128 128
5	7 128 128	3 4	128 128

Table 7. One dimensional weight–2 inputs giving a PHT of weight 4.

input bytes	input values		PHT bytes	PHT values				excepting these values of a
1	2	a -a	1 2 5 6	4a	2a	2a	a	0, 64, 128, -64
1	2	-a 2a	3 4 7 8	4a	2a	2a	a	0, 64, 128, -64
1	3	a -a	1 2 3 4	4a	2a	2a	a	0, 64, 128, -64
1	3	-a 2a	5 6 7 8	4a	2a	2a	a	0, 64, 128, -64
1	4	-a 2a	1 2 7 8	-4a	-2a	2a	a	0, 64, 128, -64
1	5	a -a	1 3 5 7	4a	2a	2a	a	0, 64, 128, -64
1	5	-a 2a	2 4 6 8	4a	2a	2a	a	0, 64, 128, -64
1	6	-a 2a	1 4 5 8	-4a	2a	-2a	a	0, 64, 128, -64
1	7	-a 2a	1 3 6 8	-4a	-2a	2a	a	0, 64, 128, -64
2	3	a -a	3 4 5 6	2a	a	-2a	-a	0, 128
2	4	a -a	1 2 3 4	2a	a	2a	a	0, 128
2	4	-a 2a	5 6 7 8	2a	a	2a	a	0, 128
2	5	a -a	2 3 6 7	-2a	2a	-a	a	0, 128
2	6	a -a	1 3 5 7	2a	2a	a	a	0, 128
2	6	-a 2a	2 4 6 8	2a	2a	a	a	0, 128
2	8	-a 2a	1 3 6 8	-2a	-2a	a	a	0, 128
3	4	a -a	1 2 5 6	2a	a	2a	a	0, 128
3	4	-a 2a	3 4 7 8	2a	a	2a	a	0, 128
3	5	a -a	2 4 5 7	-2a	-a	2a	a	0, 128
3	7	a -a	1 3 5 7	2a	a	2a	a	0, 128
3	7	-a 2a	2 4 6 8	2a	a	2a	a	0, 128
3	8	-a 2a	1 4 5 8	-2a	a	-2a	a	0, 128
4	6	a -a	2 4 5 7	-a	-a	a	a	0
4	7	a -a	2 3 6 7	-a	a	-a	a	0
4	8	a -a	1 3 5 7	a	a	a	a	0
4	8	-a 2a	2 4 6 8	a	a	a	a	0, 128
5	6	a -a	1 2 5 6	2a	2a	a	a	0, 128
5	6	-a 2a	3 4 7 8	2a	2a	a	a	0, 128
5	7	a -a	1 2 3 4	2a	2a	a	a	0, 128
5	7	-a 2a	5 6 7 8	2a	2a	a	a	0, 128
5	8	-a 2a	1 2 7 8	-2a	-2a	a	a	0, 128
6	7	a -a	3 4 5 6	a	a	-a	-a	0
6	8	a -a	1 2 3 4	a	a	a	a	0
6	8	-a 2a	5 6 7 8	a	a	a	a	0, 128
7	8	a -a	1 2 5 6	a	a	a	a	0
7	8	-a 2a	3 4 7 8	a	a	a	a	0, 128

Table 8. Weight–4 inputs giving a PHT of weight 1.

input bytes	input values	PHT byte	PHT value
1 2 3 4	128 128 128 128	2	128
1 3 5 7	128 128 128 128	3	128
1 2 5 6	128 128 128 128	5	128
1 2 3 5	-64 128 128 128	8	64
1 2 3 5	64 128 128 128	8	-64

Table 9. Weight–3 inputs giving a PHT of weight 2. (No such inputs give a PHT of weight 1.)

input bytes	input values	PHT bytes	PHT values
1 2 3	64 128 128	7 8	128 64
1 2 3	-64 128 128	7 8	128 -64
1 2 5	64 128 128	4 8	128 64
1 2 5	-64 128 128	4 8	128 -64
1 2 7	64 128 128	3 8	128 64
1 2 7	-64 128 128	3 8	128 -64
1 3 5	64 128 128	6 8	128 64
1 3 5	-64 128 128	6 8	128 -64
1 3 6	64 128 128	5 8	128 64
1 3 6	-64 128 128	5 8	128 -64
1 4 5	64 128 128	2 8	128 64
1 4 5	-64 128 128	2 8	128 -64
2 3 4	128 128 128	2 8	128 128
2 4 6	64 128 128	6 8	64 64
2 4 6	-64 128 128	6 8	-64 -64
2 5 6	128 128 128	5 8	128 128
3 4 7	64 128 128	4 8	64 64
3 4 7	-64 128 128	4 8	-64 -64
3 5 7	128 128 128	3 8	128 128
5 6 7	64 128 128	7 8	64 64
5 6 7	-64 128 128	7 8	-64 -64

Table 10. Weight–3 inputs giving a PHT also of weight 3.

input bytes	input values	PHT bytes	PHT values
1 2 3	128 128 128	4 6 8	128 128 128
1 2 4	64 64 128	4 5 6	128 128 64
1 2 4	128 128 128	2 4 8	128 128 128
1 2 4	-64 -64 128	4 5 6	128 128 -64
1 2 5	128 128 128	6 7 8	128 128 128
1 2 6	64 64 128	2 6 7	128 64 128
1 2 6	128 128 128	5 7 8	128 128 128
1 2 6	-64 -64 128	2 6 7	128 -64 128
1 3 4	64 64 128	3 4 6	128 64 128
1 3 4	128 128 128	2 6 8	128 128 128
1 3 4	-64 -64 128	3 4 6	128 -64 128
1 3 5	128 128 128	4 7 8	128 128 128
1 3 7	64 64 128	2 4 7	128 64 128
1 3 7	128 128 128	3 7 8	128 128 128
1 3 7	-64 -64 128	2 4 7	128 -64 128
1 5 6	64 64 128	3 6 7	128 128 64
1 5 6	128 128 128	5 6 8	128 128 128
1 5 6	-64 -64 128	3 6 7	128 128 -64
1 5 7	64 64 128	4 5 7	128 128 64
1 5 7	128 128 128	3 4 8	128 128 128
1 5 7	-64 -64 128	4 5 7	128 128 -64
2 3 6	64 128 128	2 6 8	128 -64 64
2 3 6	-64 128 128	2 6 8	128 64 -64
2 3 7	128 64 128	2 4 8	128 -64 64
2 3 7	128 -64 128	2 4 8	128 64 -64
2 4 5	64 128 128	5 6 8	128 -64 64
2 4 5	-64 128 128	5 6 8	128 64 -64
2 5 7	128 64 128	5 7 8	128 -64 64
2 5 7	128 -64 128	5 7 8	128 64 -64
3 4 5	64 128 128	3 4 8	128 -64 64
3 4 5	-64 128 128	3 4 8	128 64 -64
3 5 6	128 64 128	3 7 8	128 -64 64
3 5 6	128 -64 128	3 7 8	128 64 -64

B   Program for SAFER K–128

The following is a TURBO PASCAL program that implements encryption with
the cipher SAFER K–128:

PROGRAM Full_r_Rounds_max_12_of_SAFERK128_cipher;
VAR a1,a2,a3,a4,a5,a6,a7,a8, b1,b2,b3,b4,b5,b6,b7,b8, r: byte;
  k: ARRAY[1..25,1..8] OF byte; ka, kb: ARRAY[1..8] OF byte;
  i,j,flag: integer; logtab, exptab: ARRAY[0..255] OF integer;
PROCEDURE mat1(VAR a1, a2, b1, b2: byte);
BEGIN b2:= a1 + a2; b1:= b2 + a1; END; BEGIN
{The powers of the primitive element 45 of GF(257) are computed and put in
table "exptab". Logarithms are put in table "logtab".}
  logtab[1]:= 0; exptab[0]:= 1;
  FOR i:= 1 TO 255 DO
  BEGIN
   exptab[i]:= (45 * exptab[i - 1]) mod 257; logtab[exptab[i]]:= i;
  END;
  exptab[128]:= 0; logtab[0]:= 128; exptab[0]:= 1;
  flag:= 1; writeln;
  writeln('Enter number of rounds r (max. 12) then hit CR.');
  readln(r); writeln;
  REPEAT
  BEGIN
   writeln('Enter plaintext in 8 bytes (integers from 0 to 255)');
   writeln('separated by spaces, then hit CR.');
   readln(a1, a2, a3, a4, a5, a6, a7, a8);
   writeln('Enter left half of key (Ka) in 8 bytes then hit CR.');
   readln(ka[1],ka[2],ka[3],ka[4],ka[5],ka[6],ka[7],ka[8]);
   writeln('Enter right half of key (Kb) in 8 bytes then hit CR.');
   readln(kb[1],kb[2],kb[3],kb[4],kb[5],kb[6],kb[7],kb[8]); writeln;
   writeln('Key Ka is', ka[1]:4,ka[2]:4,ka[3]:4,ka[4]:4,
      ka[5]:4,ka[6]:4,ka[7]:4,ka[8]:4);
   writeln('Key Kb is', kb[1]:4,kb[2]:4,kb[3]:4,kb[4]:4,
      kb[5]:4,kb[6]:4,kb[7]:4,kb[8]:4);
   writeln('PLAINTEXT is ',a1:8,a2:4,a3:4,a4:4,a5:4,a6:4,a7:4,a8:4);
   {The next instructions implement the key schedule that derives keys
   K1, K2, ... K2r+1 from the 128 bit input key (Ka, Kb).}
   {K1 is set equal to Kb.}
   FOR j:= 1 TO 8 DO k[1,j]:= kb[j];
   {Each byte of the key Ka is right rotated by 3.}
   FOR j:= 1 TO 8 DO ka[j]:= (ka[j] shr 3) + (ka[j] shl 5);
   FOR i:= 1 TO r DO
   BEGIN
    FOR j:= 1 TO 8 DO
    BEGIN
     {Each byte of keys Ka and Kb is further left rotated by 6.}

```
    ka[j]:= (ka[j] shl 6) + (ka[j] shr 2); kb[j]:= (kb[j] shl 6) + (kb[j] shr 2);
    {The key biases are added to give the keys K2i and K2i+1.}
    k[2*i,j]:= ka[j] + exptab[exptab[18*i+j]];
    k[2*i+1,j]:= kb[j] + exptab[exptab[18*i+9+j]];
    END;
  END;
  FOR i:= 1 TO r DO {The r rounds of encryption begin here.}
  BEGIN
    {Key 2i-1 is mixed bit and byte added to the round input.}
    a1:= a1 xor k[2*i-1,1]; a2:= a2 + k[2*i-1,2];
    a3:= a3 + k[2*i-1,3]; a4:= a4 xor k[2*i-1,4];
    a5:= a5 xor k[2*i-1,5]; a6:= a6 + k[2*i-1,6];
    a7:= a7 + k[2*i-1,7]; a8:= a8 xor k[2*i-1,8];
    {The result now passes through the nonlinear layer.}
    b1:=exptab[a1];b2:=logtab[a2];b3:=logtab[a3];b4:=exptab[a4];
    b5:=exptab[a5];b6:=logtab[a6];b7:=logtab[a7];b8:=exptab[a8];
    {Key 2i is now mixed byte and bit added to the result.}
    b1:= b1 + k[2*i,1]; b2:= b2 xor k[2*i,2];
    b3:= b3 xor k[2*i,3]; b4:= b4 + k[2*i,4];
    b5:= b5 + k[2*i,5]; b6:= b6 xor k[2*i,6];
    b7:= b7 xor k[2*i,7]; b8:= b8 + k[2*i,8];
    {The PHT of the result is now computed to complete the round.}
    mat1(b1, b2, a1, a2); mat1(b3, b4, a3, a4);
    mat1(b5, b6, a5, a6); mat1(b7, b8, a7, a8);
    mat1(a1, a3, b1, b2); mat1(a5, a7, b3, b4);
    mat1(a2, a4, b5, b6); mat1(a6, a8, b7, b8);
    mat1(b1, b3, a1, a2); mat1(b5, b7, a3, a4);
    mat1(b2, b4, a5, a6); mat1(b6, b8, a7, a8);
    writeln('after round',i:2,a1:8,a2:4,a3:4,a4:4,a5:4,a6:4,a7:4,a8:4);
  END;
  {Key 2r+1 is now mixed bit and byte added to form the cryptogram.}
  a1:= a1 xor k[2*r+1,1]; a2:= a2 + k[2*r+1,2];
  a3:= a3 + k[2*r+1,3]; a4:= a4 xor k[2*r+1,4];
  a5:= a5 xor k[2*r+1,5]; a6:= a6 + k[2*r+1,6];
  a7:= a7 + k[2*r+1,7]; a8:= a8 xor k[2*r+1,8];
  writeln('CRYPTOGRAM is',a1:8,a2:4,a3:4,a4:4,a5:4,a6:4,a7:4,a8:4);writeln;
  writeln('Type 1 & CR to continue, 0 & CR to stop.');readln(flag);
  END
  UNTIL flag = 0;
END.
```

C   Examples of SAFER K–128 Encryption

Key Ka is	8	7	6	5	4	3	2	1
Key Kb is	8	7	6	5	4	3	2	1
after round 1	101	42	122	106	63	111	225	227
after round 2	102	122	66	171	75	196	228	30
after round 3	114	219	165	207	71	24	132	155
after round 4	117	53	164	99	161	204	201	48
after round 5	132	77	246	149	5	187	182	27
after round 6	199	89	95	137	71	106	55	152
after round 7	40	214	206	250	209	115	253	33
after round 8	166	126	11	244	39	244	4	61
after round 9	178	50	26	234	35	53	4	119
after round 10	107	97	193	179	197	19	126	173
after round 11	246	216	224	225	46	28	176	2
after round 12	47	211	218	110	13	45	17	209

Key Ka is	1	2	3	4	5	6	7	8
Key Kb is	8	7	6	5	4	3	2	1
after round 1	245	74	156	7	16	15	87	214
after round 2	154	238	95	247	240	190	143	127
after round 3	179	1	127	195	35	207	215	252
after round 4	25	120	166	188	225	251	99	51
after round 5	46	38	108	134	111	249	162	200
after round 6	130	171	126	19	101	109	29	199
after round 7	5	15	205	166	46	98	19	78
after round 8	37	162	212	102	129	250	124	2
after round 9	126	21	150	201	83	135	164	152
after round 10	204	215	66	130	100	178	191	96
after round 11	254	153	253	121	114	99	71	84
after round 12	224	39	89	225	161	235	19	140

References

[BIH90] E. Biham and A. Shamir, "Differential Cryptanalysis of DES-like Cryptosystems," pp. 2–21 in *Advances in Cryptology–CRYPTO '90* (Eds. A. J. Menezes and S. A. Vanstone), Lecture Notes in Computer Science No. 537. Heidelberg and New York: Springer, 1991.

[BIH93] E. Biham and A. Shamir, *Differential Cryptanalysis of the Data Encryption Standard.* New York: Springer, 1993.

[HAR95a] C. Harpes, "A Generalization of Linear Cryptanalysis Applied to SAFER," Technical Report, Signal and Info. Proc. Lab., Swiss Federal Inst. Tech., Zurich, March 9, 1995.
(http://www.isi.ee.ethz.ch/isiworld/isi/research/)

[HAR95b] C. Harpes, G. G. Kramer and J. L. Massey, "A Generalization of Linear Cryptanalysis and the Applicability of Matsui's Piling-Up Lemma," to be presented at EUROCRYPT '95.

[HUB90] K. Huber, "Neue Kryptographische Verfahren durch Kombination von Operationen in endlichen Körpern mit der schnellen Walshtransformation," unpublished manuscript, presented and distributed to participants at the Telesec Arbeitskreis Kryptosysteme, Darmstadt, Germany, Oct. 2, 1990.

[KNU95] L. R. Knudsen, "A Weakness in SAFER K– 64," manuscript submitted to CRYPTO '95, Feb. 16, 1995.

[LAI91] X. Lai, J. L. Massey and S. Murphy, "Markov Ciphers and Differential Cryptanalysis," pp. 17–38 in *Advances in Cryptology-EUROCRYPT '91* (Ed. D. W. Davies), Lecture Notes in Computer Science No. 547. Heidelberg and New York: Springer, 1991.

[LAI93] X. Lai and J.L. Massey, "Hash Functions Based on Block Ciphers," pp. 55–70 in *Advances in Cryptology-EUROCRYPT '92* (Ed. R. A. Rueppel), Lecture Notes in Computer Science No. 658. Heidelberg and New York: Springer, 1993.

[MAS94] Massey, J. L., "SAFER K–64: A Byte-Oriented Block Ciphering Algorithm," pp. 1-17 in *Fast Software Encryption* (Ed. R. Anderson), Proceedings of the Cambridge Security Workshop, Cambridge, U. K., Dec. 9–11, 1993, Lecture Notes in Computer Science No. 809. Heidelberg and New York: Springer, 1994.

[MAT93] M. Matsui, "Linear Cryptanalysis Method for DES Cipher," pp. 386-397 in *Advances in Cryptology- EUROCRYPT '93* (Ed. T. Helleseth), Lecture Notes in Computer Science No. 765. New York: Springer, 1994.

[MAT94] M. Matsui, "The First Experimental Cryptanalysis of the Data Encryption Standard," pp. 1–11 in *Advances in Cryptology-CRYPTO '94* (Ed. Y. G. Desmedt), Lecture Notes in Computer Science No. 839. Heidelberg and New York: Springer, 1994.

[PER94] S. R. Perkins, "Linear Cryptanalysis of the SAFER K–64 Block Cipher," Diploma Thesis, Signal & Info. Proc. Lab., Swiss Fed. Inst. of Tech., Zurich, 15 July 1994.

[SCH92] C. P. Schnorr, "FFT–Hash II, Efficient Cryptographic Hashing," pp. 45–54 in *Advances in Cryptology-EUROCRYPT '92* (Ed. R. A. Rueppel), Lecture Notes in Computer Science No. 658. Heidelberg and New York: Springer, 1993.

[VAU95] S. Vaudenay, "On the Need for Multipermutations: Cryptanalysis of MD4 and SAFER," pp. 286–297 in this volume.

Improved characteristics
for differential cryptanalysis
of hash functions based on block ciphers

Vincent Rijmen*   Bart Preneel**

Katholieke Universiteit Leuven
ESAT-COSIC
K. Mercierlaan 94, B-3001 Heverlee, Belgium

{bart.preneel,vincent.rijmen}@esat.kuleuven.ac.be

Abstract. In this paper we present an improvement of the differential attack on hash functions based on block ciphers. By using the specific properties of the collision attack on hash functions, we can greatly reduce the work factor to find a pair that follows the characteristic. We propose a new family of differential characteristics that is especially useful in combination with our improvement. Attacks on a hash function based on DES variants reduced to 12, 13 or 15 rounds become faster than brute force collision attacks.

1   Introduction

Hash functions are functions that compress inputs of arbitrary length to an output of fixed length n. For cryptographic applications, we impose the following properties:

1. *one-wayness:* given Y, it is difficult to find an X such that $h(X) = Y$, and given X and $h(X)$, it is difficult to find $X' \neq X$ such that $h(X') = h(X)$
2. *collision resistance:* it is difficult to find X and $X' \neq X$ such that $h(X) = h(X')$.

Most hash functions are *iterated* hash functions: the input X is divided into t blocks X_i, where each block contains b bits. The length of the input should be appended at the end. If the length of X is not a multiple of b, one uses an unambiguous *padding scheme* to extend the input. The hash function h can then be described as:

$$H_i = f(X_i, H_{i-1}) \qquad i = 1, 2, \ldots, t.$$

* N.F.W.O. research assistant, sponsored by the National Fund for Scientific Research (Belgium).
** N.F.W.O. postdoctoral researcher, sponsored by the National Fund for Scientific Research (Belgium).

Here f is called the round function, H_0 is specified together with the scheme, and H_t is the hashcode.

Hash functions can be constructed from a block cipher algorithm, e.g., the DES [FI46-77]. The main motivation for this type of construction is the minimization of design and implementation effort. An example of a well established construction is based on the following round function f:

$$f(X_i, H_{i-1}) = \text{DES}(H_{i-1}, X_i) \oplus X_i\,, \tag{1}$$

where $\text{DES}(H_{i-1}, X_i)$ denotes the DES encryption of the plaintext X_i with the key H_{i-1}. This scheme was proposed by S. Matyas, C. Meyer and J. Oseas in [MMO85], and is described in [P93a] together with eleven variants with an equivalent security level. The differential attack however works only for four of these variants, since it requires that the cryptanalyst has explicit control over the plaintext input of the block cipher and that the key is fixed.

2   Differential cryptanalysis of hash functions

Differential cryptanalysis is a powerful cryptographic tool. Its application to block ciphers (e.g., the DES) is described in [BS93]. We assume that the reader is familiar with the basic ideas of this method. Differential cryptanalysis can be applied to hash functions in the same way as to the corresponding block ciphers, but there are some important differences [P93b].

- For the case of a collision attack, we control the plaintext input. This makes the differential attack on the hash function feasible. The differential attack on a block cipher used as an encryption device, on the contrary, is only of theoretic interest.
- We know the key, and sometimes we can choose the key or choose the best alternative out of a set of possible keys. We can exploit this in several ways. First, when searching for a collision, we can select those input values that follow the characteristic with probability one in certain rounds. Precomputation of a few tables allows us to choose the inputs of four or five consecutive rounds (cf. Sect. 3). Second, the probability of some characteristics is key-dependent. If we have some control over the key, we choose it to be optimal for our characteristic, otherwise we can select a characteristic with optimal probability for the given key. Third, we can use an early abort strategy: as soon as we see that the pair is a wrong pair, we can discard it. For most inputs we need to compute only a few rounds.
- There are more restrictions on the characteristic: in block cipher analysis we only want to know the output of the most probable characteristic. For hash functions, we need a characteristic that produces a collision, i.e., the output exor of the round function f must be zero. For our example, this means that the output exor of the block cipher has to match the input exor. Moreover, the characteristic must cover all the rounds: 1R-, 2R-, or 3R-attacks do not apply. This reduces the probability of the characteristic.

– We need only one right pair to find a collision or a second preimage.

In the rest of this paper we will only consider collision attacks.

3   Choosing inputs

In a collision attack, we can choose the input values (or messages) arbitrarily. We can use this freedom to enhance our chance of success. A naive approach is to select messages that will follow the proposed characteristic in the first two rounds with probability one. In this section we present an algorithm that enables us to pass four rounds with probability one. By a very simple extension of the algorithm, it is possible to pass five rounds. Fig. 1 defines the notation for intermediate values of the hashing.

Algorithm:

Step 1: Calculate table T_1 that lists all values of R_1 that follow the characteristic in round 1. Idem for tables T_2 and T_3 that list the values of F_2 and R_3 respectively. Since in each round only a few S-boxes are active, these tables can be reduced in size by the use of "don't cares": we don't care what the inputs are of non-active S-boxes and thus we do not specify these values.

Step 2: Match these three tables and look for all possible values of (R_1, F_2, R_3).

Step 3: Calculate table T_4 with all values for R_4. For every (R_1, F_2, R_3) "invert" round two and try to match the possible values of R_2, $F(R_3)$ and R_4.

Step 4: For each match found, calculate the inputs to the first round.

By calculating an extra table we can precede these four rounds with an extra round before round one.

4   Good characteristics

It has already been observed in [P93b, Kn94] that it is a non trivial problem to find good even-round characteristics for the hash function (1). One-round iterative characteristics can have an arbitrary number of rounds, but they have a very low probability of $\frac{1}{234}$ per round (cf. Fig. 2). Two-round iterative characteristics have the highest probability for seven rounds or more [Ma94], but in hash functions they can only be applied to DES variants with an odd number of rounds. This can be concluded from Fig. 2. Each $0 \leftarrow \chi$ round has on average a probability of $\frac{1}{234}$. Dependent on the round key, this probability becomes $\frac{1}{146}$ or $\frac{1}{585}$. For 13 rounds, the attack requires the same number of encryptions as a brute-force collision attack based on the birthday paradox ($\approx 2^{32.5}$). However, since the DES has 16 rounds, any serious attack requires a characteristic with an even number of rounds.

In [P93b], B. Preneel proposed to search for an input value χ that is a good fix-point ($\chi \leftarrow \chi$) and a good building block for an iterative characteristic ($0 \leftarrow \chi$). In [Kn94], L. Knudsen shows that such a characteristic cannot have a

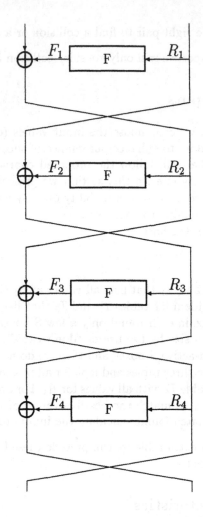

Fig. 1. Four rounds of the DES

$$
\begin{array}{ll}
(\chi \;,\; \chi) & (\chi \;,\; 0) \\
0 \leftarrow \chi & 0 \leftarrow 0 \\
0 \leftarrow \chi & 0 \leftarrow \chi \\
0 \leftarrow \chi & 0 \leftarrow 0 \\
0 \leftarrow \chi & 0 \leftarrow \chi \\
0 \leftarrow \chi & 0 \leftarrow 0 \\
(\chi \;,\; \chi) & (\chi \;,\; 0)
\end{array}
$$

Fig. 2. One-round and two-round iterative characteristics

high probability. The problem is that all χ with a good probability for $0 \leftarrow \chi$ have low probability for $\chi \leftarrow \chi$, and vice versa. L. Knudsen therefore proposes the use of an iterative characteristic based on a special four round building block (cf. Fig. 3). This building block has probability $2^{-23.6}$ averaged over all keys, and $2^{-23.0}$ for optimal keys. For a DES variant with 8 rounds the workfactor is comparable to a brute-force collision search.

$$(\text{E0000004}_x \ , \ \text{E0000004}_x)$$
$$00000002_x \leftarrow \text{E0000004}_x$$
$$00000002_x \leftarrow \text{E0000006}_x$$
$$00000002_x \leftarrow \text{E0000006}_x$$
$$00000002_x \leftarrow \text{E0000004}_x$$
$$(\text{E0000004}_x \ , \ \text{E0000004}_x)$$

Fig. 3. The 4-round iterative characteristic of L. Knudsen

Our idea is the following: we take a χ with good probability for $0 \leftarrow \chi$. Instead of inserting one $\chi \leftarrow \chi$ round, we insert five 'transient' rounds (cf. Fig. 4). These five rounds have a low probability that is however better than the fix-point construction. This is not a problem since we choose the input values of these transient rounds in such a way that they are passed with probability one. A computer search has indicated that the best transient rounds have a symmetrical pattern. For our computer search, we considered the 50 χ's with the best $0 \leftarrow \chi$ probability. For each χ all possible α's and β's were examined. The best combination is $\chi = 00196000$ and $\alpha = \beta = 04450180$. The probabilities of the different rounds are given in Table 1. Note that there exist χ's that yield a lower probability, for which the optimal α and β are different.

Table 1. The different rounds in our characteristic and their probabilities

structure	probability	comments
$0 \leftarrow \chi$	2^{-8}	key independent
$\alpha \leftarrow \chi$	$2^{-10.8}$	$2^{-9.95}$ for 50% of the keys
$\chi \oplus \alpha \leftarrow \alpha$	$2^{-20.5}$	$2^{-18.1}$ for 4.7% of the keys

The fact that the probability of these rounds depends on the key, can be exploited to reduce the work/success ratio: we can eliminate the keys that give the characteristic a low (or zero) probability. We call keys that give a non-zero probability near-optimal. For our choice of χ and α (cf. Fig 1), 4.5% of the keys are near-optimal. Stronger criteria for near-optimal keys are possible. Table 2 gives the theoretical probabilities and workfactors for DES variants with various number of rounds. The probability of the characteristic is given for near-optimal keys. The workfactors are calculated as follows: the reciprocal of the probability of the rounds where we do not choose the input values multiplied by a reduction factor that takes the early abort strategy into account. The numbers for DES variants with an odd number of rounds are obtained by choosing input values for five arbitrarily chosen consecutive rounds. The characteristic is the best 2-round iterative characteristic of [BS93].

$$
\begin{array}{ll}
(\chi \ , \ 0) & (\chi \ , \ 0) \\
0 \leftarrow 0 & 0 \leftarrow 0 \\
0 \leftarrow \chi & 0 \leftarrow \chi \\
\cdots \quad \cdots & \cdots \quad \cdots \\
\chi \leftarrow \chi & \alpha \leftarrow \chi \\
0 \leftarrow \chi & \chi \oplus \beta \leftarrow \alpha \\
\cdots \quad \cdots & \chi \oplus \alpha \leftarrow \beta \\
& \beta \leftarrow \chi \\
& 0 \leftarrow 0 \\
(\chi \ , \ 0) & (\chi \ , \ 0)
\end{array}
$$

Fig. 4. Two alternatives for the transient part of an iterative characteristic

Table 2. A survey of probabilities of the characteristics and theoretical workfactors for reduced versions of the DES

# rounds	probability $(\log_2)$	workfactor $(\log_2)$
8	–65.8	8
9	–28.8	5
12	–81.8	21.4
13	–43.2	18.9
14	–89.8	29.2
15	–50.3	25.9
16	–97.8	37.0

5   Further work

For less than seven rounds of the DES, iterative characteristics are not the best ones. Instead of building upon the iterative characteristic one could search for a characteristic of the form: optimal six round characteristic – four transient rounds (inputs chosen) – optimal six round characteristic.

With the current even-round characteristics, we make no use of the fact that we can pass five rounds. This can easily be seen on Fig. 4: the four central rounds with the worst probability are flanked by two rounds with probability one. Nothing can be gained by calculating one of these two rounds. Maybe there exist good characteristics of the form: optimal five round characteristic – five transient rounds (inputs chosen) – optimal six round characteristic. We believe this could bring the work factor of the differential attack on the variant with 16 rounds down to the value of a brute force collision attack.

References

[BS93]   E. Biham and A. Shamir, *"Differential Cryptanalysis of the Data Encryption Standard,"* Springer-Verlag, 1993.

[FI46-77] FIPS 46, *"Data Encryption Standard,"* National Bureau of Standards, 1977.

[Kn92]   L.R. Knudsen, "Iterative characteristics of DES and s^2–DES," *Advances in Cryptology, Proc. Crypto'92, LNCS 740*, E.F. Brickell, Ed., Springer-Verlag, 1993, pp. 497–511.

[Kn94]   L.R. Knudsen, *"Block Ciphers – Analysis, Design and Applications,"* PhD Thesis, Aarhus University, Denmark, DAIMI PB – 485, 1994.

[LMM91] X. Lai, J.L. Massey, and S. Murphy, "Markov ciphers and differential cryptanalysis," *Advances in Cryptology, Proc. Eurocrypt'91, LNCS 547*, D.W. Davies, Ed., Springer-Verlag, 1991, pp. 17–38.

[Ma94]   M. Matsui, "On correlation between the order of S-boxes and the strength of DES," *Advances in Cryptology, Proc. Eurocrypt'94, LNCS*, A. De Santis, Ed., Springer-Verlag, to appear.

[MMO85] S. M. Matyas, C. H. Meyer, and J. Oseas, "Generating strong one-way functions with cryptographic algorithm," *IBM Techn. Disclosure Bull.*, Vol. 27, No. 10A, 1985, pp. 5658–5659.

[P93a]   B. Preneel, R. Govaerts, and J. Vandewalle, "Hash functions based on block ciphers: a synthetic approach," *Advances in Cryptology, Proc. Crypto'93, LNCS 773*, D. Stinson, Ed., Springer-Verlag, 1994, pp. 368–378.

[P93b]   B. Preneel, "Differential cryptanalysis of hash functions based on block ciphers," *Proceedings of the 1st ACM Conference on Computer and Communications Security*, 1993, pp. 183-188.

Linear Cryptanalysis Using Multiple Approximations and FEAL

Burton S. Kaliski Jr. and M.J.B. Robshaw

RSA Laboratories
100 Marine Parkway
Redwood City, CA 94065, USA
burt@rsa.com   matt@rsa.com

Abstract. We describe the results of experiments on the use of multiple approximations in a linear cryptanalytic attack on FEAL; we pay particular attention to FEAL-8. While these attacks on FEAL are interesting in their own right, many important and intriguing issues in the use of multiple approximations are brought to light.

1 Introduction

At Crypto'94 Matsui [5] presented details on experiments with linear cryptanalysis which derived the key used for the encryption of data with DES [11]. Such attacks on DES, however, still require too much known plaintext to be considered completely practical. The technique of simultaneously using multiple linear approximations, also presented at Crypto'94 [4], might be useful in reducing the amount of plaintext required for a successful linear cryptanalytic attack.

While initial experimental evidence has demonstrated the theoretical potential of using multiple approximations [4], it remains to be seen quite how powerful it might be in practice[1]. In this paper we shall describe the results of experiments on the use of multiple approximations in a linear cryptanalytic attack on the cipher FEAL [14], in particular on the eight-round version denoted FEAL-8.

In the following section we shall describe the essential features of the technique of linear cryptanalysis together with a description of how multiple linear approximations might be used. We shall then consider the linear cryptanalysis of FEAL and provide the results of various experiments we performed.

As well as describing a very recent improvement to these attacks, two issues relating to the use of multiple approximations are examined in the section that follows.

We are interested in seeing how the advantages gained by using multiple approximations compare to those obtained by the technique of *key ranking* which was introduced by Matsui [5]. Also, we describe the seemingly contradictory

---

[1] Vaudenay has mentioned that multiple approximations can provide a factor of 64 reduction in the plaintext requirements for his work with variants of SAFER [15].

result that it is possible to improve the efficiency of an attack that uses three
linear approximations by adding a fourth derived as the algebraic sum of the
initial three. We close with our conclusions.

2 Linear cryptanalysis

2.1 Using a single approximation

Linear cryptanalysis is a technique which is proving to be very valuable in the
analysis of block ciphers. While there are fascinating comparisons [2, 7, 10] to be
made between linear cryptanalysis and the technique of differential cryptanalysis
[3], linear cryptanalysis requires known rather than chosen plaintext and, as such,
might well pose more of a practical threat to a block cipher than differential
cryptanalysis.

Gradually the technique of linear cryptanalysis has been improved [6, 5] and
in attacks on DES there is now a tantalizingly narrow gap between what can be
achieved and what would be considered a practical cryptanalytic attack.

Linear cryptanalysis requires a linear approximation to the action of the
block cipher. Such an approximation might be written as

$$P[\Gamma_1] \oplus C[\Gamma_2] = k_1$$

where $P[\Gamma_1]$ denotes the exclusive-or of specific bits of the plaintext P, $C[\Gamma_2]$
denotes the exclusive-or of certain bits of the ciphertext C and k_1 represents one
bit of key information.

Following Matsui, it has become common practice to index bits of a plaintext
or ciphertext block using zero to identify the rightmost bit; the bits of a 32-bit
block are therefore numbered $31 \ldots 0$. When linear cryptanalysis is used on DES-
like ciphers where the plaintext block is split into two halves, it is common to
describe the left and right halves as high and low, P_H and P_L. We adopt similar
notation for the ciphertext C.

A linear approximation will hold with some probability p. By taking a known
plaintext/ciphertext pair, we obtain a guess for k_1 and provided $p \neq \frac{1}{2}$ we can
use a simple algorithm (Matsui's *Algorithm 1* [6]) to decide the value of k_1. The
more data we collect, the greater our confidence that we have correctly identified
the value of k_1.

Of more practical importance are algorithms which solve for more than one
bit of key information at a time. In particular, for iterated ciphers which repeatedly use the same round function, it is possible to use techniques which recover
bits of the subkey used in either the first or last rounds of the cipher.

Typical attacks use a slightly different form of approximation to attack the
subkey K^r used in the last round of an r-round cipher:

$$P_H[\Gamma_1] \oplus P_L[\Gamma_2] \oplus C_H[\Gamma_3] \oplus C_L[\Gamma_4] \oplus f(C_L, K^r)[\Gamma_3] = k_1.$$

To use this approximation we need to predict the value of $f(C_L, K^r)[\Gamma_3]$ with
a high degree of certainty and Matsui [6, 5] has shown how this might be done.

By analyzing the form of the round function it is possible to identify which bits of the subkey K^r and which bits of the input C_L effect the value of $f(C_L, K^r)[\Gamma_3]$. Suppose there are k such subkey bits, which are termed *effective key* bits and t relevant text bits, termed *effective text* bits.

We try each of the 2^k guesses for the effective key bits with all the data we have. When we have the correct guess for the effective key bits then the value of $f(C_L, K^r)[\Gamma_3]$ will be correct and linear cryptanalysis will continue as before. When we have an incorrect key guess we assume that the value of $f(C_L, K^r)[\Gamma_3]$ is '0' or '1' roughly equally often which results in the r-round approximation having a much reduced bias. Sufficient data is then taken to ensure that the correct guess can be distinguished from among the incorrect guesses thereby identifying the correct guess for the effective key bits.

Thus there is an algorithm (often referred to as Matsui's *Algorithm 2* [6]) which requires a basic computational effort of 2^{t+k} steps and can be used to recover the value of k bits of the subkey K^r and, in attacks on DES, the value of the single bit of key information k_1.

This technique of extending a round can theoretically be used on both the first and last rounds simultaneously. Two points about this approach are worth mentioning. First, when more guessing takes place more data is required to identify the correct guess. This, however, is usually more than offset by the fact that a better approximation is being used which itself requires less data for successful cryptanalysis. Second, the number of effective text and key bits increases making the required amount of computational effort for successful cryptanalysis more substantial.

2.2   Using multiple approximations

The authors have previously presented the idea of using several linear approximations simultaneously to reduce the amount of data required to mount a successful linear cryptanalytic attack [4].

We can most easily see how multiple approximations are used in an analog to *Algorithm 1*. Here we imagine that we have n linear approximations to the same bit of key information.

$$P_H[\Gamma_1^1] \oplus P_L[\Gamma_2^1] \oplus C_H[\Gamma_3^1] \oplus C_L[\Gamma_4^1] = k_1$$
$$P_H[\Gamma_1^2] \oplus P_L[\Gamma_2^2] \oplus C_H[\Gamma_3^2] \oplus C_L[\Gamma_4^2] = k_1$$
$$\vdots$$
$$P_H[\Gamma_1^n] \oplus P_L[\Gamma_2^n] \oplus C_H[\Gamma_3^n] \oplus C_L[\Gamma_4^n] = k_1$$

Instead of getting one guess we get n guesses for the value of k_1 from each plaintext/ciphertext pair. A simple weighted sum for combining these guesses was proposed and it was shown that under certain natural assumptions, weights proportional to the absolute bias of the approximations provided the best results[2]

---

[2] We named the algorithm to do this *Algorithm 1M* thereby accentuating the fact that it is an extension of Matsui's *Algorithm 1* to allow the use of multiple approximations.

[4]. Interestingly Murphy [9] has shown that under some alternative assumptions, different weights will provide the optimal results. In practice however, since the difference between these weighting schemes involves the bias as second-order terms, the two weighting schemes are essentially the same when the approximations have very small bias.

There are now two practical issues which need to be overcome. First, we have to be sure that we can extend the use of multiple approximations to when we guess effective key bits in a round function. This is where the substantive work is done in a conventional linear cryptanalytic attack and without a similar technique for multiple approximations, the effectiveness of multiple approximations will be limited.

Second, it seems to be very unlikely that for a good block cipher, several linear approximations each with a good bias can be identified, all providing approximations to the same bit of key information. Ideally we need a technique which allows us to use several approximations simultaneously even if they involve approximations to different bits of key information.

For the first problem it is easy to draw up an analogous algorithm to Matsui's *Algorithm 2*. If the linear approximations require different effective key and text bits then when we use all n linear approximations simultaneously the set of effective key and text bits becomes equal to the union of all sets of effective key and text bits for each approximation. Thus, to ensure that there is no increase in the work effort due to an increased number of effective bits, each approximation must use the same effective key and text bits.

For the second problem we need only guess the relation between the bits of key information involved [4]. For instance, to use two linear approximations

$$P_H[\Gamma_1^1] \oplus P_L[\Gamma_2^1] \oplus C_H[\Gamma_3^1] \oplus C_L[\Gamma_4^1] = k_1$$
$$P_H[\Gamma_1^2] \oplus P_L[\Gamma_2^2] \oplus C_H[\Gamma_3^2] \oplus C_L[\Gamma_4^2] = k_2,$$

we would guess whether $k_1 = k_2$ or $k_1 \neq k_2$. We would then pursue the rest of our analysis under each assumption in turn (thereby potentially doubling the work effort for any later stages of analysis) until we are in a position to reject one of the options.

We note, however, that recent work has established a more efficient approach when multiple approximations are used to recover the effective key bits in some additional round. We shall discuss this new approach in Section 4.2 where we replace the original *Algorithm 2M* which required the use of approximations to the same bit of key information [4], with a much more general and useful algorithm.

3   Linear cryptanalysis and FEAL

Techniques that lie at the heart of linear cryptanalysis were originally used by Matsui and Yamagishi to attack small round versions of FEAL [8]. Central to these attacks is a rewriting of FEAL to give an equivalent cipher; we shall use this technique in the attack presented here.

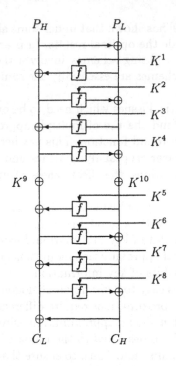

$$P_H \qquad\qquad P_L$$

$$K^9 \oplus \qquad\qquad \oplus K^{10}$$

$$C_L \qquad\qquad C_H$$

Fig. 1. Modified FEAL-8

There has been considerable recent work completed on the linear cryptanalysis of FEAL-8 [2, 1, 12, 13]. Matsui and Yamagishi [8] originally showed that there were attacks requiring 2^{28} or 2^{15} known plaintexts but with an infeasible computational workload. Biham [2] discovered a linear approximation which could be used to attack FEAL-8 with 2^{24} known plaintexts and an estimated success rate of 78% and Aoki et al. [1] describe an attack requiring 2^{25} known plaintexts for a success rate of more than 70%. We decided to closely follow the work of Aoki et al. who provide a very detailed account of their attack. In this paper we show how multiple approximations can be used very easily in two vital stages to improve the attack on FEAL-8 by almost a factor of four.

3.1   Modified FEAL

Matsui and Yamagishi [8] have shown how FEAL can be rewritten so that it is more amenable to standard linear cryptanalytic techniques.

In the original specification of FEAL [14], key material is exclusive-ored with the data that enters a Feistel network and with that leaving. However, this key material can be moved into the Feistel network provided we change the definition of the round functions to allow the introduction of additional key material. By moving key material from both ends of the Feistel network into the middle, and by assuming that all the key material is independent, FEAL-8 can be written as

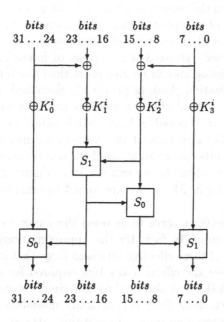

Fig. 2. The round function $f(\cdot, K^i)$ in modified FEAL. The S-boxes are defined by $S_0(x, y) = x + y \bmod 256$ and $S_1(x, y) = x + y + 1 \bmod 256$.

shown in Figure 1. The round functions for this modified version of FEAL are presented in Figure 2.

3.2 Practical issues

The attack of Aoki et al. [1] is very involved and highlights many of the complications to be found in mounting a linear cryptanalytic attack on FEAL. There are seven stages to this attack in which the first six are used to identify bits of the subkey in the first and final rounds until one round can be removed leaving the cryptanalyst with the task of attacking FEAL-7. There is an attack on FEAL-7 [8] requiring 2^{15} known plaintexts and so we find that the data requirements for the attack on FEAL-8 are dominated by those for the first six stages.

The reader is referred to the work of Aoki et al. for more detail, but in summary the requirements for the various rounds are provided in Table 1 of this paper. From the table presented there, it is clear that the most data intensive stages are the first and the third. We will see in the next section that we are able to improve the attack on FEAL-8 using multiple approximations in only two stages; fortuitously these are the first and third stages. As a consequence, we are able to lower the overall data requirements for a linear cryptanalytic attack on FEAL-8.

Before moving on to the details of using multiple approximations we mention some of the difficulties of using conventional linear cryptanalysis on FEAL; many of these issues have already been covered elsewhere [1, 13].

As we have previously stressed, the success of linear cryptanalysis almost certainly depends on being able to recover more than one bit of key information with a single approximation. And, as previously described, the usual technique is to recover what are termed effective key bits from the subkey used in either the first or last round or, if possible, both rounds simultaneously.

In DES, the parallel structure of the S-boxes ensures that the number of effective key and text bits can be minimized (to six) by considering approximations that use a single S-box. As we can see from Figure 2, in considering the output of a single S-box in FEAL, we are forced to consider at least 16 bits of subkey, usually more.

In practice these features serve to increase the number of effective text and key bits in any one round. In fact, for the approximations we currently have available, the number of these effective bits is so large that it is computationally infeasible to search over the effective key bits required for an attack which attempts to remove *both* the first and final rounds simultaneously. Thus, this very useful technique in attacking DES is practically infeasible in attacking FEAL where we can only remove one round. In addition, attacks on DES recover the value of the internal bit of key information in addition to the effective key bits in the outer rounds; with FEAL we only recover the effective key bits.

3.3   Using multiple approximations

As an example of our approach we shall closely consider the first phase of the attack due to Aoki et al. We must note that because of slight differences between FEAL and usual Feistel networks, the input to the round function used in the last round is effectively $C_H \oplus C_L$ rather than C_L.

In phase one, the following single approximation A_1, an extension of one discovered by Biham [2], is used:

$$P_L[16, 23, 25, 26, 31] \oplus C_H[31] \oplus C_L[16, 23, 25, 26]$$
$$\oplus f(C_H \oplus C_L, K^8)[23, 25, 31] = k_1.$$

The probability that this approximation holds for randomly chosen plaintext is $p = \frac{1}{2} + 2^{-11}$ and so, following Matsui [6], it might be expected that by taking $8 \times \epsilon^{-2} = 2^{25}$ known plaintexts, where the *bias* ϵ satisfies $p = \frac{1}{2} + \epsilon$, the effective key bits of K^8 can be derived with a good success rate.

In a similar experiment on eight-round DES this good success rate was 99% [6], but with FEAL-8 there is a slight degradation in the quality of the results; Aoki et al. have experimentally established that the success rate is about 78% while our own experiments (with 50 trials) reveal a success rate of 88%. This slight degradation might be due to the increased number of effective key bits we must guess in an attack on FEAL or perhaps there is some intrinsic feature of FEAL that distinguishes it from DES. Either way, we should be cautious in using algorithm performance estimates obtained for DES in estimating the effectiveness of our attacks on FEAL.

It is not difficult to identify another three linear approximations A_2, A_3, A_4 with the same bias as A_1:

$$P_H[24, 31] \oplus P_L[16, 23, 25, 26, 31] \oplus C_H[24] \oplus C_L[16, 23, 24, 25, 26, 31]$$
$$\oplus f(C_H \oplus C_L, K^8)[23, 25, 31] = k_2$$
$$P_H[22, 24] \oplus P_L[16, 23, 25, 26, 31] \oplus C_H[22, 24, 31] \oplus C_L[16, 22, 23, 24, 25, 26]$$
$$\oplus f(C_H \oplus C_L, K^8)[23, 25, 31] = k_3$$
$$P_H[22, 31] \oplus P_L[16, 23, 25, 26, 31] \oplus C_H[22] \oplus C_L[16, 22, 23, 25, 26, 31]$$
$$\oplus f(C_H \oplus C_L, K^8)[23, 25, 31] = k_1 \oplus k_2 \oplus k_3$$

There are two issues to consider when we use all four approximations. First, the bits of key information k_1, k_2 and k_3 differ only in bits of the subkey K^9. By using the technique of guessing the relation between these bits, or by using our new algorithm (Section 4.2), we can use all four simultaneously.

Second, we note that there is no increase in the number of effective key bits when we use all four of these approximations instead of one. This is because all four cases share the term $f(C_H \oplus C_L, K^8)[23, 25, 31]$ and so the same effective key and text bits are used for all four approximations.

As Aoki et al. previously discovered [1] there are many technical issues to resolve in identifying which bits of the subkey K^8 are useful to include as effective key bits and which bits of the subkey K^8 can actually be recovered with any degree of certainty.

Some bits of the subkey K^8 have little impact on the value of $f(C_H \oplus C_L, K^8)[23, 25, 31]$ and to minimize the computational effort, we ignore these bits totally. Another complication is that some bits of the subkey merely complement the value of $f(C_H \oplus C_L, K^8)[23, 25, 31]$ when they themselves are complemented; one example of this is bit 31 in the subkey K^8. With the algorithms at our disposal, we would be unable to distinguish between a guess for the effective bits of K^8 when bit 31 is set to '0' and when it is set to '1'.

In addition, there is a problem in differentiating between a set of effective key bits and a related set where every bit in the original set is complemented. To combat this we follow the example of Aoki et al. [1] and recover the value of each effective key bit exclusive-ored with bit 30 (which was chosen arbitrarily) of the subkey K^8. Note that Aoki et al. also describe the effective key bits as *explored* key bits, which are needed for successful analysis, and *detected* key bits, which are actually recovered; we do not, however, use this terminology here. Note that the effective text referred to here is the data entering the S-boxes after the initial exclusive-or in the round function.

In the following table we list the effective text and key bits we used, together with the bits[3] of subkey K^8 we were able to recover.

---

[3] We recover $12 \oplus 20$ as a bit of key information rather than $12 \oplus 20 \oplus 30$ which is recovered by Aoki et al. [1].

		Aoki et al.		multiple approximations	
phase	key attacked	plaintexts	success rate	plaintexts	success rate
1	K^8	2^{25}	79%	2^{23}	81%
2	K^1	2^{24}	100%	2^{23}	91%
3	K^1	2^{25}	91%	2^{23}	91%
4	K^1	2^{20}	100%	2^{20}	100%
5	K^8	2^{24}	100%	2^{23}	100%
6	K^8	2^{17}	100%	2^{17}	100%
7	FEAL-7	2^{15}	100%	2^{15}	100%
full FEAL-8		2^{25}	72%	2^{23}	67%

Table 1. The success rates for different stages of the attack of Aoki et al. on FEAL-8. Changes to phases one and three are the direct result of using four linear approximations; the success rate for phase one is derived experimentally and the success rate for phase three is the theoretical prediction. Success rates for phases two and five are those provided by Aoki et al. for a reduced number of plaintexts.

effective text bits	$8, 9, 10, 11, 16, 17, 18, 19, 26, 27, 28, 29, 30, 31$ and $12 \oplus 20$
effective key bits	$\{8, 9, 10, 11, 16, 17, 18, 19, 26, 27, 28, 29\} \oplus 30$ and $12 \oplus 20$
recovered key bits	$\{9, 10, 11, 17, 18, 19, 28, 29\} \oplus 30$ and $12 \oplus 20$

While we have concentrated our attention on the first phase of the attack due to Aoki et al. we note that very similar techniques can be used to devise a similar modification to the third phase. Aoki et al. use one approximation

$$P_H[7] \oplus P_L[1, 2, 8, 15] \oplus C_L[1, 2, 7, 8, 15] \oplus$$
$$\oplus f(P_H \oplus P_L, K^1)[1, 7, 15] = k_1.$$

We would use the following three linear approximations as well:

$$P_H[0] \oplus P_L[0, 1, 2, 7, 8, 15] \oplus C_L[1, 2, 7, 8, 15] \oplus C_H[0, 7]$$
$$\oplus f(P_H \oplus P_L, K^1)[1, 7, 15] = k_2,$$
$$P_H[0, 7, 14] \oplus P_L[0, 1, 2, 8, 14, 15] \oplus C_L[1, 2, 7, 8, 15] \oplus C_H[0, 14]$$
$$\oplus f(P_H \oplus P_L, K^1)[1, 7, 15] = k_3,$$
$$P_H[14] \oplus P_L[1, 2, 7, 8, 14, 15] \oplus C_L[1, 2, 7, 8, 15] \oplus C_H[7, 14]$$
$$\oplus f(P_H \oplus P_L, K^1)[1, 7, 15] = k_1 \oplus k_2 \oplus k_3.$$

3.4   Experimental results

Most of the experiments we conducted were performed on FEAL-4 with some confirmatory experiments completed using FEAL-8. All the techniques we have described for FEAL-8 can easily be converted to attacks on FEAL-4 with little or no variation except for the bias of the approximations (without regard for sign) which increases from 2^{-11} to 2^{-5}, and of course, the number of subkeys involved.

As we increased the number of linear approximations in the first phase of the analogous attack on FEAL-4 we obtained the following results. All success rates are quoted for 100 trials.

	number of plaintexts					
	$1,024$	$2,048$	$4,096$	$8,192$	$16,384$	$32,768$
A_1	3%	16%	45%	79%	100%	100%
A_1, A_2	13%	35%	82%	100%	–	–
A_1, A_2, A_3	23%	60%	96%	100%	–	–
A_1, A_2, A_3, A_4	46%	73%	99%	100%	–	–

It is easy to see that using four linear approximations allows for a factor of four reduction in the plaintext required for a successful linear cryptanalytic attack. When we turn to FEAL-8 we find that when using one linear approximation our experiments give a better success rate than that reported by Aoki et al. With 2^{25} known plaintexts we found that we could attack FEAL-8 using one linear approximation with a success rate of 88%. With 2^{26} known plaintexts this success rate increased to 98%. By using four linear approximations together, we have been able to attack FEAL-8 in experiments using 2^{23} known plaintexts with a success rate of 81% and using 2^{24} known plaintexts with a success rate of 99%.

One point we make here is that if we guess the relation between these four approximations, we force ourselves to consider various alternative results to this first phase of the attack. We would therefore continue with each of these alternatives until we can discount them. In our variant of the attack due to Aoki et al. this would mean an increased work factor of up to four times for stages two and three and of up to 16 times for stages following the third. However, results in Section 4.2 show that we can avoid this increase in the work effort *and* recover more bits of key information.

As a result of our experiments it is reasonable to conclude that the attack of Aoki et al. can be improved in terms of the amount of plaintext required using multiple approximations. Where Aoki et al. require 2^{25} known plaintexts for a 72% success rate, the use of multiple approximations should provide an attack requiring 2^{24} known plaintexts for a success rate that is very close to 99%.

Using other figures provided by Aoki et al. for the success rate of the second and fifth phases of their attack when 2^{23} known plaintexts are used, and our own experimental results for the first phase, we anticipate that FEAL-8 is vulnerable to linear cryptanalysis using 2^{23} known plaintexts with a success rate of 67%. These results are summarized in Table 1.

4   New developments and some open issues

In this section we consider some issues relevant to the general use of multiple approximations. While our observations in this section are likely to be more generally significant than those of the previous section, it is only by implementing our attack against FEAL that these issues have come to the fore. Much of our work in this section is preliminary and is still the subject of ongoing research.

4.1 Key ranking

At Crypto'94 Matsui presented the first experimental cryptanalysis of DES [5]. The innovative feature of this attack which allows an important reduction in plaintext requirements is the idea of what we shall term *key ranking*.

When attempting to identify the correct guess for a set of effective key bits, original techniques use a scoring system; the guess with the highest score is considered to be the most likely to be correct. If instead, the cryptanalyst takes the ten guesses with the highest scores (for instance) then by continuing analysis with all ten guesses the success rate will increase, or correspondingly, the same success rate will be achieved with a smaller number of known plaintexts.

This then raises a question. We have two independent techniques for reducing the amount of plaintext required for a successful attack on some cipher. Will the advantages gained by using key ranking be greater than those gained by using multiple approximations? Further, can both techniques be used together to provide an even greater advantage?

We performed some tests on FEAL-4 to see how these questions might be answered in this particular case. To compare key ranking and the use of four multiple approximations we took the four guesses with the highest score resulting from our analysis and noted how often the correct key value fell within these four[4]. The success rates for these experiments are provided below.

	number of plaintexts			
	$1,024$	$2,048$	$4,096$	$8,192$
A_1 *only*	3%	16%	45%	79%
A_1 *with ranking*	11%	31%	70%	97%
A_1, A_2, A_3, A_4	46%	73%	99%	100%
$A_1 \ldots A_4$ *with ranking*	69%	96%	100%	100%

We note that using both techniques together, multiple approximations and key ranking, we still get an improvement in performance over and above that for either method in isolation. In fact this behavior might be viewed as a natural consequence of work by Murphy et al. [10] and it suggests that any linear cryptanalytic attacks which use key ranking for enhanced performance might still be improved further when multiple approximations are used.

---

[4] Obviously we could have chosen more than four guesses with the highest score, but for comparison with multiple approximation techniques we chose four so that the same work effort would be required as when guessing the relation between the key bits in the four linear approximations we used.

4.2   Algorithm 2MG

In this section we report on a recent improvement to the technique of using multiple linear approximations to identify the effective key bits of some subkey in the outer rounds of a cipher. When we first introduced the use of several linear approximations, we stated that with approximations to different bits of key information we would have to guess the relation between the different bits in order to use all the approximations simultaneously. This would then lead to an increase in the work effort for subsequent phases of analysis.

But we have found that when we are using multiple approximations to derive the value of some set of effective key bits, it is possible to obtain the value of the bits of key information in the different approximations at the same time as deriving the value of the effective key bits.

We shall present an algorithm, *Algorithm 2MG*, to accomplish this. It is, in fact, a more general version of an algorithm presented as part of earlier work [4]. To give the algorithm in its most general form, we shall assume that we are attempting to identify the value of some effective key bits in both the first and last rounds of some cipher. We shall also assume that we are attacking a basic Feistel network. It is trivial to modify the form of the algorithm to suit our attack on FEAL. We note that there are several optimizations to the basic outline of *Algorithm 2MG* which might well be beneficial for implementation.

For an r-round Feistel cipher we approximate $(r-2)$ iterations of the round function f from the second to the $(r-1)^{\text{th}}$ round using n linear approximations while we still make guesses for the subkey bits needed to extend through the first and final rounds. Note that for practical reasons, the approximations involve the same guessed subkey bits in round one, as well as in round r. Following earlier notation we can write the ith linear approximation as follows:

$$P_H[\Gamma_1^i] \oplus P_L[\Gamma_2^i] \oplus C_H[\Gamma_3^i] \oplus C_L[\Gamma_4^i] \qquad (1)$$
$$\oplus f(P_L, K^1)[\Gamma_1^i] \oplus f(C_L, K^r)[\Gamma_3^i] = k_i.$$

We use k_i for $1 \leq i \leq n$ to denote the bit of key information in linear approximation i. We will suppose, without loss of generality, that the probability p_i that each approximation holds is greater than $\frac{1}{2}$. Recall that we define the *bias* ϵ_i of each approximation as $\epsilon_i = |p_i - \frac{1}{2}|$.

Step 1   Let $K^1[g]$ ($g = 1, 2, \ldots$) and $K^r[h]$ ($h = 1, 2, \ldots$) be possible candidates for the effective bits of subkeys K^1 and K^r respectively. Then for each pair $(K^1[g], K^r[h])$ and each linear approximation i, let $T_{g,h}^i$ be the number of plaintexts such that the left side of equation 1 is equal to 0 when K^1 is replaced by $K^1[g]$ and K^r by $K^r[h]$. Let N be the total number of plaintexts.

Step 2   Let $a_i = \epsilon_i / \sum_{i=1}^n \epsilon_i$. Define the n-tuple[5] $C = (c_1, \ldots c_n)$. Calculate

---

[5] This tuple represents the possible values for the bits of key information in each approximation.

for each g, h and each C,

$$U_{g,h}[C] = \sum_{\substack{i=1 \\ c_i=0}}^{n} a_i T_{g,h}^i + \sum_{\substack{i=1 \\ c_i=1}}^{n} a_i(N - T_{g,h}^i)$$

Step 3   Let U_{max} be the maximum value of all $U_{g,h}[C]$'s.
 − Adopt the key candidate corresponding to U_{max} and guess $k_i = c_i$ for $1 \leq i \leq n$.

Intriguingly, very little additional plaintext is required to recover these additional bits of key information. We performed experiments using two linear approximations in an attack on FEAL-4 and in an analogy to the previous method of guessing the relation between the approximated bits of key information, we derive the relation between these bits of key information.

In our experiments we get a very similar success rate with the same number of known plaintexts when we derive rather than guess the relation between the relevant bits of key information. These experiments were performed on different sets of data, explaining the slightly increased success rate for $2,048$ known plaintexts.

	number of plaintexts			
	$1,024$	$2,048$	$4,096$	$8,192$
A_1, A_2 and guessing the relation	13%	35%	82%	100%
A_1, A_2 and deriving the relation	13%	36%	75%	99%

The price we pay in using this new *Algorithm 2MG* is an increase in the amount of calculation at the time of analysis. However, there is no additional work effort for subsequent rounds. In effect, when using n linear approximations to different bits of key information, we perform linear cryptanalysis n times, but combine the results up to 2^{n-1} times, using each possible relation between the key information in the n approximations.

We then take, using a simple scoring system, the guess with the highest score from among all these possible guesses. By taking sufficient plaintext it is possible to recover both the correct guess for the effective key bits *and* the correct value of the key bits used in the approximations.

Using this enhancement in our attack on FEAL-8 we would expect the first and third phases to take four times as long as they would when using a single approximation, but there would be no increase in the work effort for subsequent phases. This gives one major advantage in the use of multiple approximations over key ranking, though we must note that this is only for this particular attack on FEAL-8.

Since we are identifying more bits of key information, which therefore require more guesses, we would expect a slight diminution in the success rate of our attack when using the same amount of known plaintext. However, as we can see in this simple case of two approximations in an attack on FEAL-4, this diminution might be very slight.

4.3   Linear independence of approximations

One very interesting development in the use of multiple approximations here is, at first sight, counter-intuitive. In our attack on FEAL-8 we used four linear approximations $A_1, \ldots, A_4$. It can easily be verified however, that A_4 is the algebraic sum of the other three approximations. In other words, the four linear approximations are not algebraically linear independent.

So why did we get an improved performance when we considered four linear approximations instead of three? Where is the additional information coming from?

We don't believe that it is a question of extra information, rather, the use of the fourth approximation with our techniques extracts more of the information that is already available when using three approximations.

When considering three linear approximations, information can be gained from each in turn, but there is other information which is revealed if we consider the approximations jointly. It is this additional information which our original techniques fail to extract.

At present, with three linear approximations, we use a rather simple scoring system to obtain the best candidate for the effective key bits. The action of adding the fourth approximation in our attack on FEAL-4, could equally have been simulated by modifying the simple scoring system used with the three approximations. And, as we saw in Section 3.4, this modified scoring system can extract more information than the previous simple one.

There is now considerable work to be done in deciding quite how we can extract the most information from the data we collect in an attack. While it may be the case that such 'optimal' techniques remain specific to the block cipher under attack, it is hoped that some general results will show the way to less data-intensive linear cryptanalytic attacks.

5   Conclusions

In this paper we have described an improvement to current linear cryptanalytic attacks on FEAL-8 by using multiple approximations. We have also confirmed experimentally some of the estimates made using current techniques and we have outlined an attack on FEAL-8 which requires 2^{23} known plaintexts for an expected success rate of 67% or 2^{24} known plaintexts for an expected success rate of 99%.

We also used FEAL-4 as a test-bed for some more interesting and, perhaps ultimately, more important experiments.

First, we showed that it is not difficult to use multiple approximations which differ in the key information they approximate. Not only that, but we have presented a new algorithm, *Algorithm 2MG*, which allows more key information to be recovered when multiple approximations are used, in addition to reducing the plaintext requirements for a successful attack.

Second, we have shown that the gains made using multiple approximations might be comparable, if not better, than those obtained using the technique of key ranking. In fact, we believe that both techniques can be used together to provide improvements over and above those available using either technique alone.

Finally, we have highlighted the somewhat counter-intuitive result that even if the linear approximations we use are not algebraically linearly independent, they might still all be used simultaneously to beneficial effect.

The full implications of all these developments are the subject of continuing research. But, under circumstances which are closely dependent on the block cipher, the use of multiple approximations in a linear cryptanalytic attack can be expected to be a valuable technique. More generally, we believe this work has demonstrated that the full potential of this relatively new type of cryptanalysis is yet to be realized.

Acknowledgement We would like to thank Sean Murphy for some interesting and helpful comments.

References

1. K. Aoki, K. Ohta, S. Araki, and M. Matsui. *Linear Cryptanalysis of FEAL-8 (Experimentation Report)*. Technical Report ISEC 94-6 (1994-05), IEICE, 1994.
2. E. Biham. On Matsui's linear cryptanalysis. In *Advances in Cryptology — Eurocrypt '94*, Springer-Verlag, to appear.
3. E. Biham and A. Shamir. *Differential Cryptanalysis of the Data Encryption Standard*. Springer-Verlag, New York, 1993.
4. B.S. Kaliski Jr. and M.J.B. Robshaw. Linear cryptanalysis using multiple approximations. In Y.G. Desmedt, editor, *Advances in Cryptology — Crypto '94*, pages 26–39, Springer Verlag, New York, 1994.
5. M. Matsui. The first experimental cryptanalysis of the Data Encryption Standard. In Y. G. Desmedt, editor, *Advances in Cryptology — Crypto '94*, pages 1–11, Springer-Verlag, New York, 1994.
6. M. Matsui. Linear cryptanalysis method for DES cipher. In T. Helleseth, editor, *Advances in Cryptology — Eurocrypt '93*, pages 386–397, Springer-Verlag, Berlin, 1994.
7. M. Matsui. On correlation between the order of the S-boxes and the strength of DES. In *Advances in Cryptology — Eurocrypt '94*, Springer-Verlag, to appear.
8. M. Matsui and A. Yamagishi. A new method for known plaintext attack of FEAL cipher. In R.A. Rueppel, editor, *Advances in Cryptology — Eurocrypt '92*, pages 81–91, Springer-Verlag, Berlin, 1992.
9. S. Murphy. August 1994. Personal communication.
10. S. Murphy, F. Piper, M. Walker and P. Wild. Likelihood estimation for block cipher keys. May 1994. Preprint.
11. National Institute of Standards and Technology (NIST). *FIPS Publication 46-2: Data Encryption Standard*. December 30, 1993.
12. K. Ohta and K. Aoki. *Linear Cryptanalysis of the Fast Data Encipherment Algorithm*. Technical Report ISEC 94-5 (1994-05), IEICE, 1994.
13. K. Ohta and K. Aoki. Linear cryptanalysis of the fast data encipherment algorithm. In Y. G. Desmedt, editor, *Advances in Cryptology — Crypto '94*, pages 12–16, Springer-Verlag, New York, 1994.

14. A. Shimizu and S. Miyaguchi.   Fast data encipherment algorithm FEAL.   In D. Chaum and W.L. Price, editors, *Advances in Cryptology — Eurocrypt '87*, pages 267–280, Springer-Verlag, Berlin, 1988.
15. S. Vaudenay.   On the need for multipermutations: Cryptanalysis of MD4 and SAFER. In these proceedings, pages 286–297.

Problems with the Linear Cryptanalysis of DES Using more than one Active S-Box per Round

Uwe Blöcher and Markus Dichtl

Siemens AG, ZFE T SN 3, D-81730 München, Germany,
E-Mail:Uwe.Bloecher@zfe.siemens.de or Markus.Dichtl@zfe.siemens.de

Abstract. Matsui introduced the concept of linear cryptanalysis. Originally only one active S-box per round was used. Later he and Biham proposed linear cryptanalysis with more than one active S-box per round. They combine equations with the Piling-up Lemma which requires independent random input variables. This requirement is not met for neighbouring S-boxes, because they share input bits. In this paper we study the error resulting from this application of the Piling-up Lemma. We give statistical evidence that the errors are severe. On the other hand we show that the Piling-up Lemma gives the correct probabilities for Matsui's Type II approximation.

1 Introduction

At Eurocrypt 1993 Matsui [3] introduced linear cryptanalysis. Matsui finds a linear equation in GF(2) between input bits, output bits, and key bits of DES which does not hold in general, but with a probability distinct from 1/2. From a sufficient number of plaintext/ciphertext pairs the attacker can derive information about key bits.

Matsui starts from linear equations for individual S-boxes which hold with a probability disjoint from 1/2. From these equations he derives equations for one round of DES. In the original publication all the bits involved in the equation for a round referred to a single S-box, the active S-box. The equations for the individual rounds have to fit together in such a way that all intermediate bits cancel, and only input bits, output bits and key bits of DES remain. To combine the probabilities of the equations Matsui applies the Piling-up Lemma:

Lemma 1 *Let X_i $(1 \leq i \leq n)$ be independent random variables whose values are in GF(2). Let p_i be the probability that $X_i = 0$. Then the probability that $X_1 + X_2 + \cdots + X_n = 0$ is $1/2 + 2^{(n-1)} \prod_{i=1}^{n} (p_i - 1/2)$.*

At Eurocrypt 1994 both Biham [1] and Matsui [4] presented a generalization of the original attack. Both allowed more than one active S-box per round and combined the equations of the active S-boxes in each round with the Piling-up Lemma. For neighbouring S-boxes this seems to be very questionable, since the Piling-up Lemma requires independent random variables whereas neighbouring

S-boxes share two input bits. As Eli Biham [1] remarks, the Piling-up Lemma holds if we average over all keys. But this is not very useful in an attack, where one wants to find the single fixed key. In this paper we study the error resulting from the application of the Piling-up Lemma on neighbouring S-boxes.

In section 2 we define the notation used. In section 3 we give some examples showing that the Piling-up Lemma gives wrong results. In section 4 we describe how the correct probability for the combination of two equations from neighbouring S-boxes is computed. In section 5 we prove that in some cases, namely for Matsui's Type II approximation, the Piling-up Lemma gives indeed the correct probabilities. In section 5 we study the error resulting from the application of the Piling-up Lemma statistically. We show that errors are frequent and severe. In section 6 conclusions are drawn.

2   Notation

Throughout this paper we use FIPS PUB-46's [5] numbering of DES bits. The input bits, key bits and output bits of the F-function, S-boxes, etc. are numbered from left to right beginning with 1. This numbering is different from Matsui's papers in which he numbers bits from right to left beginning with 0.

We use Matsui's notation in which $A[i]$ represents the i-th bit of A and $A[i_1, i_2, \ldots, i_k]$ is equal to $A[i_1] \oplus A[i_2] \oplus \ldots \oplus A[i_k]$.

We denote by X the input bits, by Y the output bits and by K the key bits of a round.

3   Examples

In this section we give some examples showing the errors resulting from applying the Piling-up Lemma.

Example 1
Combining the equations

$$X[4,7,9] \oplus Y[13,18] = K[7,10,12], \qquad p = 34/64$$
$$X[8,11,12,13] \oplus Y[6,24,30] = K[13,16,17,18], \qquad p = 28/64$$

from S-boxes S2 and S3 results in

$$X[4,7,8,9,11,12,13] \oplus Y[6,13,18,24,30] = K[7,10,12,13,16,17,18].$$

According to the Piling-up Lemma the resulting equation has probability 0.496. But there is no key for which this probability is correct! The correct probabilities for the four essentially distinct classes of keys (cf. section 4) are:

Keys	Probability
$K[11] = K[13]$ and $K[12] = K[14]$	0.461
$K[11] = K[13]$ and $K[12] \neq K[14]$	0.508
$K[11] \neq K[13]$ and $K[12] = K[14]$	0.508
$K[11] \neq K[13]$ and $K[12] \neq K[14]$	0.508

For the class of keys with probability 0.461 (25% of the keys) linear cryptanalysis will find the right key bit, because 0.461 and the Piling-up value 0.496 are on the same side of 1/2. For the remaining 75% of the keys Matsui's algorithm 1 for linear cryptanalysis will determine a wrong key bit.

Example 2

If we combine the following equations from the S-boxes S8 and S7

$$X[1, 28, 30, 32] \oplus Y[5] = K[43, 45, 47, 48], \quad p = 24/64$$
$$X[27, 29] \oplus Y[7, 12, 22, 32] = K[40, 42], \quad p = 42/64$$

we get the resulting equation

$$X[1, 27, 28, 29, 30, 32] \oplus Y[5, 7, 12, 22, 42] = K[40, 42, 43, 45, 47, 48].$$

According to the Piling-up Lemma the resulting equation has probability 0.461. But the correct probabilities for the four classes of keys are:

Keys	Probability
$K[41] = K[43]$ and $K[42] = K[44]$	0.445
$K[41] = K[43]$ and $K[42] \neq K[44]$	0.469
$K[41] \neq K[43]$ and $K[42] = K[44]$	0.430
$K[41] \neq K[43]$ and $K[42] \neq K[44]$	1/2

For 25% of the keys the probability is 1/2. For these keys linear cryptanalysis will give no information about the key bits.

Example 3

Combining the equations

$$X[4] \oplus Y[9, 17, 31] = K[5], \quad p = 30/64$$
$$X[4] \oplus Y[2, 18, 28] = K[7], \quad p = 32/64 = 1/2$$

from the S-boxes S1 and S2 results in

$$Y[2, 9, 17, 18, 28, 31] = K[5, 7].$$

Because the probability of one equation is 1/2 the resulting equation must have probability 1/2 according to the Piling-up Lemma! But the actual probabilities for the four classes of keys are:

Keys	Probability
$K[5] = K[7]$ and $K[6] = K[8]$	0.512
$K[5] = K[7]$ and $K[6] \neq K[8]$	0.488
$K[5] \neq K[7]$ and $K[6] = K[8]$	0.512
$K[5] \neq K[7]$ and $K[6] \neq K[8]$	0.488

Example 3 shows that it is probably not sufficient to restrict linear cryptanalysis to those equations for one S-box which have probability unequal to 1/2.

Example 4

Combining the equations

$$X[1,4] \oplus Y[9,17,23,31] = K[5], \qquad\qquad p = 22/64$$
$$X[4,5,6,7,8,9] \oplus Y[13] = K[7,8,9,10,11,12], \quad p = 30/64$$

from the S-boxes S1 and S2 results in

$$X[1,5,6,7,8,9] \oplus Y[9,13,17,23,31] = K[5,7,8,9,10,11,12].$$

According to the Piling-up Lemma the resulting equation has probability 0.5098. But the correct probabilities for the four classes of keys are:

Keys	Probability
$K[5] = K[7]$ and $K[6] = K[8]$	0.375000
$K[5] = K[7]$ and $K[6] \neq K[8]$	0.656250
$K[5] \neq K[7]$ and $K[6] = K[8]$	0.398438
$K[5] \neq K[7]$ and $K[6] \neq K[8]$	0.609375

These probabilities deviate largely from the Piling-up probability.

4   How to find the correct probabilities for a given key

In this paper we restrict ourselves to two active S-boxes per round. We state that for a given key the combination of two equations of neighbouring S-boxes in one round with the Piling-up Lemma often gives a wrong probability. The correct probability for a given key can be computed if we regard two neighbouring DES S-boxes as one bigger S-box. The expansion mapping E, the permutation P and the round key bits are taken into account. The bigger S-box therefore has 8 output bits and 10 input bits (2 of them are doubled), which are xored with 12 round key bits. Figure 1 shows the S-box combined of S1 and S2.

For a given equation and given round key bits the probability can be computed by testing all 2^{10} inputs and counting the number of inputs for which the equation holds. The fast Walsh Transform can be used to speed up this computation.

It is obvious that only the four round key bits which are xored to the doubled input bits can affect the probability. (The other key bits only change the order of counting.) These four round key bits decide whether the doubled input bits remain equal after xoring the round key or not. So we have four classes of round keys.

5   A case for which the Piling-up Lemma holds

The following theorem shows that the Piling-up Lemma gives the correct probability for Matsui's Type II approximation of DES ([4], [2]). Matsui gives the example where the two equations from S7 and S8

$$X[28,29] \oplus Y[7,12,22,32] = K[41,42], \quad p = 40/64$$
$$X[28,29] \oplus Y[5,21,27] = K[43,44], \quad p = 20/64$$

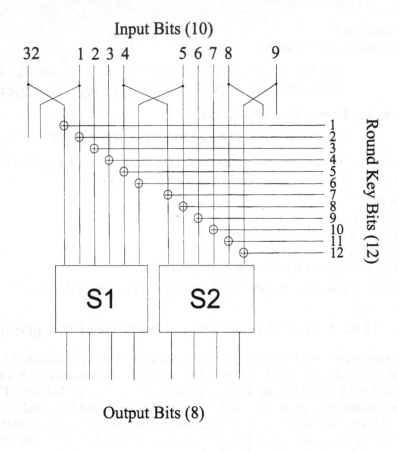

Fig. 1. Combined S-box of S1 and S2.

are combined to the equation

$$Y[5, 7, 12, 21, 22, 27, 32] = K[41, 42, 43, 44].$$

in which all input bits are cancelled out. The probability 0.453 computed with the Piling-up Lemma is correct for all keys.

Theorem 1 *Let E_1 and E_2 be linear equations for adjoining DES S-boxes. Let E_1 and E_2 be such that the two input bits of the DES F-function which go to both S-boxes are terms of both equations, and that no other input bits are used in the equations. Let p_i be the probability of equation E_i (i=1,2). Then for each value of the round key the probability of the combined equation of E_1 and E_2 is $\frac{1}{2} + 2 \cdot (p_1 - \frac{1}{2}) \cdot (p_2 - \frac{1}{2})$, the Piling-up Lemma can be applied.*

Proof of Theorem 1: Let B_1 and B_2 be two adjoining DES S-Boxes. Let $b_{i1}, \ldots, b_{i6}$ be the input bits of B_i ($i = 1, 2$). Let $y_{i1}, \ldots, y_{i4}$ be the output bits of B_i ($i = 1, 2$). We have $b_{11} = k_1 + x_1$, $b_{12} = k_2 + x_2$, $b_{13} = k_3 + x_3$,

$b_{14} = k_4 + x_4$, $b_{15} = k_5 + x_5$, $b_{16} = k_6 + x_6$, $b_{21} = k_7 + x_5$, $b_{22} = k_8 + x_6$, $b_{23} = k_9 + x_7$, $b_{24} = k_{10} + x_8$, $b_{25} = k_{11} + x_9$, $b_{26} = k_{12} + x_{10}$. The k_j are key bits, the x_j are input bits of the F-function.

We consider the two equations

$$x_5 + x_6 + \sum_{i \in I_1} y_{1i} = k_5 + k_6$$

and

$$x_5 + x_6 + \sum_{i \in I_2} y_{2i} = k_7 + k_8$$

with $I_1 \subseteq \{1, \ldots, 4\}$ and $I_2 \subseteq \{5, \ldots, 8\}$

We prove that the number of 10-tuples $(x_1, \ldots, x_{10})$ for which the combined equation

$$\sum_{i \in I_1} y_{1i} + \sum_{i \in I_2} y_{2i} = k_5 + k_6 + k_7 + k_8 \tag{1}$$

holds does not depend on the values of k_5, k_6, k_7, and k_8.

The Piling-up Lemma gives the probability of the combined equation averaged over all possible (k_5, k_6, k_7, k_8). When the probability does not depend on the key bits, the Piling-up Lemma gives a correct result also for fixed key bits.

We prove now that changing the key bit k_5 does not change the number of 10-tuples $(x_1, \ldots, x_{10})$ for which equation (1) holds.

We have to consider the 10-tuples for which $\sum_{i \in I_1} y_{1i} + \sum_{i \in I_2} y_{2i}$ does not change when k_5 is toggled. In these cases $\sum_{i \in I_1} y_{1i}$ does not depend on k_5. Each of those 10-tuples falls into one of two classes:

Class A : Those 10-tuples for which $\sum_{i \in I_2} y_{2i}$ does not depend on the key bit k_7. Then for one of the 10-tuples $(x_1, \ldots, x_4, x_5, x_6 \ldots, x_{10})$ and $(x_1, \ldots, x_4, 1 + x_5, x_6 \ldots, x_{10})$ the sum $\sum_{i \in I_1} y_{1i} + \sum_{i \in I_2} y_{2i}$ takes the value 0 , and for the other 10-tuple the value 1 is taken.

Class B : Those 10-tuples for which $\sum_{i \in I_2} y_{2i}$ depends on the key bit k_7. Here we make use of the special form of DES S-boxes, namely that the input bits b_1 and b_6 select from four permutations. As a consequence $\sum_{i \in I_1} y_{1i}$ takes as many zeros as ones when we consider as inputs all the 10-tuples $(x_1, \ldots, x_{10})$ where $x_1, \ldots, x_5$ run through all possible 2^5 values whereas $x_6, \ldots x_{10}$ remain fixed. As a consequence of the balanced zeros and ones for each pair of such 10-tuples $(x_1, \ldots, x_4, x_5, x_6, \ldots, x_{10})$ and $(x_1, \ldots, x_4, 1 + x_5, x_6, \ldots x_{10})$ where $\sum_{i \in I_1} y_{1i}$ takes the value z for both, there is another pair $(x'_1, \ldots, x'_4, x'_5, x_6, \ldots x_{10})$ and $(x'_1, \ldots, x'_4, 1 + x'_5, x_6, \ldots x_{10})$ where the sum takes the value 1+z for both.

In both classes there are as many 10-tuples $(x_1, \ldots, x_{10})$ where $\sum_{i \in I_1} y_{1i} + \sum_{i \in I_2} y_{2i}$ takes the value 0 as there are with the value 1. So the number of those 10-tuples for which equation 1 holds does not change when k_5 is toggled.

The argument for k_7 is very similar, k_6 and k_8 follow by symmetry. □

6   Statistics

We compare the probabilities given by the Piling up Lemma to the correct ones which were computed as described in section 4. When they are different, three things can happen:

a) We may get wrong key bits from the attack. This happens when the computed probability and the actual probability for the key are on different sides of 1/2.

b) We may get no information at all. This happens when the actual probability for the key is equal to 1/2.

c) Our estimations of the number of plaintext ciphertext pairs required for the attack may be wrong.

We studied the effects a), b), and c) statistically. We started from the set of all 5507 equations for single DES S-boxes with probability distinct from 1/2 and 1. We considered pairs of such equations which refer to neighbouring S-boxes. (Of course the S-boxes S1 and S8 are considered as neighbouring.) We took a random sample of 1000 from these pairs. For each pair there are four essentially different types of round keys.

For each type of round key we computed the probability of the combined equation. So we had 4000 cases for comparing the correct probability and the probability from the application of the Piling-up Lemma.

The results are shown in table 1.

Case	Percentage
Piling-up Lemma holds	14.1%
a) Wrong key bits	22.2%
b) No information	9.8%
c) Wrong estimation of number of plain-/ciphertext pairs	54.0%

Table 1. Statistics of 1000 samples of equations referring to neighbouring S-boxes. The equations were chosen from all 5507 equations.

The Piling-up Lemma gave the correct result in 562 cases, which makes 14.1% of 4000. 886 cases or 22.2% belong to case a), an attacker which relies on the Piling-up Lemma will get wrong key bits. 393 cases or 9.8% belong to case b), the attacker does not get information.

For the 2721 cases which do not belong to a) or b) the deviation of the result computed by the Piling-up Lemma from the correct value was analysed. The number of plaintext/ciphertext pairs required for linear cryptanalysis is proportional to $(p - 1/2)^{-2}$ where p is the probability of the equation used for the attack. We use p_l to denote probabilities computed with the Piling-up

Lemma and p_c for the correct probabilities. The work factor $f = \frac{(p_l - 1/2)^{-2}}{(p_c - 1/2)^{-2}}$ is the factor by which the number of plaintext/ciphertext pairs according to the Piling-up Lemma deviates from the correct value. In 20% of the the cases f is above 10, linear cryptanalysis in these cases is more than an order of magnitude easier than suggested by the Piling-up Lemma. For 53% of the cases, f is above 2. For 12% of the cases f is below 0.5. Figure 2 shows the distribution of f.

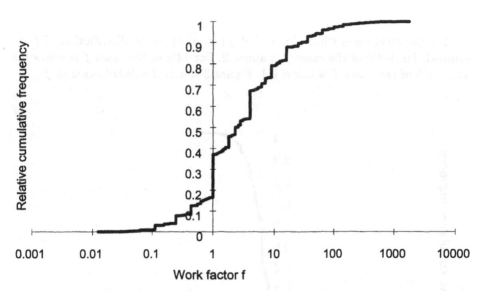

Fig. 2. Relative error in the number of required plaintext/ciphertext pairs caused by the Piling-up Lemma for all equations.

The effect of the Piling-up Lemma on "good" equations was also studied. Of course "good" equations are preferable for the attack, but in order to find equation which fit, one has to accept "bad" ones as well. The best set of equations for the linear cryptanalysis of DES [3] contains three times an equation with probability 30/64, the worst probability possible.

For the "good" equations we used the same method as above, but took only those equations for whose probabilities p holds $| p - 1/2 | \geq 8/64$. These 669 are the best 12% of the equations. Again we considered 1000 pairs of equations from neighbouring S-boxes, which makes 4000 cases.

The results are shown in table 2.

The Piling-up Lemma gave the correct result in 901 cases, which makes 22.5% of 4000. 37 cases or 0.9% belong to case a), an attacker which relies on the Piling-up Lemma will get wrong key bits. 19 cases or 0.4% belong to case b), the attacker does not get information.

Case	Percentage
Piling-up Lemma holds	22.5%
a) Wrong key bits	0.9%
b) No information	0.4%
c) Wrong estimation of number of plain-/ciphertext pairs	76.1%

Table 2. Statistics of 1000 samples of equations referring to neighbouring S-boxes. The equations were chosen from the best 12% of the equations.

For the 3944 cases which do not belong to a) or b) the distribution of f was studied. For 9.4% of the cases f is above 2. For 14% of the cases f is below 0.5. For 2.7% of the cases f is below 0.1. Figure 3 shows the distribution of f.

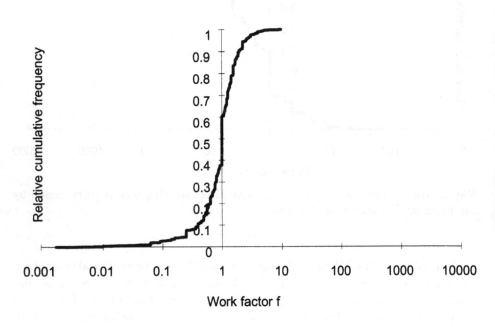

Fig. 3. Relative error in the number of required plaintext/ciphertext pairs caused by the Piling-up Lemma for the best 12% of the equations.

7 Conclusions

We have shown that the extension of Matsui's linear cryptanalysis as suggested by Matsui and Biham does not have a sound theoretical basis. Our statistical results suggest that this method leads to significant errors in practical attacks. This goes as far as the computation of wrong key bits.

On the other hand we have proved that Matsui' s approximations of Type II are valid for DES under the assumption of independent round keys.

But in general, the extension of linear cryptanalysis to more than one active S-box per round has, to the best of our knowledge, to be considered as an open problem.

References

1. Eli Biham. On Matsui's linear cryptanalysis. In *Pre-proceedings of Eurocrypt '94*, pages 349 – 361, 1994.
2. Mitsuru Matsui. Linear cryptanalysis of DES cipher (I) (Version 1.03). Preprint.
3. Mitsuru Matsui. Linear cryptanalysis method for DES cipher. In *Advances in Cryptology - Eurocrypt '93*, number 765 in Lecture Notes in Computer Science, pages 386 – 397. Springer-Verlag, 1993.
4. Mitsuru Matsui. On correlation between the order of S-boxes and the strength of DES. In *Pre-proceedings of Eurocrypt '94*, pages 377 – 387, 1994.
5. National Bureau of Standards. Data Encryption Standard. FIPS Publ. 46, Washington, DC, 1977.

Correlation Matrices

Joan Daemen, René Govaerts, and Joos Vandewalle

Katholieke Universiteit Leuven
ESAT-COSIC
K. Mercierlaan 94, B-3001 Heverlee, Belgium

joan.daemen@esat.kuleuven.ac.be

Abstract. In this paper we introduce the *correlation matrix* of a Boolean mapping, a useful concept in demonstrating and proving properties of Boolean functions and mappings. It is argued that correlation matrices are the "natural" representation for the proper understanding and description of the mechanisms of linear cryptanalysis [4]. It is also shown that the difference propagation probabilities and the table consisting of the squared elements of the correlation matrix are linked by a scaled Walsh-Hadamard transform.

Key Words: Boolean Mappings, Linear Cryptanalysis, Correlation Matrices.

1 Introduction

Most components in encryption schemes are Boolean mappings. In this paper, we establish a relation between Boolean mappings and specific linear mappings over real vector spaces. The matrices that describe these mappings are called *correlation matrices*. The elements of these matrices consist of the correlation coefficients associated with linear combinations of input bits and linear combinations of output bits.

Correlation matrices describe correlation properties of Boolean mappings in a direct way and are therefore the natural representation for the description and understanding of the mechanisms of linear cryptanalysis [4]. Moreover, they provide a useful tool for theoretical derivations and proofs.

After giving some preliminary definitions, we describe the Walsh-Hadamard transform of Boolean functions. Subsequently, we introduce the concept of correlation matrices and show how to calculate elements of this matrix for some particular types of mappings. This is followed by a treatment of the correlation properties of iterated transformations. We conclude with deriving the relations between the table of difference propagation probabilities of a mapping and its correlation matrix. For a more thorough treatment of difference propagation and additional properties of correlation matrices we refer to [6].

2 Preliminaries

A binary *vector* consists of an array of binary-valued components, that are indexed starting from 0. A binary vector a with *dimension* (or equivalently *length*) n has components $a_0, a_1, \ldots, a_{n-1}$. The set of all binary vectors with dimension n is denoted by $\mathbb{Z}_2^n$.

A Boolean function $f(a)$ is a two-valued function with domain $\mathbb{Z}_2^n$ for some n. A Boolean mapping $h(a)$ maps $\mathbb{Z}_2^n$ to $\mathbb{Z}_2^m$ for some n, m and can be seen as the parallel application of m Boolean functions: $(h_1(a), h_2(a), \ldots, h_{m-1}(a))$. If $m = n$, the Boolean mapping is called a transformation of $\mathbb{Z}_2^m$. This transformation is called *invertible* if it is a bijection.

The addition modulo 2 of two binary variables α and β is denoted by $\alpha + \beta$. Hence $\alpha + \beta$ is 0 if $\alpha = \beta$ and 1 otherwise. The bitwise addition, sum or difference of two binary vectors a and b is denoted by $a + b$ and consists of a vector c with components $c_i = a_i + b_i$. If the plus sign is used to denote arithmetic addition, it will be clear form the context. A Boolean mapping h is *linear* (with respect to bitwise addition) if $h(a + b) = h(a) + h(b)$ for all $a, b \in \mathbb{Z}_2^n$.

3 The Walsh-Hadamard transform

Linear cryptanalysis can be seen as the exploitation of *correlations* between linear combinations of bits of different intermediate encryption values in a block cipher calculation. The correlation between two Boolean function can be expressed by a *correlation coefficient* that ranges between -1 and 1:

Definition 1. The correlation coefficient associated with a pair of Boolean functions $f(a)$ and $g(a)$ is denoted by $\mathrm{C}(f, g)$ and given by

$$\mathrm{C}(f, g) = 2 \cdot \mathrm{prob}(f(a) = g(a)) - 1 .$$

From this definition it follows that $\mathrm{C}(f, g) = \mathrm{C}(g, f)$. If the correlation coefficient is different from zero the functions are said to be *correlated*.

A selection vector w is a binary vector that *selects* all components i of a vector that have $w_i = 1$. Analogous to the inner product of vectors in linear algebra, the linear combination of the components of a vector a selected by w can be expressed as $w^t a$ where the t suffix denotes transposition of the vector w. A linear Boolean function $w^t a$ is completely specified by its corresponding selection vector w.

Let $\hat{f}(a)$ be a real-valued function that is -1 for $f(a) = 1$ and $+1$ for $f(a) = 0$. This can be expressed by $\hat{f}(a) = (-1)^{f(a)}$. In this notation the real-valued function corresponding to a linear Boolean function $w^t a$ becomes $(-1)^{w^t a}$. The bitwise sum of two Boolean functions corresponds to the bitwise product of their real-valued counterparts, i.e.,

$$\widehat{f(a) + g(a)} = \hat{f}(a)\hat{g}(a) . \tag{1}$$

We define an *inner product* for real-valued functions, not to be confused with the inner product of *vectors*, by

$$< \hat{f}(a), \hat{g}(a) >= \sum_a \hat{f}(a) \hat{g}(a) \ , \tag{2}$$

It can easily be shown that

$$C(f(a), g(a)) = 2^{-n} < \hat{f}(a), \hat{g}(a) > \ . \tag{3}$$

The real-valued functions corresponding to the linear Boolean functions form an orthogonal basis with respect to the defined inner product:

$$< (-1)^{u^t a}, (-1)^{v^t a} >= 2^n \delta(u + v) \ , \tag{4}$$

with $\delta(w)$ the real-valued function that is equal to 1 if w is the zero vector and 0 otherwise. The representation of a Boolean function with respect to this basis is called its Walsh-Hadamard transform [5, 1]. The link between the Walsh-Hadamard transform of a Boolean function and its correlation with linear Boolean functions was first established in [2]. If the correlation coefficients $C(f(a), w^t a)$ are denoted by $\hat{F}(w)$ we have

$$\hat{f}(a) = \sum_w \hat{F}(w)(-1)^{w^t a} \tag{5}$$

and dually

$$\hat{F}(w) = 2^{-n} \sum_a \hat{f}(a)(-1)^{w^t a} \ , \tag{6}$$

summarized by

$$\hat{F}(w) = \mathcal{W}(f(a)) \ . \tag{7}$$

Hence a Boolean function is completely specified by the set of correlation coefficients with all linear functions.

The Walsh-Hadamard transform of the sum of two Boolean functions $f(a) + g(a)$ can be derived using (5). If $h = f + g$, we have

$$\hat{H}(w) = \sum_v \hat{F}(v + w)\hat{G}(v) \ . \tag{8}$$

Hence, addition modulo 2 in the Boolean domain corresponds to convolution in the transform domain. If the convolution operation is denoted by $\otimes$ this is expressed by

$$\mathcal{W}(f + g) = \mathcal{W}(f) \otimes \mathcal{W}(g) \ . \tag{9}$$

The subspace of $\mathbb{Z}_2^n$ generated by the vectors w such that $\hat{F}(w) \neq 0$ is called its *support space* $\mathcal{V}_f$. The support space of the sum of two Boolean functions is a subspace of the (vector) sum of their corresponding support spaces: $\mathcal{V}_{f+g} \subseteq \mathcal{V}_f + \mathcal{V}_g$. This follows directly from the convolution property. Two Boolean functions are called *disjunct* if their support spaces are disjunct, i.e., if the intersection of their support spaces only contains the origin. A vector $v \in \mathcal{V}_{f+g}$ with f and g

disjunct, can be decomposed in only one way into a component $u \in V_f$ and a component $w \in V_g$. In this case the transform values of $h = f + g$ are given by

$$\hat{H}(v) = \hat{F}(u)\hat{G}(w) \text{ with } v = u + w \text{ and } u \in V_f, w \in V_g .\tag{10}$$

Pairs of Boolean functions that depend on non-overlapping sets of input bits are a special case of disjunct functions.

4   Correlation matrices

A mapping $h : \mathbb{Z}_2^n \mapsto \mathbb{Z}_2^m$ can be decomposed into m *component* Boolean functions: $(h_0, h_1, \ldots, h_{m-1})$. Each of these component functions h_i has a Walsh-Hadamard transform $\hat{H}_i$. The vector function with components $\hat{H}_i$ is denoted by $\hat{H}$ and can be considered the Walsh-Hadamard transform of the mapping h. As in the case of Boolean functions, $\hat{H}$ completely determines the Boolean transformation h. The Walsh-Hadamard transform of any linear combination of components of h is specified by a simple extension of (9):

$$\mathcal{W}(u^t h) = \bigotimes_{u_i=1} \hat{H}_i .\tag{11}$$

All correlation coefficients between linear combinations of input bits and that of output bits of the mapping h can be arranged in a $2^m \times 2^n$ *correlation matrix* C^h. The element C_{uw} in row u and column w is equal to $C(u^t h(a), w^t a)$. The rows of this matrix can be interpreted as

$$(-1)^{u^t h(a)} = \sum_w C_{uw}^h (-1)^{w^t a} .\tag{12}$$

A matrix C^h defines a linear mapping with domain $\mathbb{R}^{2^n}$ and range $\mathbb{R}^{2^m}$. Let $\mathcal{R}$ be a mapping from the space of binary vectors to the space of real vectors, where a binary vector of dimension n is depicted onto a real vector with dimension 2^n. $\mathcal{R}$ is defined by

$$\mathcal{R} : \mathbb{Z}_2^n \mapsto \mathbb{R}^{2^n} : \alpha = \mathcal{R}(a) : \alpha_u = (-1)^{u^t a} .\tag{13}$$

Since $\mathcal{R}(a+b) = \mathcal{R}(a)\mathcal{R}(b)$, $\mathcal{R}$ is a group-homomorphism from $< \mathbb{Z}_2^n, + >$ to $< \mathbb{R}^{2^n}, \cdot >$, with $\cdot$ denoting the componentwise product. From (12) it can easily be seen that

$$C^h \mathcal{R}(a) = \mathcal{R}(h(a)) .\tag{14}$$

Consider the composition of two Boolean mappings $h = h_2 \circ h_1$ or $h(a) = h_2(h_1(a))$, with h_1 mapping n-dimensional vectors to p-dimensional vectors and with h_2 mapping p-dimensional vectors to m-dimensional vectors. The correlation matrix of h is determined by the correlation matrices of the component mappings. We have

$$(-1)^{u^t h(a)} = \sum_v C_{uv}^{h_2} (-1)^{v^t h_1(a)}$$

$$= \sum_v C_{uv}^{h_2} \sum_w C_{vw}^{h_1} (-1)^{w^t a}$$

$$= \sum_w (\sum_v C_{uv}^{h_2} C_{vw}^{h_1})(-1)^{w^t a} \ .$$

Hence,

$$C^{h_2 \circ h_1} = C^{h_2} \times C^{h_1} \ , \tag{15}$$

with $\times$ denoting the matrix product. The input-output correlations of $h = h_2 \circ h_1$ are given by

$$C(u^t h(a), w^t a) = \sum_v C(u^t h_1(a), v^t a) C(v^t h_2(a), w^t a) \ . \tag{16}$$

If h is an invertible transformation in $\mathbb{Z}_2^n$, we have (with $b = h^{-1}(a)$)

$$C(u^t h^{-1}(a), w^t a) = C(u^t b, w^t h(b)) = C(w^t h(b), u^t b) \ . \tag{17}$$

Using this fact and $C^h \times C^{(h^{-1})} = C^{h \circ h^{-1}} = I = C^h \times (C^h)^{-1}$ we obtain

$$(C^h)^{-1} = C^{(h^{-1})} = (C^h)^t \ , \tag{18}$$

hence, C^h is an orthogonal matrix.

This can be used to give an elegant proof of the following proposition:

Proposition 2. *Every linear combination of output bits of an invertible transformation is a balanced Boolean function of its input bits.*

Proof : If h is an invertible transformation, its correlation matrix C is orthogonal. Since $C_{00} = 1$ and all rows and columns have norm 1, it follows that there are no other elements in row 0 or column 0 different from 0. Hence, $C(u^t h(a), 0) = \delta(u)$ or equivalently, $u^t h(a)$ is balanced for all $u \neq 0$. $\square$

A mapping from $\mathbb{Z}_2^n$ to $\mathbb{Z}_2^m$ is converted into a mapping from $\mathbb{Z}_2^{n-1}$ to $\mathbb{Z}_2^m$ by fixing a single component of the input. More generally, a component of the input can be set equal to a linear combination of other input components, possibly complemented. Such a restriction is of the type

$$v^t a = \epsilon \ , \tag{19}$$

with $\epsilon \in \mathbb{Z}_2$. Assume that $v_s \neq 0$. The restriction can be seen as the result of a mapping $a' = h_r(a)$ from $\mathbb{Z}_2^{n-1}$ to $\mathbb{Z}_2^n$ specified by $a_i' = a_i$ for $i \neq s$ and $a_s' = \epsilon + v^t a + a_s$. The nonzero elements of the correlation matrix of h_r are

$$C_{ww}^{h_r} = 1 \text{ and } C_{(v+w)w}^{h_r} = (-1)^\epsilon \text{ for all } w \text{ with } w_s = 0 \ . \tag{20}$$

It can be seen that columns indexed by w with $w_s = 0$ have exactly two nonzero entries with magnitude 1 and those with $w_s = 1$ are all-zero. Omitting the latter gives a $2^n \times 2^{n-1}$ correlation matrix C^{hr} with only columns indexed by the vectors with $w_s = 0$.

The transformation restricted to the specified subset of inputs can be seen as the consecutive application of h_r and the transformation itself. Hence, its correlation matrix C' is given by $C \times C^{hr}$. The elements of this matrix are

$$C'_{uw} = C_{uw} + (-1)^\epsilon C_{u(w+v)} , \tag{21}$$

if $w_s = 0$ and 0 if $w_s = 1$.

5   Specific types of mappings

Consider the transformation that consists of the bitwise addition of a constant vector k: $h(a) = a + k$. Since $u^t h(a) = u^t a + u^t k$ the correlation matrix is a diagonal matrix with

$$C_{uu} = (-1)^{u^t k} . \tag{22}$$

Therefore the effect of bitwise addition of a constant vector before (or after) a mapping h on its correlation matrix is a multiplication of some columns (or rows) by -1.

Consider a linear transformation $h(a) = Ma$ with M a $m \times n$ binary matrix. Since $u^t h(a) = u^t M a = (M^t u)^t a$ the elements of the corresponding correlation matrix are given by

$$C_{uw} = \delta(M^t u + w) . \tag{23}$$

If M is an invertible square matrix, the correlation matrix is a permutation matrix. The single nonzero element in row u is in column $M^t u$. The effect of applying an invertible linear transformation before (or after) a transformation h on the correlation matrix is only a permutation of its columns (or rows).

Consider a mapping from $\mathbb{Z}_2^n$ to $\mathbb{Z}_2^m$ that consists of the parallel application of ℓ component mappings (S-boxes) from $\mathbb{Z}_2^{n_i}$ to $\mathbb{Z}_2^{m_i}$ with $\sum_i n_i = n$ and $\sum_i m_i = m$. We will call such a mapping a *boxed* mapping. We have $a = (a_{(0)}, a_{(1)}, \ldots, a_{(\ell-1)})$ and $b = (b_{(0)}, b_{(1)}, \ldots, b_{(\ell-1)})$ with the $a_{(i)}$ vectors of dimension n_i and the $b_{(i)}$ vectors with dimension m_i. The mapping $b = h(a)$ is defined by $b_{(i)} = h_{(i)}(a_{(i)})$ for $0 \le i < \ell$. With every S-box $h_{(i)}$ is associated a $2^{n_i} \times 2^{m_i}$ correlation matrix denoted by $C^{(i)}$. Since the $h_{(i)}$ are disjunct, (10) can be applied and the elements of the correlation matrix of h are given by

$$C_{uw} = \prod_i C^{(i)}_{u_{(i)} w_{(i)}} . \tag{24}$$

with $u = (u_{(0)}, u_{(1)}, \ldots, u_{(\ell-1)})$ and $w = (w_{(0)}, w_{(1)}, \ldots, w_{(\ell-1)})$. In words this can be expressed as: the correlation coefficient associated with input selection w and output selection u is the product of its corresponding S-box input-output correlations $C^{(i)}_{u_{(i)} w_{(i)}}$.

6   Application to iterated transformations

Correlation matrices can be easily applied to express correlations in iterated transformations such as most block ciphers. The studied transformation is

$$\beta = \rho_q \circ \ldots \circ \rho_2 \circ \rho_1 \ , \tag{25}$$

with the ρ_i selected from a set of invertible transformations $\{\rho[b] | b \in \mathbf{Z}_2^{n_b}\}$ by round keys $\kappa^{(i)}$: $\rho_i = \rho[\kappa^{(i)}]$ The round keys $\kappa^{(i)}$ are derived from the cipher key κ by the key schedule.

6.1   Fixed key

In the transform domain, a fixed succession of round transformations corresponds to a $2^n \times 2^n$ correlation matrix that is the product of the correlation matrices corresponding to the round transformations. We have

$$C = C^{\rho_q} \times \ldots \times C^{\rho_2} \times C^{\rho_1} \ . \tag{26}$$

Linear cryptanalysis exploits the occurrence of large elements in product matrices corresponding to all but a few rounds of a block cipher.

A q-round *linear trail* Ω, denoted by

$$\Omega = (\omega_0 \triangleleft \rho_1 \triangleright \omega_1 \triangleleft \rho_2 \triangleright \omega_2 \triangleleft \ldots \triangleright \omega_{q-1} \triangleleft \rho_1 \triangleright \omega_q) \ , \tag{27}$$

is obtained by chaining q single-round correlations $C(\omega_i{}^t \rho_i(a), \omega_{i-1}{}^t a)$. With this linear trail is associated a *correlation contribution coefficient* C_p ranging between -1 and $+1$.

$$C_p(\Omega) = \prod_i C^{\rho_i}_{\omega_i \omega_{i-1}} \ . \tag{28}$$

From this definition and (26) we have

$$C(u^t \beta(a), w^t a) = \sum_{\omega_0 = w, \omega_q = u} C_p(\Omega) \tag{29}$$

Hence the correlation between $u^t \beta(a)$ and $w^t a$ is the sum of the correlation contribution coefficients of all q-round linear trails Ω with initial selection vector w and terminal selection vector u.

6.2   Variable key

In cryptanalysis, the succession of round transformations is not known in advance but is governed by an unknown key or some input-dependent value. In general, the elements of the correlation matrix of ρ_i depend on the specific value of the round key $\kappa^{(i)}$.

For some block ciphers the strong round-key dependence of the correlation and propagation properties of the round transformation have been cited as a design criterion. The analysis of correlation or difference propagation would have to be repeated for every specific value of the cipher key, making linear and differential analysis infeasible. A typical problem with this approach is that the *quality* of the round transformation with respect to LC or DC strongly depends on the specific value of the round key. While the resistance against LC and DC may be very good on the average, specific classes of cipher keys can exhibit linear trails with excessive correlation contribution coefficients.

These complications can be avoided by designing the round transformation in such a way that the amplitudes of the elements of its correlation matrix are independent of the specific value of the round key. As was shown in Sect. 4, this is the case if the round transformation consists of a fixed transformation ρ followed (or preceded) by the bitwise addition of the round key $\kappa^{(i)}$.

The correlation matrix C^ρ is determined by the fixed transformation ρ. The correlation contribution coefficient of the linear trail Ω becomes

$$C_{\mathrm{p}}(\Omega) = \prod_i (-1)^{\omega_i^t \kappa^{(i)}} C_{\omega_i \omega_{i-1}}^\rho = (-1)^{d_\Omega + \sum_i \omega_i^t \kappa^{(i)}} |C_{\mathrm{p}}(\Omega)| . \qquad (30)$$

with d_Ω equal to 1 if $\prod_i C_{\omega_i \omega_{i-1}}^\rho$ is negative and 0 otherwise. $|C_{\mathrm{p}}(\Omega)|$ is independent of the round keys, and hence only the sign of the correlation contribution coefficient depends on the round keys.

The correlation coefficient between $u^t \beta(a)$ and $w^t a$ can be expressed in terms of the correlation contribution coefficients of linear trails:

$$C(u^t \beta(a), w^t a) = \sum_{\omega_0 = w, \omega_q = u} (-1)^{d_\Omega + \sum_i \omega_i^t \kappa^{(i)}} |C_{\mathrm{p}}(\Omega)| . \qquad (31)$$

The amplitude of this correlation coefficient is no longer independent of the round keys since the terms are added or subtracted depending on the value of the round keys.

6.3   Matsui's linear cryptanalysis of DES

The multiple-round linear expressions described in [4] correspond with what we call linear trails. The probability p that such an expression holds corresponds with $\frac{1}{2}(1 + C_{\mathrm{p}}(\Omega))$, with $C_{\mathrm{p}}(\Omega)$ the correlation contribution coefficient of the corresponding linear trail. This implies that the considered correlation coefficient is assumed to be dominated by a single linear trail. This assumption is valid because of the large amplitude of the described correlation coefficients on the one hand and the structure of the DES round transformation on the other.

The correlation contribution coefficient of the linear trail is independent of the key and consists of the product of the correlation coefficients of its single-round components. In general, the elements of the correlation matrix of the DES round transformation are not independent of the round keys. In the linear trails described in [4] the independence is caused by the fact that the single-round correlations of the described linear trail only involve bits of a single S-box.

7 Difference propagation

Say we have two n-dimensional vectors a and a^* with bitwise difference $a + a^* = a'$. Let $b = h(a), b^* = h(a^*)$ and $b' = b + b^*$. Hence, the difference a' propagates to the difference b' through h. This is denoted by $(a' \dashv h \vdash b')$, In general b' is not determined by a' but depends on the value of a (or a^*).

Definition 3. The prop ratio R_p of a difference propagation $(a' \dashv h \vdash b')$ is given by

$$R_p(a' \dashv h \vdash b') = 2^{-n} \sum_a \delta(b' + h(a + a') + h(a)) . \tag{32}$$

The prop ratio ranges between 0 and 1 and must be an integer multiple of 2^{1-n}. The difference propagation $(a' \dashv h \vdash b')$ restricts the values of a to a fraction of all possible inputs. This fraction is given by $R_p(a' \dashv h \vdash b')$. It can easily be seen that

$$\sum_{b'} R_p(a' \dashv h \vdash b') = 1 . \tag{33}$$

Differential cryptanalysis [3] can be seen as the exploitation of large prop ratios.

The prop ratios of the difference propagations of Boolean functions and mappings can be expressed respectively in terms of their Walsh-Hadamard transform values and their correlation matrix elements. Analogous with (8), it can be shown that the components of the inverse transform of the componentwise product of two spectra $\hat{c}_{fg} = \mathcal{W}^{-1}(FG)$ are given by

$$\hat{c}_{fg}(b) = 2^{-n} \sum_a \hat{f}(a)\hat{g}(a + b) = 2^{-n} \sum_a (-1)^{f(a)+g(a+b)} . \tag{34}$$

$\hat{c}_{fg}(b)$ is not a Boolean function. It is generally referred to as the *cross correlation function* of f and g. If $g = f$ it is called the autocorrelation function of f and denoted by $\hat{r}_f$. The components of the spectrum of the autocorrelation function consist of the squares of the spectrum of f, i.e.,

$$\hat{F}(w)^2 = \mathcal{W}(\hat{r}_f(a)) . \tag{35}$$

This is generally referred to as the Wiener-Khintchine theorem [5].

The difference propagation in a Boolean function f can be expressed easily in terms of the autocorrelation function. The prop ratio of difference propagation $(a' \dashv f \vdash 0)$ is given by

$$R_p(a' \dashv f \vdash 0) = 2^{-n} \sum_a \delta(f(a) + f(a + a'))$$

$$= 2^{-n} \sum_a \frac{1}{2}(1 + \hat{f}(a)\hat{f}(a + a'))$$

$$= \frac{1}{2}(1 + \hat{r}_f(a'))$$

$$= \frac{1}{2}(1 + \sum_w (-1)^{w^t a'} \hat{F}^2(w)) . \tag{36}$$

The component of the autocorrelation function $\hat{r}_f(a')$ corresponds to the amount that $R_p(a' \dashv f \vdash 0)$ deviates from $1/2$.

For mappings from $\mathbb{Z}_2^n$ to $\mathbb{Z}_2^m$, let the autocorrelation function of $u^t h(a)$ be denoted by $\hat{r}_u(a')$, i.e.,

$$\hat{r}_u(a') = 2^{-n} \sum_a (-1)^{u^t h(a) + u^t h(a+a')} \ . \tag{37}$$

The prop ratio of difference propagation $(a' \dashv h \vdash b')$ is given by

$$\begin{aligned}
R_p(a' \dashv h \vdash b') &= 2^{-n} \sum_a \delta(h(a) + h(a + a') + b') \\
&= 2^{-n} \sum_a \prod_i \frac{1}{2}((-1)^{h_i(a) + h_i(a+a') + b'_i} + 1) \\
&= 2^{-n} \sum_a 2^{-m} \sum_u (-1)^{u^t h(a) + u^t h(a+a') + u^t b'} \\
&= 2^{-m} \sum_u (-1)^{u^t b'} 2^{-n} \sum_a (-1)^{u^t h(a) + u^t h(a+a')} \\
&= 2^{-m} \sum_u (-1)^{u^t b'} \hat{r}_u(a') \\
&= 2^{-m} \sum_u (-1)^{u^t b'} \sum_w (-1)^{w^t a'} C_{uw}^2 \\
&= 2^{-m} \sum_{u,w} (-1)^{w^t a' + u^t b'} C_{uw}^2 \ . \tag{38}
\end{aligned}$$

Hence the array containing the prop ratios is the (scaled) two-dimensional Walsh-Hadamard transform of the array that contains the squares of the elements of the correlation matrix. Inverting the transform gives the dual expression:

$$C_{uw}^2 = 2^{-n} \sum_{a',b'} (-1)^{w^t a' + u^t b'} R_p(a' \dashv h \vdash b') \ . \tag{39}$$

8   Conclusions

The correlation matrix of a Boolean mapping is an alternative representation that reveals properties of a more global nature. Correlation matrices are the "natural" representation for the description and understanding of linear crypt-analysis.

References

1. S. W. Golomb, *Shift Register Sequences*, Holden–Day Inc., San Francisco, 1967.
2. G.Z. Xiao, J.L. Massey, A Spectral Characterization of Correlation-Immune Functions, *IEEE Trans. Inform. Theory*, Vol. 34, No. 3, 1988, pp. 569–571

3. E. Biham and A. Shamir, *Differential Cryptanalysis of of the Data Encryption Standard*, Springer-Verlag, 1993.
4. M. Matsui, Linear Cryptanalysis Method for DES Cipher, *Advances in Cryptology – Proceedings of Eurocrypt '93, LNCS 765*, T. Helleseth, Ed., Springer-Verlag, 1993, pp. 386–397.
5. B. Preneel, *Analysis and Design of Cryptographic Hash Functions*, Doct. Dissertation K.U.Leuven, January 1993.
6. J. Daemen, *Cipher and Hash Function Design. Strategies Based on Linear and Differential Cryptanalysis*, Doct. Dissertation K.U.Leuven, March 1995.

On the Need for Multipermutations: Cryptanalysis of MD4 and SAFER

Serge Vaudenay[*]

Ecole Normale Supérieure — DMI
45, rue d'Ulm
75230 Paris Cedex 5 France
Serge.Vaudenay@ens.fr

Abstract. Cryptographic primitives are usually based on a network with boxes. At EUROCRYPT'94, Schnorr and the author of this paper claimed that all boxes should be multipermutations. Here, we investigate a few combinatorial properties of multipermutations. We argue that boxes which fail to be multipermutations can open the way to unsuspected attacks. We illustrate this statement with two examples.

Firstly, we show how to construct collisions to MD4 restricted to its first two rounds. This allows one to forge digests close to each other using the full compression function of MD4. Secondly, we show that variants of SAFER are subject to attack faster than exhaustive search in 6.1% cases. This attack can be implemented if we decrease the number of rounds from 6 to 4.

In [18], *multipermutations* are introduced as formalization of perfect diffusion. The aim of this paper is to show that the concept of multipermutation is a basic tool in the design of dedicated cryptographic functions, as functions that do not realize perfect diffusion may be subject to some clever cryptanalysis in which the flow of information is controlled throughout the computation network. We give two cases of such an analysis.

Firstly, we show how to build collisions for MD4 restricted to its first two rounds[2]. MD4 is a three rounds hash function proposed by Rivest[17]. Den Boer and Bosselaers[2] have described an attack on MD4 restricted to its last two rounds. Another unpublished attack on the first two rounds has been found by Merkle (see the introduction of [2]). Here, we present a new attack which is based on the fact that an inert function is not a multipermutation. This attack requires less than one tenth of a second on a SUN workstation. Moreover, the same attack applied to the full MD4 compression function produces two different digests close to each other (according to the Hamming distance). This proves the compression function is not correlation-free in the sense of Anderson[1].

---

[*] *Laboratoire d'Informatique de l'Ecole Normale Supérieure*, research group affiliated with the CNRS

[2] This part of research has been supported by the CELAR.

Secondly, we show how to develop a known plaintext attack to a variant of SAFER K-64, in which we replace the permutation $\exp_{45}$ by a (weaker) one. SAFER is a six rounds encryption function introduced by Massey[9]. It uses a byte-permutation (namely, $\exp_{45}$ in the group of nonzero integers modulo 257) for confusion. If we replace $\exp_{45}$ by a random permutation P (and $\log_{45}$ by P^{-1}), we show that in 6.1% of the cases, there exists a known plaintext attack faster than exhaustive search. Furthermore, this attack can be implemented for the function restricted to 4 rounds. This attack is based on the linear cryptanalysis introduced by Matsui[10] and recently gave way to the first experimental attack of the full DES function[11].

1   Multipermutations

In [18], multipermutations with 2 inputs and 2 outputs are introduced. Here, we propose to generalize to any number of inputs and outputs.

Definition. A (r, n)-multipermutation over an alphabet Z is a function f from Z^r to Z^n such that two different $(r + n)$-tuples of the form $(x, f(x))$ cannot collide in any r positions.

Thus, a $(1, n)$-multipermutation is nothing but a vector of n permutations over Z. A $(2, 1)$-multipermutation is equivalent to a *latin square*[3]. A $(2, n)$-multipermutation is equivalent to a set of n two-wise *orthogonal* latin squares[4]. Latin squares are widely studied by Dénes and Keedwell in [3].

An equivalent definition says that the set of all $(r + n)$-tuples of the form $(x, f(x))$ is an error correcting code with minimal distance $n + 1$, which is the maximal possible. In the case of a linear function f, this is the definition of MDS codes: codes which reach Singleton's bound (for more details about MDS codes, see [15]). More generally, a (r, n)-multipermutation is equivalent to a $((\#Z)^r, r + n, \#Z, r)$-orthogonal array[5].

A multipermutation performs a *perfect diffusion* in the sense that changing t of the inputs changes at least $n - t + 1$ of the outputs. In fact, it corresponds to the notion of *perfect local randomizer* introduced by Maurer and Massey[13] with the optimal parameter. If a function is not a multipermutation, one can find several values such that both few inputs and few outputs are changed. Those values can be used in cryptanalysis as is shown in two examples below. This motivates the use of multipermutations in cryptographic functions.

---

[3] a latin square over a finite set of k elements is a $k \times k$ matrix with entries from this set such that all elements are represented in each column and each row.

[4] two latin squares A and B are orthogonal if the mapping $(i, j) \mapsto (A_{i,j}, B_{i,j})$ gets all possible couples.

[5] a $(M, r + n, q, r)$-orthogonal array is a $M \times (r + n)$ matrix with entries from a set of q elements such that any set of r columns contains all q^r possible rows exactly $\frac{M}{q^r}$ times.

The design of multipermutations over a large alphabet is a very difficult problem, as the design of two-wise orthogonal latin squares is a well-known difficult one. The only powerful method seems to use an MDS code combined with several permutations at each coordinate.

In the particular case of 2 inputs, it is attractive to choose latin squares based on a group law: if we have a group structure over Z, we can seek permutations α, β, γ, δ, ϵ and ζ such that

$$(x, y) \mapsto (\alpha[\beta(x).\gamma(y)], \delta[\epsilon(x).\zeta(y)])$$

is a permutation, as it will be sufficient to get a multipermutation. Unfortunately, it is possible to prove that such permutations exist only when the 2-Sylow subgroup of Z is not cyclic[6], using a theorem from Hall and Paige[7]. More precisely, they do not exist when the 2-Sylow subgroup is cyclic. They are known to exist in all solvable groups in which the 2-Sylow subgroup is not cyclic, but the existence in the general case is still a conjecture. Hence, Z should not have a cyclic group structure. For instance, we can use the $\mathbb{Z}_2^n$ group structure for $n > 1$. Such multipermutations are proposed in [18].

In MD4, the group structure of $\mathbb{Z}_2^{32}$ is used, but some functions are not multipermutations. On the other hand, in SAFER, the group structure of $\mathbb{Z}_{256}$, which is cyclic, is used, so without multipermutations.

2   Cryptanalysis of MD4

2.1   Description of MD4

MD4 is a hash function dedicated to 32-bit microprocessors. It hashes any bit string into a 128-bit digest. The input is padded following the Merkle-Damgård scheme[4, 12] and cut into 512-bit blocks. Then, each block is processed iteratively using the Davies-Meyer scheme[5, 14] i.e. with an encryption function C in a feedforward mode: if $B_1, \ldots, B_n$ is the sequence of blocks (the padded message), the hash value is

$$h_{B_n}(\ldots h_{B_1}(v_i) \ldots)$$

where v_i is an Initial Value, and $h_x(v)$ is $C_x(v) + v$ (x is the key and v is the message to encrypt).

Here we intend to build a single block collision to $h(v_i)$, that is to say two blocks x and x' such that $C_x(v_i) = C_{x'}(v_i)$. It is obvious that this can be used to build collisions to the hash function. So, we only have to recall the definition of the function $C_x(v)$.

---

[6] we agree the trivial group is not cyclic. Actually, $x \mapsto x^2$ is an orthomorphism in all groups with odd order, in which the 2-Sylow subgroup is trivial.

The value v is represented as 4 integers a, b, c and d (coded with 32 bits), and the key x is represented as 16 integers $x_1, \ldots, x_{16}$. The initial definition of C uses three rounds $i = 1, 2, 3$. The figure 1 shows the computational graph of a single round i. It uses a permutation σ_i and some boxes B_i^j. B_i^j is fed with a main input, a block integer $x_{\sigma_i(j)}$ and three side inputs. If p is the main input and q, r and s are the side inputs (from top to bottom), the output is

$$R^{\alpha_{i,j}}(p + f_i(q, r, s) + x_{\sigma_i(j)} + k_i)$$

where R is the right circular rotation, $\alpha_{i,j}$ and k_i are constants and f_i is a particular function. In the following, we just have to know that f_2 is the bit-wise majority function, σ_1 is the identity permutation, and

$$\sigma_2 = \begin{pmatrix} 1 & 2 & 3 & 4 & 5 & 6 & 7 & 8 & 9 & 10 & 11 & 12 & 13 & 14 & 15 & 16 \\ 1 & 5 & 9 & 13 & 2 & 6 & 10 & 14 & 3 & 7 & 11 & 15 & 4 & 8 & 12 & 16 \end{pmatrix}$$

2.2   Attack on the first two rounds

If we ignore the third round of C, it is very easy to build collisions. We notice that no B_2^j are multipermutations: if $p = 0$, $x = -k_2$ and two of the three integers q, r and s are set to zero, then $B_2^j(x, p, q, r, s)$ remains zero (the same remark holds with -1 instead of 0). So, we can imagine an attack where two blocks differ only in x_{16}, the other integers are almost all set to $-k_2$ and such that almost all the outputs of the first round are zero. This performs a kind of corridor where the modified values are controlled until the final collision.

More precisely, let $x_1, \ldots, x_{11}$ equal $-k_2$, x_{12} be an arbitrary integer (your phone number for instance) and x_{13}, x_{14} and x_{15} be such that the outputs a, c and d of the first round are zero. The computation of x_{13}, x_{14} and x_{15} is very easy from the computational graph. Thanks to the previous remark, we can show that the outputs a, c and d of the second round do not depend on x_{16} as the modified information in x_{16} is constrained in the register b. Thus, modifying x_{16} does not modify a, c and d.

Letting the b output be a function of x_{16}, we just have to find a collision to a 32 bits to 32 bits function. This can be done very efficiently using the birthday paradox or the ρ method. An implementation on a Sparc Station uses one tenth of second.

If we use the same attack on the full-MD4 function, since $\sigma_3^{-1}(16) = 16$, the only modified x occurs in the very last computation in the third round. So, if this round is fed with a collision, it produces a collision on the a, c and d output. The digests differ only in the second integer b. Hence, the average Hamming distance between both digests is 16. This proves the compression function of MD4 is not correlation-free, according to Andersons's definition[1].

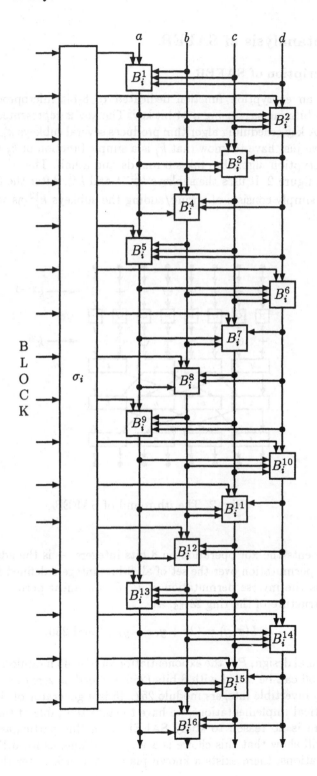

Fig. 1. One round of C

3 Cryptanalysis of SAFER

3.1 Description of SAFER

SAFER is an encryption function dedicated to 8-bit microprocessors. It encrypts a 64 bits message using a 64-bit key. The key is represented as 8 integers $k_1, \ldots, k_8$. A key scheduling algorithm produces several subkeys $k_1^i, \ldots, k_8^i$. In the following, we just have to know that k_j^i is a simple function of k_j (and $k_j^1 = k_j$).

The encryption algorithm takes 6 rounds and a half. The ith round is summarized on figure 2. It uses the subkeys k_j^{2i-1} and k_j^{2i}. After the 6th round, the half round simply consists of xoring/adding the subkeys k_j^{13} as we would do in a 7th round.

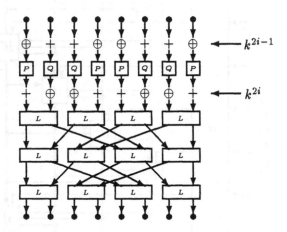

Fig. 2. The ith round of SAFER

$\oplus$ represents the xor operation on 8 bits integers. $+$ is the addition modulo 256. P is a permutation over the set of all 8-bits integers defined in the SAFER design. Q is the inverse permutation of P. L is a linear permutation over the algebraic structure of the ring $\mathbb{Z}_{256}$, as

$$L(x, y) = (2x + y, x + y) \pmod{256}.$$

In the original design, P is the exponentiation in base 45 modulo 257: all integers from 1 to 256 can be coded with 8 bits (256 is coded as zero) and represent the group of all invertible integers modulo 257. 45 is a generator of this group.

In practical implementations, we have to store the table of the permutation P. So, there is no reason to study SAFER with this particular permutation. Here, we will show that this choice is a very good one, as for 6.1% of all possible permutations, there exists a known plaintext attack faster than exhaustive search.

3.2 Linear cryptanalysis of SAFER

*I have many times used the discrete exponential or the discrete logarithm
as nonlinear cryptographic functions and they have never let me down.*
James Massey

The permutation L is not a multipermutation, as we have

$$L_1(x + 128, y) = L_1(x, y)$$

for all x and y (where L_1 denotes the first output of L). So, we have pairs
of 4-tuples $(x, y, L(x, y))$ at Hamming distance 2. Actually, there are no $(2, 2)$-
multipermutations which are linear over $\mathbb{Z}_{256}$ as its 2-Sylow subgroup is cyclic
(it is itself here). We can use this property of L_1 by a dual point of view noticing
that some information about $L_1(x, y)$ only depends on y. Namely, we have

$$L_1(x, y) \cdot 1 = y \cdot 1$$

where $\cdot$ is the inner product over $(\mathbb{Z}_2)^8$, so, $y \cdot 1$ is the least significant bit of y.
Similarly, we have

$$(L_1(x, y) \cdot 1) \oplus (L_2(x, y) \cdot 1) = x \cdot 1$$

Let F denote the function defined by the three bottom layers on figure 2
(layers which uses L in a round). If $x_1, \dots, x_8$ are the inputs of a round, the
outputs are $F(y_1, \dots, y_8)$ where $y_1 = P(x_1 \oplus k_1^1) + k_1^2$, ... We notice that if
$F(y_1, \dots, y_8) = (z_1, \dots, z_8)$, we have a 2-2 *linear characteristic*

$$(z_3 \cdot 1) \oplus (z_4 \cdot 1) = (y_3 \cdot 1) \oplus (y_4 \cdot 1)$$

(this means there is a linear dependence using 2 inputs and 2 outputs of F).
There are 5 other 2-2 linear characteristics:

$$(z_2 \cdot 1) \oplus (z_6 \cdot 1) = (y_2 \cdot 1) \oplus (y_6 \cdot 1)$$
$$(z_5 \cdot 1) \oplus (z_7 \cdot 1) = (y_5 \cdot 1) \oplus (y_7 \cdot 1)$$
$$(z_3 \cdot 1) \oplus (z_7 \cdot 1) = (y_5 \cdot 1) \oplus (y_6 \cdot 1)$$
$$(z_5 \cdot 1) \oplus (z_6 \cdot 1) = (y_2 \cdot 1) \oplus (y_4 \cdot 1)$$
$$(z_2 \cdot 1) \oplus (z_4 \cdot 1) = (y_3 \cdot 1) \oplus (y_7 \cdot 1).$$

If L were a multipermutation, the smallest characteristics would be a-b ones such
that $a+b = 6$. This property is similar to the well-known Heisenberg's inequality
which states we cannot have any precise information on both the input and the
output of the Fourier transform. This means more information would be required
in a cryptanalysis.

Let q denote $\mathrm{Prob}_x[x \cdot 1 = P(x) \cdot 1] - \frac{1}{2}$, the bias which measures the dependence between the least significant bits of $P(x)$ and x. We get the same bias with Q in place of P. If $(x_1, \ldots, x_8)$ is a plaintext, if $y_1 = P(x_1 \oplus k_1)$, $y_2 = Q(x_2 + k_2)$, ..., $y_8 = P(x_8 \oplus k_8)$, and if $(z_1, \ldots, z_8)$ is the ciphertext, let us write

$$b(x, z) = (y_3 \cdot 1) \oplus (y_4 \cdot 1) \oplus (z_3 \cdot 1) \oplus (z_4 \cdot 1).$$

Lemma 1 in appendix A states that $b(x, z) = \phi(k)$ with probability $\frac{1}{2}(1 + (2q)^{10})$ where $\phi(k)$ denotes the exclusive or of all least significant bits of k_3^i and k_4^i for $i = 2, \ldots, 13$. For a given (x, z), to compute $b(x, z)$, we only have to know k_3 and k_4. Lemma 1 states that it occurs with probability roughly equal to $\frac{1}{2}$ (the difference with $\frac{1}{2}$ is negligible against $(2q)^{10}$) when wrong k_3 and k_4 are used in the computation of $b(x, z)$. Thus, trying all the possible (k_3, k_4), it is possible to distinguish the good one from the other candidates by a statistical measure.

Let us recall the central limit theorem (see [6] for instance):

Theorem. *If B is the arithmetic mean of N independent random variables with the same probability distribution with average μ and standard deviation σ, we have*

$$\mathrm{Prob}\left[(B - \mu)\frac{\sqrt{N}}{\sigma} \in [a, b]\right] \to \frac{1}{\sqrt{2\pi}} \int_a^b e^{-\frac{t^2}{2}} dt.$$

Let $B(k_3, k_4)$ be the average of $b(x, z)$ over all the N available couples (x, z). Lemma 1 proves that the standard deviation of $b(x, z)$ is close to $\frac{1}{2}$. Let

$$\lambda_1 + \lambda_2 = \sqrt{N}(2q)^{10}.$$

The central limit theorem states that if (k_3, k_4) is wrong,

$$\mathrm{Prob}\left[\left|B(k_3, k_4) - \frac{1}{2}\right| < \frac{\lambda_1}{2\sqrt{N}}\right] \to \frac{1}{\sqrt{2\pi}} \int_{-\lambda_1}^{\lambda_1} e^{-\frac{t^2}{2}} dt;$$

and if (k_3, k_4) is good,

$$\mathrm{Prob}\left[\left|B(k_3, k_4) - \frac{1}{2}\right| < \frac{\lambda_1}{2\sqrt{N}}\right] \to \frac{1}{\sqrt{2\pi}} \int_{\lambda_2}^{\lambda_2 + 2\lambda_1} e^{-\frac{t^2}{2}} dt.$$

The statistical test consists in accepting any (k_3, k_4) such that

$$\mathrm{Test}(k_3, k_4): \qquad \left|B(k_3, k_4) - \frac{1}{2}\right| > \frac{\lambda_1}{2\sqrt{N}}.$$

If $\lambda_1 = \lambda_2 = 2$ the good (k_3, k_4) is accepted with probability 98% and the bad ones are rejected with probability 95%. So, the number of plaintexts/ciphertexts required to distinguish the good (k_3, k_4) is

$$N \sim \frac{16}{(2q)^{20}}.$$

If $|q|$ is greater than 2^{-4}, this is faster than exhaustive search.

For only 4 rounds in SAFER, we have $N \sim \frac{16}{(2q)^{12}}$. So, for all permutations P which are biased ($q \neq 0$), this attack is faster than exhaustive search. For $|q| \geq 2^{-4}$, the attack can be implemented.

The analysis of the distribution of q shows that we have $|q| \geq 2^{-4}$ for 6.1% of the possible permutations P (see appendix B). We have $q = 0$ for only 9.9% of the permutations. Unfortunately (or fortunately), for the P chosen by Massey, we have $q = 0$, so, the weakness of the diffusion phase is balanced by the strength of the confusion phase. Actually, $q = 0$ is a property of all exponentiations which are permutations (see appendix C). This analysis illustrates how Massey was right in context with his quotation.

Further analysis can improve this attack. It is possible to use tighter computations. We can look for a better tradeoff between the workload and the probability of success. It is also possible to use several characteristics to decrease N (for more details, see [8]). At least, it is possible to decrease N by a factor of 64. Actually, we believe it is possible to improve successfully this attack for all the 90.1% biased permutations.

Conclusion

In MD4, we have shown that the fact that f_2 is not a multipermutation allows one to mount an attack. Similarly, in SAFER, the diffusion function is not a multipermutation. This allows one to imagine another attack. This shows that we do need multipermutations in the design of cryptographic primitives. Research in this area should be motivated by this general statement.

Acknowledgments

I would like to thank Antoon Bosselaers, Ross Anderson, James Massey and Carlo Harpes for helpful information. I thank the CELAR for having motivated the research on MD4. I also thank Jacques Stern and Hervé Brönnimann for their help.

References

1. R. J. Anderson. The Classification of Hash Functions. In *Proceedings of the 4th IMA Conference on Cryptography and Coding*, Cirencester, United Kingdom, pp. 83–95, Oxford University Press, 1995.
2. B. den Boer, A. Bosselaers. An Attack on the last two Rounds of MD4. In *Advances in Cryptology CRYPTO'91*, Santa Barbara, California, U.S.A., Lectures Notes in Computer Science 576, pp. 194–203, Springer-Verlag, 1992.
3. J. Dénes, A. D. Keedwell. *Latin Squares and their Applications*, Akadémiai Kiadó, Budapest, 1974.
4. I. B. Damgård. A Design Principle for Hash Functions. In *Advances in Cryptology CRYPTO'89*, Santa Barbara, California, U.S.A., Lectures Notes in Computer Science 435, pp. 416–427, Springer-Verlag, 1990.
5. R. W. Davies, W. L. Price. Digital Signature – an Update. In *Proceedings of the International Conference on Computer Communications*, Sydney, pp. 843–847, North-Holland, 1985.

6. W. Feller. *An Introduction to Probability Theory and its Applications*, vol. 1, Wiley, 1957.

7. M. Hall, L. J. Paige. Complete Mappings of Finite Groups. In *Pacific Journal of Mathematics*, vol. 5, pp. 541–549, 1955.

8. B. R. Kaliski Jr., M. J. B. Robshaw. Linear Cryptanalysis using Multiple Approximations. In *Advances in Cryptology CRYPTO'94*, Santa Barbara, California, U.S.A., Lectures Notes in Computer Science 839, pp. 26–39, Springer-Verlag, 1994.

9. J. L. Massey. SAFER K-64: a Byte-oriented Block-ciphering Algorithm. In *Fast Software Encryption – Proceedings of the Cambridge Security Workshop*, Cambridge, United Kingdom, Lectures Notes in Computer Science 809, pp. 1–17, Springer-Verlag, 1994.

10. M. Matsui. Linear Cryptanalysis Method for DES Cipher. In *Advances in Cryptology EUROCRYPT'93*, Lofthus, Norway, Lectures Notes in Computer Science 765, pp. 386–397, Springer-Verlag, 1994.

11. M. Matsui. The first Experimental Cryptanalysis of the Data Encryption Standard. In *Advances in Cryptology CRYPTO'94*, Santa Barbara, California, U.S.A., Lectures Notes in Computer Science 839, pp. 1–11, Springer-Verlag, 1994.

12. R. C. Merkle. One way Hash Functions and DES. In *Advances in Cryptology CRYPTO'89*, Santa Barbara, California, U.S.A., Lectures Notes in Computer Science 435, pp. 416–427, Springer-Verlag, 1990.

13. U. M. Maurer, J. L. Massey. Local Randomness in Pseudorandom Sequences. In *Journal of Cryptology*, vol. 4, pp. 135–149, 1991.

14. S. M. Matyas, C. H. Meyer, J. Oseas. Generating Strong One-way Functions with Cryptographic Algorithm. *IBM Technical Disclosure Bulletin*, vol. 27, pp. 5658–5659, 1985.

15. F. J. McWilliams, N. J. A. Sloane. *The Theory of Error-correcting Codes*, North-Holland, 1977.

16. L. O'Connor. Properties of linear approximation tables. In these proceedings, pp. 131–136.

17. R. Rivest. The MD4 Message Digest Algorithm. In *Advances in Cryptology CRYPTO'90*, Santa Barbara, California, U.S.A., Lectures Notes in Computer Science 537, pp. 303–311, Springer-Verlag, 1991.

18. C.-P. Schnorr, S. Vaudenay. Black box Cryptanalysis of Hash Networks based on Multipermutations. To appear in *Advances in Cryptology EUROCRYPT'94*.

A   Linear characteristic

Lemma 1. *If $\phi(k)$ denotes the least significant bit of the sum of all k_3^i and k_4^i for $i = 2, \ldots, 13$, let us denote $y_3 = Q(x_3 + k_3)$, $y_4 = P(x_4 \oplus k_4)$ and*

$$b(x, z) = (y_3 \cdot 1) \oplus (y_4 \cdot 1) \oplus (z_3 \cdot 1) \oplus (z_4 \cdot 1)$$

where z is the encrypted message of x using an unknown key. $b(x, z) = \phi(k)$ holds with probability

$$\frac{1}{2}(1 \pm (2q)^{10+e})$$

where e is the number of wrong integers in (k_3, k_4) ($e = 0$ if both are good and $e = 2$ in most of cases). The standard deviation of $b(x, z)$ is

$$\frac{1}{2}\sqrt{1 - (2q)^{20+2e}}.$$

Proof. Thanks to the property of the linear characteristic, if we denote by t_j^i the xor of the least significant bit of the input and the output of the P/Q box in position j in round #i, it is easy to se that

$$b(x, z) = (y_3 \cdot 1) + (y_4 \cdot 1) + (y_3' \cdot 1) + (y_4' \cdot 1) + \sum_{i=2}^{6} \sum_{i=3}^{4} t_j^i + \phi(k') \quad (\text{mod } 2)$$

where $\phi(k')$ denotes the real $\phi(k)$ and y_3' (resp. y_4') denotes the real y_3 (resp. y_4). Under the heuristic assumption that all inputs to P/Q boxes are uniformly distributed and independent, it is easy to prove that

$$\text{Prob}\left[\sum_{i=2}^{6} \sum_{i=3}^{4} t_j^i = 0\right] = \frac{1}{2}(1 + (2q)^{10})$$

using the piling-up lemma pointed out by Matsui[10]. This finishes the case where k_3 and k_4 are good.

If k_3 or k_4 are wrong, let us denote $e = 2$ if both are bad, and $e = 1$ if only one is bad. Assume k_3 is bad without loss of generality. We have

$$\text{Prob}\left[(y_3 \cdot 1) \oplus (y_3' \cdot 1) = 0\right] = \frac{1}{2}(1 + 2q).$$

The $\pm$ comes from whether $\phi(k) = \phi(k')$ or not. This finishes the computation of the probability.

The standard deviation comes from the following formula which holds for all 0/1 random variables :

$$\sigma(b) = \sqrt{E(b)(1 - E(b))}.$$

$\square$

B Distribution of the bias

Lemma 2. *If* $q = Prob[x \cdot 1 = P(x) \cdot 1] - \frac{1}{2}$ *where* P *is a permutation over* $\{0, \ldots, n - 1\}$ *(we assume that* n *is a multiple of 4), nq is always an even integer and for all integers k*

$$\text{Prob}\left[q = \frac{2k}{n}\right] = \frac{\left((\frac{n}{2})!\right)^4}{n! \left((\frac{n}{4} - k)!\right)^2 \left((\frac{n}{4} + k)!\right)^2}$$

for a uniformly distributed permutation Pd.

All those kind of distribution has been studied by O'Connor, but we give here and independant study in this particular case [16].

Proof. If $k + \frac{n}{4}$ denotes the number of even integers x such that $P(x)$ is even, we have $q = \frac{2k}{n}$. So, we just have to enumerate the number of permutations for a given $k + \frac{n}{4}$.

We have to choose 4 sets with $k + \frac{n}{4}$ elements in sets with $\frac{n}{2}$ elements: the set of even integers which are mapped on even integers, the set of their images, the set of odd integers which are mapped on odd integers and the set of their images. We also have to choose 2 permutations over a set of $k + \frac{n}{4}$ integers (how to connect even to even integers and odd to odd integers) and 2 permutations over a set of $-k + \frac{n}{4}$ integers (how to connect even to odd integers and odd to even integers). So, the number of permutations is

$$\left(\begin{array}{c} \frac{n}{2} \\ \frac{n}{4} + k \end{array} \right)^4 \times \left(\left(\frac{n}{4} + k \right)! \right)^2 \times \left(\left(\frac{n}{4} - k \right)! \right)^2.$$

$\square$

This allows one to compute

$$\text{Prob}\left[|q| \geq 2^{-4} \right] \simeq 6.1\%$$

for $n = 256$ and

$$\text{Prob}\left[q = 0 \right] \simeq 9.9\%.$$

C   Bias of the exponentiation

Lemma 3. *For any generator g of $\mathbb{Z}_{257}^*$, the permutation $x \mapsto g^x$ is unbiased (i.e. $q = 0$).*

Proof. We have $(g^{128})^2 \equiv g^{256} \equiv 1 \pmod{257}$ so g^{128} is 1 or -1. As the exponentiation in base g is a permutation and $g^0 = 1$, we have $g^{128} \equiv -1 \pmod{257}$.

We have $g^{x+128} \equiv -g^x \equiv 257 - g^x \pmod{257}$, so, we can partition all the integers into pairs $\{x, x + 128\}$ of integers with the same least significant bit. The image of this pair by the exponentiation has two different least significant bits, so the bias q is 0. $\square$

How to Exploit the Intractability of Exact TSP for Cryptography

Stefan Lucks

Institut für Numerische und Angewandte Mathematik
Georg–August–Universität Göttingen
Lotzestr. 16–18, D–37083 Göttingen, Germany
(email: lucks@namu01.gwdg.de)

Abstract. We outline constructions for both pseudo-random generators and one-way hash functions. These constructions are based on the exact TSP (XTSP), a special variant of the well known traveling salesperson problem. We prove that these constructions are secure if the XTSP is infeasible. Our constructions are easy to implement, appear to be fast, but require a large amount of memory.

1 Introduction

In some fields of modern cryptography, e.g. public key cryptography, it is common to base the security of cryptosystems on the intractability of well known mathematical problems. Examples for these problems are the factorization of integers and the discrete logarithm. But for the fields of one-way hash functions and secret key cryptography there seems to be no scheme, which has a simple mathematical description, is provably secure under a reasonable assumption *and* is fast.

Merkle and Hellman [8] were the first to suggest an NP-hard problem for cryptography. They couldn't prove their public key cipher to be as secure as the underlying problem, and later their cryptosystem was broken. In fact it is regarded to be very unlikely, that one can prove the equivalence of breaking a public key cipher and computing some NP-hard problem.

Shamir [13] suggested an identification scheme based on the NP-hard permuted kernel problem (PKP). For a discussion of the hardness of the PKP see [1], [5] and [11]. Other identification schemes based on NP-hard problems were due to Stern, see [14] and [15].

What about secret key cryptography? Theoretical constructions are known which are as hard to break as any one-way function, though these constructions are too inefficient for practical applications. Furthermore Impagliazzo and Naor [6] did discuss constructions for pseudo-random bit generators and universal one-way hash functions, which are as secure as the Subset Sum problem. This is NP-hard.

A one-way hash function very similar to Impagliazzo's and Naor's scheme was suggested the same year by Damgård [4]. This was broken by Camion and Patarin [2], using essentially brute force and applying the birthday paradox. We conclude that Damgård's scheme did not fall due to an inherent feasibility of the Subset Sum problem.

Anyway the Subset Sum problem *may* be too easy for cryptography. The whole theory of NP-hardness is based on worst-case complexity. Therefore NP$\neq$P would only imply that Subset Sum is infeasible in the worst case. But even being infeasible on average might be not enough.

Consider a problem P and assume for simplicity, that all instances of size n are equally likely. Let P be hard on average. In other words, the expected runtime for every probabilistic algorithm to solve P, when an instance of size n is randomly chosen, is exponential (in n). There may exist an algorithm, such that, say, for 50% of all instances of size n we need exponential time to solve them, and for the remaining instances we need time n^c. The expected runtime for this algorithm is exponential, but a randomly chosen instance is easy with probability $1/2$.

In cryptography we demand for any attacker the probability of a successful attack to be negligible. Consequently we also demand that a randomly generated instance of the underlying problem is hard with overwhelming probability—not only hard in the worst case or hard on average.

Empirical results on heuristics for the Subset Sum problem—e.g. by Schnorr and Euchner [12]—raise serious doubts on the security of Impagliazzo's and Naor's schemes. Nevertheless other NP-complete problems can be good cryptographic one-way functions.

The traveling salesperson problem (TSP) is among the oldest and most prominent problems in algorithm and computational complexity theory. It is unsolved if we only regard an efficient algorithm as a valid solution. It has been studied long before the theory of NP-hardness was developed, see [7] for details.

2   The Exact TSP (XTSP)

Essentially the XTSP is a variant of the TSP, where we are looking for a Hamiltonian path of a given length—not for the shortest one.

In the following $A = (a_{i,j})$ is an $n \times n$-matrix with $a_{i,i} = 0$ and for $i \neq j$ $a_{i,j}$ randomly chosen from $\{0, \dots, 2^{l(n)} - 1\}$. We think of A as the distance matrix for distances in the complete directed Graph G_n with n vertices. Therefore the XTSP is actually a family of problems depending on the parameter $l(n)$.

In this paper we only deal with directed graphs, but our results can easily be adopted to the undirected case too.

We regard the numbers $a_{i,j}$ as fixed public constants like, say, the S-boxes of DES.

Any Hamiltonian cycle X for G_n can be coded as an integer with

$$\lceil \log_2((n-1)!) \rceil \quad \text{Bits.}$$

By $\text{Length}_A(X)$ we mean the length of X with respect to A. Given a number B, the XTSP is to find a Hamiltonian cycle X with

$$\text{Length}_A(X) = B.$$

It is easy to prove the NP-hardness of the XTSP and the NP-completeness of the corresponding existence problem.

By the following theorem we find a relationship between different members of our problem family.

Theorem 1. *Let*

$$l'(n) \ll l(n) \ll \log_2((n-1)!) \quad or \quad l'(n) \gg l(n) \gg \log_2((n-1)!).$$

Then the XTSP with number length $l'(n)$ can—with respect to probabilistic algorithms and except for a polynomial factor in computation time—be no harder than the XTSP with $l(n)$.

Sketch of proof: For $l'(n) \ll l(n) \ll \log_2((n-1)!)$ we regard the difference matrices $A = (a_{i,j})$ and $A' = (a'_{i,j})$ with $a'_{i,j} \equiv a_{i,j} \pmod{2^{l'(n)}}$. For any random X and $B = \text{Length}_{A'}(X)$ there exists with overwhelming probability a Y with $\text{Length}_A(Y) = B$. If Y exists, it is a solution with respect to A' too.

The proof for $l'(n) \gg l(n) \gg \log_2((n-1)!)$ uses the fact, that any solution with respect to the number length $l(n)$ is unique with overwhelming probability. Then any solution for the number length $l'(n)$ is—if existing at all—the same as for $l(n)$.   □

Thus $l(n) \approx \log_2((n-1)!)$ describes the most secure cases. It seems recommendable to bound $l(n)$ by $\frac{1}{c}\log_2((n-1)!) \leq l(n) \leq c\log_2((n-1)!)$ for some $c > 1$, e.g. $c = 2$.

To get a "more uniform" output we define the modular XTSP. Given the matrix A, the number B and the number length $l(n)$, the problem is to find an X with

$$\text{Length}_A(X) \equiv B \left(\text{mod } 2^{l(n)}\right).$$

Theorem 2. *With respect to probabilistic algorithms, and except for a polynomial factor in computation time, the modular XTSP is as hard as the XTSP itself.*

Sketch of proof: Let X be random, $B = \text{Length}_A(X)$. A random Y with $\text{Length}_A(Y) \equiv B$ modulo $2^{l(n)}$ satisfies the equation $\text{Length}_A(Y) = B$ with probability $1/\Theta(n)$, since B is the sum of only n numbers, hence $B < 2^{l(n)} + n$.   □

In the following we only regard the modular XTSP, thus all computations of the function Length_A are done modulo $2^{l(n)}$, where $l(n)$ is the number length of A.

3   One-Way Hash Functions

One-way hash functions are useful for electronic fingerprints. For $m > k(m)$ the functions $f_m : \{0,1\}^m \longrightarrow \{0,1\}^{k(m)}$ are *one-way hash functions*, if it is easy to compute $f_m(x)$ when given m and $x \in \{0,1\}^m$, but infeasible to find a $y \neq x$ in $\{0,1\}^m$ with $f_m(x) = f_m(y)$ for random x.

Theorem 3. *Let* $l(n) < m = \lfloor \log_2((n-1)!) \rfloor$*; if* Length_A *is a one-way function then* $f_m(X) = \text{Length}_A(X)$ *is a one-way hash function.*

Sketch of proof: If, given a random X, we can efficiently find some $Y \neq X$ with $\text{Length}_A(Y) = \text{Length}_A(X)$, then we can invert the function "Length_A". Let a target number B be given, then

- we chose a random Hamiltonian cycle X and let $B' = \text{Length}_A(X)$,
- we chose a random edge $i^* \to j^*$ of X and compute the matrix $A' = (a'_{i,j})$ with $a'_{i,j} = a_{i,j}$ for $(i \neq i^*$ or $j \neq j^*)$, $a'_{i^*,j^*} = a_{i^*,j^*} + B - B'$, consequently $\text{Length}_{A'}(X) = B$,
- and finally we compute a $Y \neq X$ with $\text{Length}_{A'}(X) = \text{Length}_{A'}(Y)$.

With nonnegligible probability the edge $i^* \to j^*$ is no part of the Hamiltonian cycle Y. In this the case $\text{Length}_A(Y) = B$.   □

Universal one-way hash function were defined by Naor and Yung [9], who also outlined their application to digital signatures and fingerprints. For $m > k(m)$ the collections F_m of functions $f_{m,i} : \{0,1\}^m \longrightarrow \{0,1\}^{k(m)}$ constitute *families of universal one-way hash functions*, if given m, for any $x \in \{0,1\}^m$ and randomly chosen $f_{m,i} \in F_m$, it is easy to compute $f_{m,i}(x)$, but with overwhelming probability infeasible to find a $y \neq x$ in $\{0,1\}^m$ with $f_{m,i}(x) = f_{m,i}(y)$. Note that x is not random, but it's choice does not depend on $f_{m,i}$.

The Length_A-functions are families of universal one-way hash functions. Let X^* be fixed, A a random matrix and X a random Hamiltonian cycle. A is a distance matrix of the complete directed graph G_n, and by renaming (i.e. permuting) the vertices of G_n we can compute the distance matrix A^*, such that X and X^* "are the same cycle". If we can compute a Y^* with $\text{Length}_{A^*}(X^*) = \text{Length}_{A^*}(Y^*)$, this directly leads us to a Y with $\text{Length}_A(X) = \text{Length}_A(Y)$.

4   A Pseudo-Random Generator (PRG)

A PRG uses a short, "really random" input to generate a longer, "randomly looking" bit string S. For a *cryptographic PRG* it must be infeasible to distinguish between S and a "really random" bit string S', where each bit is generated independently according to the uniform distribution. I.e. there must be no probabilistic polynomial time algorithm, to distinguish between S and S' with probability significantly greater than 0.5.

Cryptographic PRGs are highly useful for many cryptographic applications. It is straightforward to use them as (secret key) stream ciphers.

Theorem 4. *Let $l(n) > 1 + log_2((n - 1)!)$; if Length$_A$ is a one-way function, then $g(X) = $ Length$_A(X)$ is a cryptographic PRG.*

Sketch of proof: Assume there exists a polynomial time algorithm D' to distinguish between S and S' with probability $\frac{1}{2} + \frac{1}{p(n)}$, $p(n)$ a polynomial in n. Then there also exists a polynomial time algorithm D to distinguish with overwhelming probability. We will use D to find out for any B, if there is an X with Length$_A(X) = B$.

Since $l(n) > 1 + log_2((n - 1)!)$, x is unique with nonnegligible probability. For any i, j, $i \neq j$ we randomly change $a_{i,j}$. As in the proof of theorem 3 we get a new distance matrix A'. We have Length$_{A'}(X) = B$ if and only if the edge $i \rightarrow j$ is no part of the Hamiltonian cycle X. In the other case with significant probability there is no Y with Length$_{A'}(Y) = B$. Thus we can use D to find the edges of X.   □

5   On the Choice of A

Before one can apply our schemes, the coefficients $a_{i,j}$ of the matrix A must be fixed. We consider two natural ways to do this:

1. Generate n random points in a finite plane (or some higher dimensional space) and compute $a_{i,j}$ as the (e.g. Euclidean) distance between the points i and j. In order to save space one might store the coordinates of the points and compute the distances on demand.
2. Generate the $a_{i,j}$ as independent random numbers from $\{0, 2^{l(n)} - 1\}$, according to the uniform distribution.

Note that the first option leads to an undirected graph, i.e. $a_{i,j} = a_{j,i}$. Though we could cope with this, the first option is not recommendable. It is well known (cf. [3], section 37.2), that, if the triangle inequality holds for a TSP, there is a good deterministic approximation algorithm. But no such algorithm can exist for general TSPs, if NP$\neq$P. There is no obvious way to make use of the triangle inequality for solving the XTSP. Nevertheless such inherent structures in the matrix A should be considered as possible weaknesses.

The second option forces us to generate and store a large number of random bits. A very convincing way to solve the generation problem is to use the first $l(n)n^2$ bits of the binary representation of $\pi - 3$. Other mathematical constants would do as well, if the resulting bits appear to be uniformly distributed.

6   How Infeasible is the XTSP?

As outlined in the introduction, we should not trust in the infeasibility of any problem simply because it is NP-complete. The "classical" TSP minimization problem is feasible for dimensions $n = $"several thousand", see Padberg and Rinaldi [10]. This is alarming!

On the other hand there are some reasons to believe that algorithms like Padberg's and Rinaldi's—which seams to be typical for all approaches to solve large-scale TSPs—are of few help for attacks against our cryptographic schemes:

1. The triangle inequality does hold for all solved problems.
2. The number representation for distances is limited. Padberg and Rinaldi used at most 64 bits; these were floating point numbers.
3. The results are achieved by branch–and–cut or branch–and–bound algorithms. The basic branch–and–... principle can roughly be described as follows:
   - Let a solution space S be given.
   - Divide S into subsets $S_1, S_2, \ldots, S_k$.
   - Compute lower and/or upper bounds for all solutions in S_i.
   - For all S_i do:
     - If the lower bound is too large (or the upper bound too low) then discard S_i
     - else apply branch–and–... on the solution space S_i.

Clearly this is efficient if we can discard many subsets S_i at a high level of the recursion tree. In our case the target number B is computed as the length of some random Hamiltonian cycle X. Hence we can expect a nonnegligible fraction of all Hamiltonian cycles to be shorter than X and a nonnegligible fraction to be longer. Thus nearly all large subsets S_i of the solution space will contain both longer and shorter Hamiltonian cycles and we can discard almost none.

So the XTSP appears to be a variant of the TSP, where branch–and–... works exceptionally bad.

Nevertheless some further research is necessary before we can suggest "probably secure" values for the parameters n and $l(n)$. The size of these parameters is, of course, essential for the speed of our schemes and for the size of the required memory.

Much more research is necessary before we can recommend our schemes for practical use. Any effort in attacking the schemes is appreciated!

References

1. T. BARITAUD, M. CAMPANA, P. CHAUVAUD, H. GILBERT, *On the security of the Permuted Kernel Identification Scheme*, in: Proc. Crypto '92, Springer LNCS 760, 305-311.
2. P. CAMION, J. PATARIN, *The Knapsack Hash Function proposed at Crypto '89 can be broken*, in: Proc. EuroCrypt '91, Springer LNCS 547, 39-53.
3. T. H. CORMEN, C. E. LEISERSON, R. L. RIVEST, *Introduction to Algorithms*, McGraw-Hill, 1990.
4. I. DAMGÅRD, *Design Principles for Hash Functions*, in: Proc. Crypto '89, Springer LNCS 435, 416-427.
5. J. GEORGIADES, *Some Remarks on the Security of the Identification Scheme Based on Permuted Kernels*, in: J. Cryptology (1992), Vol. 5, 133-137.

6. R. IMPAGLIAZZO, M. NAOR, *Efficient Cryptographic Schemes Provably as Secure as Subset Sum*, in: Proc. FOCS '89, 236-241.
7. E. L. LAWLER, J. K. LENSTRA, A. H. G. RINNOY KAN, D. B. SHMOYS (eds.), *The Traveling Salesman Problem*, Wiley, 1985.
8. R. C. MERKLE, M. HELLMAN, *Hiding information and Signature in Trapdoor Knapsack*, in: IEEE Trans. on Inf. Theory 24 (1978), 525-530.
9. M. NAOR, M. YUNG, *Universal One Way Hash Functions and Their Cryptographic Applications*, in: Proc. STOC '89, 33-43.
10. M. PADBERG, G. RINALDI, *A Branch–and–Cut Algorithm for the Resolution of Large-Scale Symmetric Traveling Salesman Problems*, in: Siam Review 33, No. 1 (1991), 60-100.
11. J. PATARIN, P. CHAUVAUD *Improved Algorithms for the Permuted Kernel Problem*, in: Proc. Crypto '93, Springer LNCS 773, 391-402.
12. C. P. SCHNORR, M. EUCHNER, *Lattice Basis Reduction: Improved Practical Algorithms and Solving Subset Sum Problems*, in: Proc. FCT '91, 68-85.
13. A. SHAMIR, *An Identification Scheme based on Permuted Kernels*, in: Proc. Crypto '89, Springer LNCS 435, 606-609.
14. J. STERN, *A new identification scheme based on syndrome decoding*, in: Proc. Crypto '93, Springer LNCS 773, 13-20.
15. J. STERN, *Designing Identification Schemes with Keys of Short Size.* in: Proc. Crypto '94, Springer LNCS 839, 164-173.

How to Reverse Engineer an EES Device

Michael Roe

Cambridge University Computer Laboratory,
Pembroke Street, Cambridge CB2 3QG, UK
Email: mrr@cl.cam.ac.uk

1 Introduction

In April 1993, the Clinton Administration announced a new encryption system
for the protection of government and civilian telephone conversations. This proposal, called the Escrowed Encryption Standard (EES), has a number of features
which led to widespread public outcry [9, 14]:

- It contains special features to enable government officials to listen in on
  civilian telephone conversations.
- Many of the technical details of how it will work are secret.

In this paper we will not consider the moral issues of whether it is right for a
government to monitor its citizens in this way; instead, we will concentrate on
the technical means which the Clinton Administration is going to use to do it.

The details of the mathematical operations used by the EES are classified. While
this would be fine for a miliary cryptographic device, it is extremely undesirable
in a product intended for the general commercial market (e.g. one which is to
be placed inside every telephone in America):

- It is unnecessary.
- It greatly reduces public confidence in the scheme.
- The secret details are unlikely to remain secret for long.

Clearly, *something* in the system has to be secret. The goal of EES is that
government agencies will be able to tap telephone calls, but no-one else will.
To achieve this, the government must have, or know, something that no-one
else does. This something is the *Unit Key* (KU), a cryptographic key which is
different for every telephone. The basic idea is that if you know a phone's unit
key you can tap it, and if you don't you can't.

In a well-designed system, this would indeed be how it worked. Everything except
the actual value of the unit keys would be public, and it would be clear to
everyone that phones could only be tapped by someone who had obtained a unit
key through the proper channels.

Unfortunately, EES is not quite like that. As well as the unit keys, most of the technical information about how the system works is secret. This leaves a lingering doubt in people's minds that the undisclosed technical details contain a "back door" which enables phones to be tapped by someone who has not obtained proper legal authorisation.

We will describe a number of ways in which the (classified) internal workings could have been constructed so as to allow government agencies to by-pass the procedural controls, and hence decrypt EES-protected conversations without an authorising court order.

In support of the third claim (that the secret details are unlikely to remain secret for long), we will describe a number of experiments which can used to discover some of the classified internal details of an EES device.

2   Technical Overview

An overview of the EES system is shown in figure 1. Two telephone subscribers are engaged in a telephone conversation. Meanwhile, an FBI agent is intercepting the call. The possibility exists that some other person (who is not a government agent) might also have physical access to the wires that carry the telephone conversation. Briefly, the goal of EES is that the FBI agent will be able to listen in on the call, but the other attacker will not.

To prevent this other attacker from monitoring the call, the voice signal is digitised and encrypted. This encryption is carried out by a "tamper-proof" device within each telephone. Henceforth, we will refer to this "tamper-proof" encryption device as an "EES Device".

Several different models of EES device have been manufactured. The first such device, code-named "Clipper", was manufactured by Mykotronx Ltd. of Torrance, California. Unfortunately, "Clipper" was already a registered trademark of the Intergraph corporation. The Intergraph product of the same name is entirely unrelated, and has nothing to do with telephone tapping. In order to avoid perpetuating this source of confusion, we will use the term "EES device" rather than the code-name.

A block diagram of the internals of an EES device is shown in figure 2. The information used to construct this diagram came from the Escrow and Encryption Standard, FIPS 185 [16]. FIPS 185 acknowledges that the description it contains is incomplete, and asserts that "The complete specifications are classified".

Stored within the device are the unit key (KU) and the family key (KF). The unit key is different for every device, while the family key is the same in all interoperable devices.

The EES devices are used in two phases. In the first phase, the two communicating EES devices are both loaded with the same value of the session key (KS), and they agree upon an *Initialisation Vector* (IV). The session key is generated

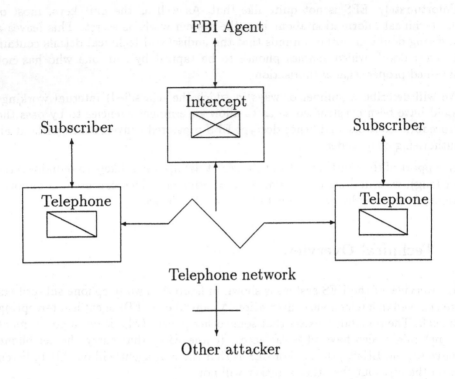

Fig. 1. Overview

by some other electronics elsewhere in the telephone; the Initialisation Vector is generated by one of the EES devices. In the second phase, the two EES devices encrypt and decrypt digitised voice using the session key and Initialisation Vector.

The procedure used in the first phase is asymmetrical between the two EES devices. A choice is made as to which of the two will generate the Initialisation Vector. We will call the device which generates the Initialisation Vector the "initiator" and the device which does not generate the Initialisation Vector key the "responder".

The "initiator" EES device is given as input an 80-bit session key (KS), and provides as output a 64-bit Initialisation Vector (IV) and a 128-bit *Law Enforcement Access Field* (LEAF). According to FIPS 185, the initiator calculates the LEAF in the following way:

1. The session key (KS) is enciphered under the unit key (KU). The mode of operation which is used for this encipherment has not been disclosed. Note, however, that it cannot be ECB mode, as ECB mode operates on 64-bit blocks and the session key is 80 bits long.
2. An Escrow Authenticator (EA) is computed. The Escrow Authenticator is

a form of cryptographic checkvalue. The size of the Escrow Authenticator, the algorithm used to compute it, and the inputs to this algorithm have not been disclosed.

3. The enciphered session key, the Device Identifier (DID) and the Escrow Authenticator are enciphered under the family key (KF) to form the LEAF. The mode of operation which is used for this encipherment has not been disclosed.

The "responder" EES device is given as input the session key (KS), the Initialisation Vector (IV), and the Law Enforcement Access Field (LEAF). The responder decrypts the LEAF value it is given, using the family key (KF) and the function f_3^{-1}. Part of this decrypted result is the Escrow Authenticator (EA). The responder computes the value the Escrow Authenticator *should* have using the function f_2. If the two values are not the same, the responder refuses to function (i.e. it will neither encrypt nor decrypt data).

Important details which are missing from FIPS 185 include:

- The size (in bits) of the Device Identifier (DID) and the Escrow Authenticator (EA).
- The functions f_1, f_2, f_3.
- The inputs to the function f_2.
- The inputs to the function f_4.

On reading FIPS 185, it is not apparent that there is any connection between the Initialisation Vector (IV) and the Law Enforcement Access Field (LEAF). Experiments with actual devices reveal that the value of the IV is used in the computation of the LEAF. As we shall explain later, this connection between the IV and the LEAF is fundamental to the operation of EES.

To complete the picture, the process used to generate the Unit Key and the Family Key is shown in figure 3. (This figure is derived from an article by Dorothy Denning [7]). The first escrow agent supplies KS_1, RS_1 ("Random Seed"), AI_1 ("Arbitrary Input") and KFC_1 ("Family Key Component") , while the second escrow agent supplies KS_2, RS_2, AI_2 and KFC_2. The small squares represent exclusive-or operations. The rectangles labelled $E(KCK)$ represent encipherment using KCK as a key. The key generation unit has two outputs (shown on the right of the figure). One of these outputs is given to each of the two escrow agents.

The process by which KU and KC_1 are generated are not shown on this figure. The "interim" key escrow system which is currently operational uses a variant of the key generation algorithm from annex C of ANSI X9.17 [7, 1]. In the next phase of the deployment of EES, a classified algorithm will be used for key generation.

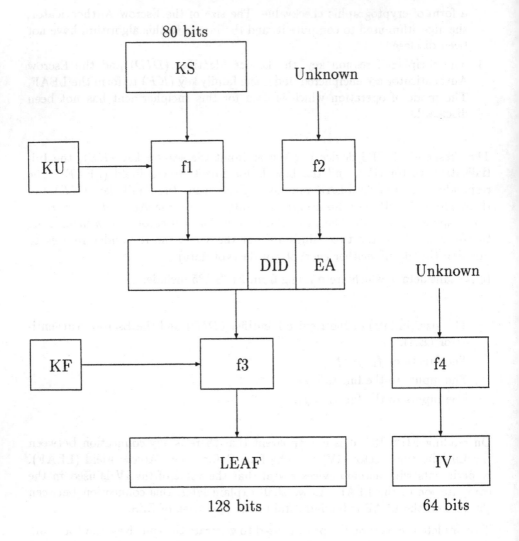

Fig. 2. Leaf Creation Method 1 (LCM1)

3 Correct use of an EES Device

An EES device does not, by itself, constitute a secure communication system. A manufacturer incorporating an EES device into a product (e.g. a telephone) must supply additional components before it will work. The design of EES makes a number of assumptions about how an EES device will be incorporated into a product, but FIPS 185 does not make these explicit. To summarise:

- Some additional means must be provided to transmit the session key (KS) between the communicating parties. The session key must be protected

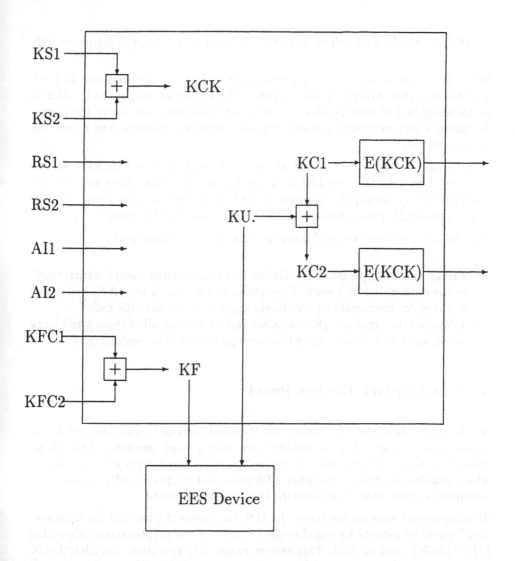

Fig. 3. EES Key Generation

against wiretapping e.g. by encrypting it. Furthermore, the EES encryption algorithm (SKIPJACK) will typically not be used for this encryption; the EES device will not encrypt any data with SKIPJACK until a session key has been exchanged, and you can't exchange a session key until you have some means of encrypting it. To break this deadlock, some other form of cryptography can be used. For example, the "Capstone" EES device provides yet another classified cryptographic algorithm, the NIST Key Exchange Algorithm (KEA) for this purpose.

- The LEAF and IV must also be transmitted between the communicating parties. For EES to work as its designers intended, these must *not* be encrypted.

If they are, the FBI will be unable to decipher an intercepted telephone call.

When it comes to incorporating cryptographic devices into products, it is often the case that Murphy's Law applies: "What can go wrong, will". This is particularly true of devices whose internals are classified, and where the person designing a system around the device is not permitted to know how to use the device correctly.

It is alleged that some of the first designers of products based on EES devices were not even told that the FBI-intercept feature was there; they were simply told that the cryptographic algorithm had a 192-bit Initialisation Vector (in fact, it has a 64-bit IV plus a 128-bit LEAF, giving a total of 192 bits).

The two most obvious ways of using an EES device will not work:

- Taking all the cryptographic variables and transmitting them (unencrypted) to the other side won't work. This results in the session key (KS) being sent in the clear, thus enabling absolutely anyone to decipher the call.
- Taking all the cryptographic variables and encrypting all of them won't work either, as it will prevent the FBI-intercept feature from operating.

4   The Skipjack Review Panel

As the internals of the EES device are classified, and as it was designed by an organisation whose activities include intercepting large numbers of telephone calls, it is reasonable to suspect the existence of some form of "back door" which enables the NSA to decipher EES-protected telephone calls without first obtaining a court order and following the proper procedures.

To allay public fears on the issue, the U.S. Government convened an "independent" panel of experts to examine the security of the cryptographic algorithm (SKIPJACK) used in EES. This review panel only examined the SKIPJACK algorithm itself [4], and did not examine the way it is used by an EES device [1]. As we will show later, even if the SKIPJACK algorithm itself is perfectly good, the way in which it is used by the EES device could introduce many security loopholes. On top of this, the results of the experiments carried out by the review panel fall far short of being a whole-hearted endorsement of SKIPJACK.

The most significant experiment carried out by the SKIPJACK review panel is the *cyclic closure test* (CCT). The objective of this test is to discover how many different possible keys there are. If the number of keys is too small, an attacker can break the algorithm by trying all possible keys. The key input to the SKIPJACK algorithm is 80 bits long, so it is clear that there are at most 2^{80} possible keys. This is a very large number, and should provide adequate security

---

[1] The review panel originally intended to produce a second report, which would examine the security of the way the algorithms were used. This report was never published.

for most purposes. However, the possibility remains that the number of different keys is much smaller than 2^{80}. For example, devices that implement the Data Encryption Standard (DES) [11] have a 64 bit key input, but only 56 of these bits are used (the remaining 8 bits are used as a parity check). Thus, DES has at most 2^{56} different keys, even though it has a 64 bit key input. If you are given a black box that implements DES, it is easy to determine that it has at most 2^{56} keys; a simple experiment will show that the values of 8 of the input bits have no effect on the output. Determining effective key size is not always so simple; the IBM Commercial Data Masking Facility (CDMF) [8] has only 2^{40} possible keys, and yet changing each of its 64 key input bits changes the output. In these more complex cases, the cyclic closure test is used to determine the true key size.

The cyclic closure test is a statistical test. You first choose a key size k, and then run the test several times to test the hypothesis that the number of different keys is at least 2^k. As this is a statistical test, it is very important that it is run multiple times. This is analogous to testing a coin for fairness; if it comes up heads once, this means nothing; if it comes up heads twice in row, this could be a co-incidence; if it keeps on coming up heads, something is seriously wrong.

The Skipjack review panel ran the CCT 1,000 times to test the hypothesis that there are at least 2^{40} distinct keys. These 1,000 samples gave a strong indication that SKIPJACK really does have at least 2^{40} keys. However, 2^{40} is not big enough to provide adequate security. Algorithms such as CDMF have 2^{40} keys, and they are known to be reasonably easy to break by exhaustive search.

The more interesting test is for the hypothesis that SKIPJACK has at least 2^{56} distinct keys. That is, does it have at least as many keys as DES, the 1970's algorithm which it is intended to replace. Unfortunately, the review panel only published the result of one run of the test for this hypothesis. As the CCT is a statistical test, a single run is almost meaningless. However, it is worth noting that for the one run they published, the measured cycle length (28,767,197) was shorter than the expected cycle length (168,216,976). If this happens consistently, then there is strong evidence that SKIPJACK has fewer than 2^{56} keys, rather than the 2^{80} which its proponents claim. In fact, the review team did run the test again, although they did not publish the second result [5]. The second time, the measured cycle length was also shorter than it should be.

5   Experimental Apparatus

The experiments described in the next section require access to an EES device. One way to obtain such a device would be to obtain an EES-protected telephone and dismantle it. The EES devices used in telephones are often of the surface-mounted "Quad flat pack" type and are soldered to the printed circuit board. This makes it troublesome to remove the device without destroying it. However, most university electronics laboratories will have equipment for doing this.

An alternative means of obtaining an EES device is to use the identification smart-cards (code-name "Tessera" [2]) used in the U.S. Defense Message System. The Tessera card has the great advantage that it is designed to be plugged into a computer (via the PCMCIA socket on an IBM-compatible laptop). This means that it can be used to perform the experiments described in this section without needing the use of a soldering iron.

There are some significant differences between the facilities provided by a telephone-oriented EES device such as "Clipper" and a computer-oriented EES device such as the Tessera card:

1. Some of the devices intended for use in telephones only support one cryptographic mode of operation (typically, OFB mode). The Tessera card supports all the FIPS 81 modes [12], including Electronic Codebook Mode.
2. The Tessera card imposes an additional layer of key management protocols on top of EES. Functionally, the Tessera card can be regarded as two different devices on one chip: an EES device, and another processor which mediates all accesses to the EES device. It has been alleged that this second processor is based on an ARM 600 series macrocell. No EES device lets the user choose the Initialisation Vector; the Tessera card does not let the user choose (or even know) the session key either.

   Tessera provides two basic means of setting up a session key: a shared key can be established between two cards using the NIST KEA algorithm, or a previously saved session key can be reloaded. The operation to reload a key takes as input an 80 bit key and 16 bits of redundancy, all encrypted under a master key which is unique to the device. The Tessera card will only reload a key if the 16 bits of redundancy take on the correct value. Furthermore, the device master key is different for each card, and is not revealed to the owner of the card. Note that this makes it impossible for the owner of the device to either choose or to determine the value of a session key.

After I had started work on this paper, the Tessera cards were withdrawn and replaced with a new computer-based EES device, which was code-named "Fortezza". The change in name was because "Tessera" (like "Clipper") was already a registered trademark of another company. The Fortezza card also incorporated a technical change which was intended to prevent the LEAF-forging attack described by Matt Blaze [3]. After it has rejected several false LEAFs in succession, the Fortezza card goes into a state in which it does nothing until it is reset.

This modification does not actually prevent the LEAF-forging attack: it merely makes it slower. For example, the LEAF-forger can just reset the card each time it goes into this state. However, this is probably enough to prevent people from using LEAF-forging as a practical way to defeat the key escrow system. It was doubtful whether anyone would do this with the Tessera card, as 40 minutes is

---

[2] *Tessera* is the Latin word for a clay tile. Such tiles were used as primitive identity cards in the Roman empire.

a long time to wait every time you send a message. The Fortezza card makes LEAF-forging take even longer than 40 minutes, and hence it is very unlikely that anyone would use this a means to defeat the escrow system.

However, there are other reasons for creating forged LEAFs. Some of the experiments described in this paper depend upon the ability to create forged LEAFs. These experiments are still possible with the Fortezza card, but take longer. They could be redesigned to take account of the different performance characteristics of the Fortezza card. For example, in many cases there are two possible experiments, one which involves generating a large number of real LEAFs, and another which involves verifying a large number of fake LEAFs. With the Tessera card, there is no reason to choose one of these experiments over the other. With the Fortezza card, experiments which involve generating LEAFs are much quicker than experiments which involve verifying fake LEAFs.

6 Experiments

6.1 Re-Run Cyclic Closure Test

Rationale As described in the previous section, the SKIPJACK review panel decided to stop running the cyclic closure test on the SKIPJACK algorithm just as the results began to look interesting (i.e. began to provide evidence of a weakness in the algorithm). It would be interesting to continue the test, to find out if the results obtained by the SKIPJACK review panel were a statistical fluke or a real sign of weakness.

The cyclic closure test is a "black box" test; it does not require use of any information about the internals of the algorithm under test. Thus it is often possible to perform the cyclic closure test using only a hardware implementation of a secret algorithm. Unfortunately, it is not possible to perform this test using either the Tessera card or the telephone-oriented EES devices.

This experiment (or rather, non-experiment!) is included for two reasons:

- The fact that this experiment can't be performed is very significant. If the Skipjack algorithm was actually a weak algorithm, it would be very hard for users to discover this.
- This experiment might be possible with EES devices other than the Tessera card or Clipper.

Method The cyclic closure test involves finding a cycle in the following iteration:

$$x_{i+1} = E(d, x_i)$$

Where $E(d, k)$ is the result of encrypting d with key k in Electronic Codebook Mode, and d is a data block chosen at random.

The telephone-oriented EES devices *cannot* be used for this test, as they are only capable of OFB-mode encipherment with an internally generated IV. A device which supported OFB-mode with an externally-supplied IV could still be used for the cyclic closure test (simply keep the IV constant). However, the telephone-oriented devices generate a new IV at random each time they are loaded with a new key. This ruins the cyclic closure test, as the ever-changing IV will prevent the iteration from settling down into a repeating cycle.

The Tessera card also cannot be used for this test, as it does not allow the user to choose a session key. It is interesting to note that this test would be possible if the Tessera card did not put any redundancy in saved session keys. The test would be run as follows: encrypt d under the current session key, pretend that the result is a saved session key, "reload" it, and repeat. This gives the following iteration:

$$x_{i+1} = E(d, D(x_i, K_M))$$

Where K_M is the (unknown) device master key. Note that encryption and decryption are both reversible operations, so decrypting a key with K_M permutes the key space without changing its size. Thus, signs of weakness found with this modified iteration are just as valid as those found using the standard cyclic closure test.

However, the 16 bits of redundancy prevent us from using this trick; the device will usually detect that the input to the reload operation is not a genuine saved key (it will be fooled one time in 2^{16}, but this isn't often enough to be useful for the CCT).

Resources Needed This estimate of resources is somewhat hypothetical, as this experiment isn't possible for the reasons outlined above. However, if an EES device without these restrictions was obtained, the resources required would be approximately as follows.

Assuming that the true key size is 56 bits (rather than 80 as is claimed) the cyclic closure test will take on average 2×2^{28} iterations to find a cycle. Assuming that each key change takes 50ms (the Tessera card is rather slow at changing keys) then the cyclic closure test will take 310 days. To give statistically significant results, this test must be run at least three times. Thus, the total resources needed to perform this test are about 3 machine-years.

Note that if the true key size is greater than 56 bits, this test will not need any more time. If a cycle is not found after the expected number of iterations, then the run should be terminated. If all three runs are terminated without finding a cycle, then it is possible to reject the hypothesis that the the key size is 2^{56} or smaller.

Opportunities for parallelism Clearly, each of the runs of the cyclic closure test can be run in parallel. Hence this experiment could be performed in one year using three devices rather than three years using one device. If a very large number of devices are available, and it is desired to perform the test quickly, then it is possible to use alternative algorithms which permit a greater degree of parallelism. Some algorithms of this type are described in a paper by Michael Wiener [17].

6.2   Search for Equivalent Keys

Rationale As the cyclic closure test cannot be performed with the Tessera card, it is necessary to use some other test for effective key size. The test described in this section tests the size of the key space by searching for pairs of keys which have the same effect on all plaintexts. If the true key size is less than the claimed 2^{80} there will be many such pairs. Finding even one such pair would reveal interesting information about the internal structure of the SKIPJACK algorithm (analagous to the existence of weak keys in DES [10]). Finding many such pairs would be proof of a serious weakness in the algorithm.

This test is not quite as satisfactory as the cyclic closure test, for at least the following reasons:

– It requires at least $O(2^{28})$ 64-bit words of disc space. The cyclic closure test needs only a very small abqut of disc space and memory.
– We have a reason to believe that SKIPJACK will fail the cyclic closure test (the results of the Skipjack review panel). We have no reason to suppose that this alternative test will be as good at detecting SKIPJACK's weaknesses.

Method Repeat the following at least 2^{28} times:

1. Get the Tessera card to generate a random session key. (It can do this, even though it won't reveal the value of the key it has generated).
2. Use the "save key" transaction to save this key (encrypted under the device master key!) to external storage.
3. Encrypt a fixed test pattern (0, say) in ECB mode using the session key, and save the result.

The results of encrypting 0 under different keys can then be sorted in order to find matches. A match can be caused in any of the following three ways:

1. The Tessera card generated the same session key on more than one occasion. This case can be recognised by comparing the saved session keys (which are in enciphered form). If the two session keys are equal, then their enciphered forms will also be equal, as the enciphered form does not contain any randomness to prevent comparison.

2. The two keys have different effects in general, but happen to give the same result when enciphering 0.

   This case can be recognised by reloading the saved session keys, and using them to encipher other values (e.g. 1, 2, ...) The existence of such pairs of keys is to be expected, and is not a sign of weakness in the algorithm.

3. The two keys have the same effect on all (or nearly all) inputs.

   This case can be recognised by re-loading the saved session keys, and using them to encipher some randomly-chosen values. If the two keys have the same effect on a large number of randomly-chosen inputs, then this is strong evidence that they have the same effect on most inputs. If this case is encountered, it is a sign of weakness in the algorithm.

6.3 Check that IV is used as an Initialisation Vector

Rationale This experiment confirms that the "IV" parameter which is exchanged between EES devices is the same as the Initialisation Vector that the device actually uses when it encrypts or decrypts. An alternative possibility would be that "IV" is the Initialisation Vector enciphered under a key common to all EES devices.

Method One EES device (the "initiator") is used to generate an IV and LEAF for session key k_1. The initiating device is then used to encipher a sample message in OFB mode. A second EES device (the "responder") is used to decipher this message twice, in two different ways:

- Firstly, the responder is put into OFB mode and used to decipher the message.
- Secondly, the responder is put into ECB mode. The responder is used to decipher the message, with the OFB mode chaining being implemented in software external to the device. This is possible because OFB mode is defined in terms of invocations of a "black box" which provides ECB mode.

If the "IV" input to the device was not the same as the Initialisation Vector used by the device in OFB-mode chaining, then these two decipherments would give different answers.

Results This experiment has been carried out with a Tessera card by Matt Blaze [2]; both decipherments are the same. A similar experiment using CBC mode instead of OFB mode also showed that on-chip chaining and software chaining gave the same answer.

Conclusions The "IV" input really is the Initialisation Vector, and it is not enciphered. Furthermore, we can conclude that when OFB mode is selected, the device really does encipher and decipher in OFB mode (rather than some other keystream mode).

Further Remarks We can also conclude that the mode of operation is not one of the variables which are used to compute the Escrow Authenticator (EA). If the mode of operation were an input to EA, it would be impossible to use a LEAF generated by an initiator in one mode with a responder in a different mode.

However, even if the mode of operation was an input to the Escrow Authenticator, it would still be possible to perform this experiment. The reason for this is that the Initialisation Vector has no effect in ECB mode. A modified version of this experiment would involve two LEAFs, one (L_1) for OFB mode, session key k_1 and Initialisation Vector V_1, and the other (L_2) for ECB mode, session key k_1 and Initialisation Vector V_2. As the device chooses its own IV when generating a LEAF, $V_1 \neq V_2$. The software emulation of OFB mode would input k_1, L_2 and V_2 to the EES device, but would use V_1 as the real Initialisation Vector in the software emulation of chaining.

6.4 Determine whether device checks session key equals enciphered session key

Rationale A "responding" EES device is supplied with the session key via two different paths:

- Via a key management mechanism external to the device. We shall call this the "clear text session key".
- Via the LEAF, which contains the session key enciphered under the initiator's unit key and the shared family key. We shall call this the "enciphered session key".

Given these inputs, a device might do one of the following things:

- Use the clear text session key, and ignore the enciphered session key.
- Use the enciphered session key, and ignore the clear text session key.
- Check that the two versions of the key are equal, and signal an error condition if they are not equal.

If there is no "back door" in the key escrow mechanism, the responding device should not be able to use either the second or the third of these possible methods. The Unit Key is supposed to be secret and unique to every device. Hence the responding EES device should not know the unit key of the initiating device, and so should not be able to extract the session key from the LEAF.

However, the question still remains. It is theoretically possible that some of the claims made about the device are false, that in fact there is no unit key, and the responding device does check that the two values of the key are equal. It is desirable to have an experiment to show that the device does not perform this check.

Method The first EES device (the "initiator") is used to generate an IV and LEAF for a particular session key (k_1). This IV and LEAF are loaded into a second EES device (the "responder"), but with a different session key (k_2). A data block of 64 bits is encrypted using the initiating device, and the resulting ciphertext is decrypted using the responding device. For the purposes of this experiment, it does not matter which of the supported modes of operation is used for this encryption and decryption (although initiator and responder must use the same mode).

Results Most of the time, the responding device will detect that the session key has been modified. The Tessera card reveals that it has detected the tampering by refusing to perform encipherment or decipherment with the modified session key.

Some of the time (1 in 2^{16} for the Tessera card [3]) the responding device accepts the modified session key. In these cases it is possible to continue the experiment and to attempt to decipher the block of ciphertext. The result of this decipherment is nearly always different from the original input plaintext.

Conclusions It is possible to reject the hypothesis that the responding device only uses the enciphered session key. If the responder only used the enciphered session key, it would decrypt the ciphertext block with k_1, and hence recover the original plaintext. Instead, the responder decrypts the ciphertext block with k_2; as this ciphertext was produced by enciphering with k_1 (not k_2) this results in a block which is almost always different from the original input.

The observed behaviour can be explained by any of the following hypotheses:

- The responder does not check that the session keys are equal; it only checks the value of the Escrow Authenticator (EA).
- It does check that the session keys are equal, but never acts upon the result. This is equivalent to the above!
- The responder extracts the session key from the LEAF, computes a hash of the key which is *different* from the Escrow Authenticator (EA), and uses this other hash to check if the two keys are equal. While this is theoretically possible, it makes no sense!

Given the above experiment, and the claim that the unit keys are unique to each device, it seems likely that the responder does not check the encrypted session key in the LEAF. Furthermore, we can conclude that the size of the Escrow Authenticator is at least 16 bits. If the Escrow Authenticator was smaller, then modified session keys would be accepted more frequently.

Further Remarks It is theoretically possible that the Escrow Authenticator is larger than 16 bits, but the responding device only checks 16 bits of it. Why would the device be designed to do this? One possible reason is as follows: by rejecting LEAFs with a bad EA, the responding device is revealing valuable information about the secret family key, KF. An attacker who does not know KF can still use a responding EES device as an oracle to provide answers to certain questions about the internals of a LEAF. Indeed, most of the experiments described in this paper are different methods of using a responder as an oracle. To prevent this type of attack on the family key, it would be better if a responding EES device always accepted a LEAF, regardless of the value of EA. However, if responding devices never checked EA, then it would be very easy to by-pass the key escrow mechanism (e.g. by not transmitting the LEAF at all).

A compromise between these positions would be for the responding devices to only check part of the EA. The devices would check enough of the EA to make it hard to bypass the key escrow system, but not enough of it to give an attacker perfect information about the LEAF.

How big should the LEAF be, and how many bits of it should be checked? Enough bits should be checked to make it inconvenient to forge an acceptable LEAF. If t is the time taken for a device to generate a single LEAF, and k is the number of bits that are checked, $t \times 2^k$ should be an inconveniently long time. A value of 50 ms for t and 16 for k results in LEAF forgery requiring 55 minutes, which is certainly inconvenient.

The Escrow Authenticator should contain enough extra bits to make it reasonably likely that a forged LEAF will not be a perfect forgery. However, the EA cannot be made very large because there is limited amount of space in the LEAF (128 bits). A LEAF size of 18 bits (with 16 of them checked) would mean that a single forged LEAF has a 75% chance of not being perfect. An 18-bit EA would ensure that anyone who forges LEAFs regularly will almost certainly create some imperfect forgeries.

If the responding devices do not check these extra bits, does it matter that they are there? It does, because there might be specialised equipment (such as the FBI intercept processors) that does check the additional bits. This would enable government officials (who have the special equipment) to detect attempts at LEAF forgery. Note that for this to work, the extra two bits do not even need to be the output of a strong hash function; it will even work if they are always zero. This would allow this function to be concealed even from people who have access to the (classified) description of the internals of the device; the specification could say "these two bits always zero (reserved for future use)" without arousing suspicion as to their true purpose. Indeed, it would be sufficient to make the serial number two bits longer, and to arrange that no EES device is ever manufactured with the top two bits of its serial number non-zero. Then, if an intercept processor sees a LEAF containing a serial number with either of the top two bits set, it has strong evidence that the LEAF has been forged.

It is worth noting that in Dorothy Denning's original description of Clipper [6], the serial number is described as being 30 bits long. As the LEAF is 128 bits long, and the encrypted session key is 80 bits long, this would leave 18 bits for the Escrow Authenticator (EA). Subsequent descriptions of Clipper take care to avoid mentioning the length of EA [16].

6.5 Determine whether the session key affects EA

Method The first EES device (the "initiator") is used to generate an IV and LEAF for a particular session key (k_1). This IV and LEAF are loaded into a second EES device (the "responder"), but with a different session key (k_2).

Results This experiment was first performed by Matt Blaze [3]. Most of the time, the responding device will refuse to encrypt using the modified session key. Given that the responder does not make use of the session key inside the LEAF, this means that the value of the session key affects the value of EA.

6.6 Determine whether the IV affects EA

Method The first EES device (the "initiator") is used to generate an IV and LEAF for a particular session key (k_1). This session key and LEAF are loaded into a second EES device (the "responder"), with a different IV.

Results Most of the time, the responding device will refuse to encrypt using the modified IV.

Conclusions There are two possible explanations for this behaviour:

1. The value of the IV is an input to the function (f_2) used to compute EA.
   If the IV is modified, the responder will compute a different value of the Escrow Authenticator from that contained in the LEAF, and so will reject the LEAF.
2. The IV is an input to the function (f_3) used to encipher the LEAF. That is, IV is used as an Initialisation Vector for two different encipherments: the encipherment of the user-supplied data with the session key KS, and the encipherment of the LEAF with the family key KF.
   If the IV is modified, the responder will incorrectly decipher the LEAF, the EA contained within the incorrectly deciphered LEAF will not match the recomputed EA, and so the responder will reject the LEAF.

Further Remarks Encrypting two different messages with the same key and the same IV is usually a bad idea, as it makes some additional cryptanalytical attacks possible. However, what may be happening with EES is that two different messages (the LEAF and the user data) are enciphered with the same IV but *different* keys. This is certainly unusual, but does not seem to be cryptographically weak.

6.7   Determine whether f_2 is a checksum

Rationale If the function f_2 (used to compute EA) was one of the commonly-used error detecting codes (such as a simple checksum, a CRC, or Fletcher's algorithm), then it would be possible to produce false LEAF values in a much more efficient way that that proposed by Matt Blaze [3]. Hence, it is likely that the designers of EES made the function f_2 a cryptographically strong hash function. This experiment determines whether or not this is the case.

Method Use an "initiating" EES device to produce a LEAF (L_1) and IV (V_1) for key k_1. For each of the common error detecting codes, calculate values of the key (k_2) and IV (V_2) which give the same error detection code as k_1 and V_1. For some codes, doing this requires knowledge of the order in which k and V occur in the input to f_2; simply try all possible orderings.

Are the modified key and IV, together with the original LEAF, accepted by a "responding" EES device? If they are, repeat the experiment with different values of k_1. If the fake k_2 and V_2 are always accepted, then the function f_2 has been found.

Further Remarks With the Tessera card, it is impossible for the user to select the session key, so this experiment is restricted to changing the Initialisation Vector. Furthermore, if the Initialisation Vector is also used as the IV in the decipherment of the LEAF (as suggested in section 6.6) then it will not be possible to derive useful information about f_2 by changing IV. This experiment is only really informative with EES devices that allow the user to select a session key.

6.8   Determine whether DID or encrypted session keys affects EA

Method With a fixed value of the session key (k_1) and a fixed value of the IV (V_1), try random values of the LEAF until two LEAF values (L_1 and L_2) are found that will be accepted by a responding device. Then the following two equations hold:

$$f_3^{-1}(L_1) = (E_1', D_1', f_2(k_1, V_1, E_1', D_1', \ldots))$$

$$f_3^{-1}(L_2) = (E_2', D_2', f_2(k_1, V_1, E_2', D_2', \dots))$$

One in 2^{16} LEAF values will be accepted, so this can be done in a few hours. It is highly likely that these two fake LEAF values will contain different encrypted session keys. That is, $E_1' \neq E_2'$.

With one fake LEAF (say L_1), keep the IV fixed at V_1 and find other values of the session key that will be accepted by a responding device. Call these k_i'.

$$\forall i : f_3^{-1}(L_1) = (E_1', D_1', f_2(k_i', V_1, E_1', D_1', \dots))$$

Check to see whether these k_i are also accepted by a device given the other fake LEAF. If they are always accepted, then the following equation holds:

$$\forall i : f_3^{-1}(L_2) = (E_2', D_1', f_2(k_i', V_1, E_2', D_2', \dots))$$

That is, changing the values of the Device ID (DID) and the encrypted session key from (E_1', D_1') to (E_1', D_2') does not change the set of keys (k') for which the value of f_2 is constant.

If f_2 were a simple function, this fact would not allow us to draw any conclusions about its inputs. For example, f_2 might be a checksum of all the inputs (including E and D), with $E_1' + D_1' = E_2' + D_2'$. However, if f_2 is a collision-free hash function, it would follow that E and D could not be inputs to f_2!

Note that this experiment will still work even if the IV is used in the decipherment of the LEAF (as suggested in section 6.6).

6.9   Determine whether encrypted session keys affects EA

Method With two different session keys (k_1 and k_2), use an initiating device to generate (IV, LEAF) pairs until an IV collision is found. That is, we have two LEAFs generated by the same device (with the same DID) with the same IV but different session keys.

Repeat this process until many such pairs are found. Eventually, a pair will be found where k_1 is accepted with the second LEAF as well as the first. Call these two LEAFs L_1 and L_2.

Then find session keys k_i such that k_i is accepted with L_1 and V_1. If these keys are also accepted with L_2 and V_2, then the enciphered session key is probably not used in the computation of EA. If these keys are usually rejected, then the enciphered session key is used in the computation of EA.

Note that this experiment will still work even if the IV is used in the decipherment of the LEAF (as suggested in section 6.6).

Resource Requirements Assuming that IVs are randomly generated, about 2^{32} IVs will need to be generated before an IV collision is found. If it takes significantly fewer or more attempts to find a collision, this is very interesting:

- If it takes longer, then the initiating device must have some memory; perhaps the IV is the result of encrypting a counter with a key stored inside the device. If this is the case, then this experiment cannot be completed.
- If a collision is found sooner, then the IV is not as random as it should be. Perhaps a few bits of the IV are being used to leak part of the session key (or even the unit key!) to an eavesdropper? This would be very bad!

Assume that it does indeed take 2^{32} attempts to find a collision. One in 2^{16} of these collisions will have the additional property that k_1 is accepted with both IVs. Thus the total number of trials needed is less than 2^{48}. (It is less because there are economies of scale when searching for multiple collisions).

This is undoubtedly an expensive and time-consuming experiment. However, it only needs to be done once *for the algorithm*. The discussions of DES breaking machines usually hypothesise an attacker being prepared to do an $O(2^{56})$ search for each key she wishes to break. Here, it is only necessary to do an $O(2^{48})$ search *once*; this will reveal a (classified) fact about every EES device in existence.

7   Other Observations

As far as we know, there has never been a good description of the internals of an EES device in the unclassified literature. Attempts to reverse engineer these devices have led to a better description of their internal workings than has previously been available. This in turn led to some new observations on the security of the scheme [13].

7.1   Forgery

The telephone-oriented EES devices only support Output Feedback Mode. This mode does not provide integrity or authentication, although someone who had not examined the scheme in detail might be fooled into thinking that it does. The fact that OFB mode is unsuitable for integrity or authentication has been known for at least ten years [15]. The new observation is that a workable key escrow system would need to have authentication as well as confidentiality.

Consider the following scenario. A person is on trial for a criminal offence, and the only evidence for the prosecution is an intercept of a telephone call that was protected by an EES device. The prosecution shows that the ciphertext can be decrypted into intelligible speech by an intercept processor loaded with the escrowed copy of the defendant's telephone's unit key. The content of the telephone conversation is clearly incriminating. However, the defendant denies that they made the telephone call. The compression algorithms used in EES telephones distort speech, so it is hard to recognise the voice. The prosecution argues that since only the defendant's escrowed key will decipher the call, it must be the case that it was made from the defendant's telephone.

The defence argues that the call has been forged in the following manner. Given the ciphertext (from wiretapping) and the plaintext (either as the output from an intercept processor, or from a bug in the same room as the telephone) it is possible to exclusive-or them together to recover the key stream. This key stream can then be exclusive-ored with an entirely different plaintext to produce a forged ciphertext. This forged ciphertext will be converted into the forged plaintext when fed into the intercept processor.

Based on this argument, juries may well be persuaded to disregard evidence from EES intercepted telephone calls. As the stated purpose of EES is to aid law-enforcement, this is a serious blow to its credibility.

Alternatively, all this "cryptographic magic" may convince a jury that EES provides absolute proof that a person made a particular call. This carries a grave risk of innocent people being convicted on the basis of falsified evidence.

Note that access to the escrowed keys is *not* needed to make a forgery. Anyone with access to the output from an intercept processor can make such a forgery, regardless of whatever physical and procedural controls are used to restrict access to the actual keys.

7.2   Masquerade

Another attack on the EES protocols has been found by Moti Yung and Yair Frankel. Their attack has some similarities with the forgery attack described in the previous section, in that it shows the need for authentication as well as confidentiality in an escrow system. Their attack proceeds as follows. Suppose Alice initiates a conversation with Bob, and Bob later initiates a conversation with Chris. If Bob gets to choose his session key (KS) with Chris, then Bob can re-use the session key and IV he shared with Alice. Furthermore, Bob can give Chris a copy of Alice's LEAF, rather than a new LEAF created by Bob. This has two unfortunate consequences:

- If they only have a wiretap warrant for Bob (and not Alice or Chris), the FBI may experience a certain amount of difficulty in tapping the call between Bob and Alice; the enciphered session key inside the LEAF Bob sends will not be decipherable with Bob's escrowed unit key.
- Bob may be able to incriminate Alice. Bob (knowing that Chris's phone is tapped) could say something incriminating in his call to Chris. When the FBI decipher the LEAF Bob sent (to determine who called Chris), they will find Alice's Device Identifier. Furthermore, when they have obtained a wiretap warrant for Alice and an escrowed copy of Alice's unit key, they will discover

that decipherment of the contents of the LEAF with Alice's unit key yields the session key, and that decipherment of the message with this session key yields the incriminating plaintext. This can then be used to convince a jury that Alice is guilty.

- Alternatively, Alice actually makes the incriminating telephone calls to Chris. When Alice is caught and prosecuted, Alice's lawyer claims that Alice is an innocent victim of an attack carried out by Bob (as described above).

While the forgery attack can be prevented by using a different mode of operation, this attack works against all modes. However, the key management system used to exchange session keys can be chosen so as to prevent this attack (e.g. by preventing either party from choosing what the session key will be). The Escrowed Encryption Standard (FIPS 185) makes no mention of how session keys are to be exchanged, so a manufacturer can make a product which is vulnerable to this attack whilst still being compliant with FIPS 185.

7.3   The Role of the IV

Suppose that Alice and Bob frequently talk to each other using EES telephones, and that their conversations are reasonably short (a few minutes, say). Furthermore, suppose that Alice wishes to defeat the key escrow mechanism without Bob knowing. To do this, Alice can build a modified EES telephone that uses Matt Blaze's LEAF-forging attack [3]. This attack has the disadvantage that it takes Alice a long time to forge a LEAF for each call — 40 minutes with a Tessera card, and even longer with a Fortezza card.

Alice might try to optimise this attack by computing a forged LEAF once, and re-using this forged LEAF for each of her telephone conversations with Bob. What are the problems with this optimisation?

If she uses the same key too often, Alice increases her vulnerability to attacks based on differential or linear cryptanalysis. However, if SKIPJACK is safe against these attacks for one long phone call, then it is also safe against these attacks for a series of short phone calls with the same key. So this isn't a problem.

It's also true that if this re-used session key is ever compromised (e.g. by physical means, such as a bug inside Bob's telephone), then *all* of Alice and Bob's conversations are compromised. However, it is a design assumption of EES that the FBI can no longer afford to do this sort of close-in monitoring, and so Alice feels safe from this too.

What really stops Alice from using this optimisation is that the Initialisation Vector affects the value of the LEAF. To re-use a forged LEAF, Alice must re-use the Initialisation Vector as well as the session key, as these are both cryptographically bound to the LEAF. Re-using the same key and IV renders Alice vulnerable to easy cryptanalytical attacks, particularly as the telephones use

OFB mode for encipherment. The exclusive-or of two ciphertext conversations will be the exclusive-or of the plaintexts (the keystream cancels out), and there is enough redundancy in digitised speech to enable both plaintexts to be recovered from their exclusive-or.

Note that this is very dependent on the mode of operation that is used. If CBC mode was used instead of OFB mode, Alice could successfully re-use a LEAF. To do this, Alice would tell Bob the same IV each time, but would actually use a different IV. As CBC mode is resynchronising, this works; Bob hears a few milliseconds of noise at the start of a call (too little to arouse suspicion), and then the call proceeds normally. OFB mode is not resynchronising. If Alice tries this trick with OFB mode, Bob will hear nothing but static.

Hence, we observe that tying the IV to the LEAF increases the security of the system against attempts to use unescrowed keys; but that this additional protection is only obtained with some modes of operation (e.g. OFB mode). However, OFB mode was a bad choice of mode for reasons explained in the previous section. We conclude that a version of EES which worked (in so far as such a thing is possible at all!) would have had to use a mode of operation which supports non-repudiation and isn't resynchronising.

Acknowledgements

I would like to thank Steve Kent for explaining the physical tamper-proofing used by the Clipper Chip; Dorothy Denning for providing additional information on the experiments carried out by the SKIPJACK Review Panel; Matt Blaze for providing additional information on his experiments with the Tessera card; and Robert Morris for practical advice.

References

1. American Bankers' Association. *ANSI X9.17-1985: Financial Institution Key Management (Wholesale)*, 1985.
2. M. Blaze. Personal communication, December 1994.
3. M. Blaze. Protocol failure in the escrowed encryption standard. In *Second ACM Conference on Computer and Communications Security*, pages 59 – 67, November 1994.
4. E. F. Brickell, D. E. Denning, S. T. Kent, D. P. Maher, and W. Tuchman. *SKIP-JACK Review — Interim Report — The SKIPJACK Algorithm*, June 1993.
5. D. E. Denning. Personal communication, August 1993.
6. D. E. Denning. *The Clipper Chip : A Technical Summary*, April 1993.
7. D. E. Denning and M. Smid. Key escrowing today. *IEEE Communications Magazine*, 32(9), September 1994.
8. D. Johnson, S. Matyas, A. Le, and J. Wilkins. Design of the commercial data masking facility data privacy algorithm. In *Proceedings of the first ACM Conference on Computer and Communications Security*, November 1993.

9. S. Landau. *Codes, Keys and Conflicts: Issues in U.S. Crypto Policy.* ACM U.S. Public Policy Committee, June 1994.

10. C. H. Meyer and S. M. Matyas. *Cryptography: a new dimension in computer data security.* John Wiley and Sons, 1982.

11. National Bureau of Standards. *Federal Information Processing Standard — Publication 46: Data Encryption Standard,* 1977.

12. National Bureau of Standards. *Federal Information Processing Standard — Publication 81: DES Modes of Operation,* 1977.

13. M. Roe and M. Lomas. Forging a clipper message. *Communications of the ACM,* 37(12):12, December 1994.

14. William Safire. Sink the clipper chip. In *New York Times,* 14th February 1994.

15. (UK) Government Communications Headquarters. *Review of DEA-1,* October 1985.

16. (U.S.) National Institute of Standards and Technology. *Federal Information Processing Standards Publication 185 — Specifications for the Escrowed Encryption Standard,* February 1994.

17. Michael Wiener and Paul van Oorschot. Parallel collision search with application to hash functions and discrete logarithms. In *Second ACM Conference on Computer and Communications Security,* pages 210 – 217, November 1994.

A Fast Homophonic Coding Algorithm Based on Arithmetic Coding

W T Penzhorn

Department of Electrical and Electronic Engineering
University of Pretoria, 0002 Pretoria, South Africa
e-mail: walter.penzhorn@ee.up.ac.za

Abstract. We present a practical algorithm for the homophonic coding of a message source, as required for cryptographic purposes. The purpose of homophonic coding is to transform the output of a non-uniformly distributed message source into a sequence of uniformly distributed symbols. This is achieved by randomly mapping each source symbol into one of a set of homophones. The selected homophones are then encoded by means of arithmetic coding, after which they can be encrypted with a suitable cryptographic algorithm. The advantage of homophonic coding above source coding is that source coding merely protects against a ciphertext-only attack, whereas homophonic coding provides additional protection against known-plaintext and chosen-plaintext attacks. This paper introduces a fast algorithm for homophonic coding based on arithmetic coding, termed the *shift-and-add* algorithm, which makes use of the fact that the set of homophones are chosen according to a dyadic probability distribution. This leads to a particularly simple, efficient implementation, requiring no multiplications but only shifts and additions. The usefulness of the algorithm is demonstrated by the homophonic coding of an ASCII textfile. The simulation results show that homophonic coding increases the entropy by less than 2 bits per symbol, and also provides source encoding (data compression).

1 Introduction

The history of cryptology shows that most secret-key cipher systems were broken by exploiting the fact that plaintext characters are not uniformly distributed. The technique of *homophonic substitution*, or *homophonic coding*, (sometimes also called *multiple substitution*) is a well-known method for converting a given plaintext sequence into a (more) random sequence. Such a coding scheme was, for example, used by the Duke of Mantua in his correspondence with Simeone de Crema and was also used in the well-known Beale ciphers [6].

In *simple substitution* each plaintext source symbol is replaced with a corresponding codeword by means of a *one-to-one* mapping. *Homophonic coding* is similar, except that the mapping is *one-to-many*, as each source symbol is mapped into a *set of codewords* referred to as *homophones*. The term *homophonic* means to *"sound the same"* which indicates that different codewords refer to the same source symbol. Source symbols which occur frequently are mapped into more than one codeword, so as to flatten the frequency distribution of the resulting codewords. An important attribute of homophonic coding is that in the one-to-many mapping, a codeword is picked at *random* from the set of homophones to represent the given source symbol, thereby introducing external randomness into a given message.

2   Illustration of Homophonic Coding Based on Huffman Source Coding

To illustrate the main idea behind homophonic coding, consider the example in Fig. 1, due to Massey [9] and Günther [4]. The latter introduced *variable-length homophonic coding*, which was subsequently put into a more general framework in [5]. In this example, the homophonic coder consists of the following three elements: a memoryless binary message source (BMS), a homophonic channel and a binary Huffman source encoder. The message source U produces symbols from the two-letter alphabet $U = \{u_1, u_2\}$, with probability $P(u_1) = P(U_i = u_1) = 1/4$ and $P(u_2) = P(U_i = u_2) = 3/4$.

The (memoryless) homophonic coding channel substitutes the source symbols U_i for the homophones taken from the alphabet $V = \{v_1, v_2, v_3\}$. It is characterized by the set of transition probabilities $P(V = v_j | U = u_i)$ such that for each j there is exactly one i such that $P(V = v_j | U = u_i) \neq 0$. Those v_j for which $P(V = v_j | U = u_i) \neq 0$ are the *homophones* for u_i. For the example in Fig. 1 we have $P(v_1|u_1) = 1$, $P(v_2|u_2) = 2/3$ and $P(v_3|u_2) = 1/3$ from which the probability of occurrence of the homophones readily follows as $P(v_1) = 1/4$, $P(v_2) = 1/2$ and $P(v_3) = 1/4$. Finally, the selected homophones are encoded as binary codewords in the alphabet $X = \{x_1, x_2\}$ by means of Huffman source coding.

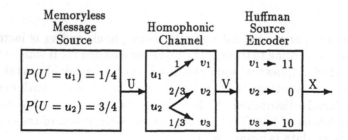

Fig. 1. Variable-length homophonic coder with Huffman source coding.

The operation of the homophonic coder is as follows: When the message source produces the symbol u_1, the homophonic channel outputs the codeword v_1, and

the resulting binary codeword produced by the Huffman coder is $x_1 = 11$. Note that this is a deterministic mapping, since u_1 is always represented as v_1. However, source symbol u_2 is represented as either v_2 or v_3, selected randomly: v_2 is chosen with probability $2/3$, and v_3 with probability $1/3$, as indicated in Fig. 1. Note that the resulting binary codewords produced by the Huffman coder are of unequal length ($l_2 = 1$ and $l_1 = l_3 = 2$ respectively). It is easily verified that the resulting bit string output by the Huffman source coder is uniformly distributed, and that the average codeword length $E[w]$ is $3/2$. Fig. 1 also introduces the general form of homophonic coding, or multiple substitution, which consists of a *message source*, a *reversible memoryless mapping* and a *source encoder*. Conceptually, the required mapping is performed by the *homophonic channel*, whose output is then "compressed" by a suitable source encoder. The homophonic channel and source encoder will be jointly referred to as the *homophonic coder*. Furthermore, in Fig. 1 it is implicitly assumed that an external *randomiser* provides the appropriate random numbers for the selection of the homophones.

In spite of the elegance and simplicity of this scheme, its main drawback is the fact that it is not easily adaptable to changing source statistics, as was also noted in [10]. For this reason we propose the use of arithmetic source coding in homophonic coding.

3   Information-theoretic Analysis of Homophonic Coding

In this section we briefly review some important information theoretical results from [5] and [10]. In Fig. 2 the block diagram of a general homophonic coding scheme is shown. The *message source* U produces a sequence of random variables $U_1, U_2, U_3 \ldots$. The U_i are taken from the L-ary alphabet $U = \{u_1, u_2, \ldots, u_L\}$, $2 \leq L < \infty$, according to the probability distribution $P(u_1), P(u_2), \ldots, P(u_L)$. For simplicity we shall assume that the source is *memoryless* and *stationary*, so that the coding problem reduces to the encoding of the single variable U_i.

The homophonic coder, which consists of a memoryless homophonic channel and a source encoder, maps the output of the message source into the sequence $X_1, X_2, X_3 \ldots$. The randomizer provides the random numbers needed for the selection of the homophones. In [5] an elegant method is introduced for the generation of the required "awkward" random numbers by means of a random binary symmetric source (BSS). For the purpose of this analysis we shall regard the output of the homophonic coder as the *plaintext* sequence to be encrypted.

Let X^n and Y^n denote the *finite* plaintext sequence $\{X_1, X_2, \ldots, X_n\}$ and ciphertext sequence $\{Y_1, Y_2, \ldots, Y_n\}$, respectively. As is customary, and as Fig. 2 suggests, we assume that the secret key Z is statistically independent from the plaintext sequence X^n for all n. A cipher system is called *non-expanding* if the plaintext digits and ciphertext digits take values in the same D-ary alphabet and there is an increasing infinite sequence of positive integers $n_1, n_2, n_3, \ldots$ such that, when Z is known, X^n and Y^n uniquely determine one another for all $n \in S = \{n_1, n_2, n_3, \ldots\}$. We shall also call a sequence of L-ary random

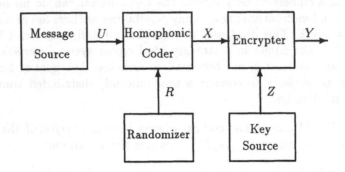

Fig. 2. Schematic diagram of a general homophonic coding scheme.

variables *completely random* if each of its digits is statistically independent of the preceding digits and is equally likely to take on any of the L possible values. Shannon [12] defined the *key equivocation*, or *conditional entropy*, of the secret key Z, given the first n digits of ciphertext as

$$H(Z|Y_1, Y_2, \ldots, Y_n) = H(Z|Y^n) \ .$$

Since $H(Z|Y^n)$ can only decrease as n increases, Shannon called a cipher system *ideal* if the key equivocation approaches a non-zero value as n tends toward infinity, and *strongly ideal* if $H(Z|Y^n)$ is constant, i.e.

$$H(Z|Y^n) = H(Z) \quad \forall n$$

which is equivalent to the statement that the ciphertext sequence is statistically independent of the secret key. The following important results were proved in [5, 10] and are quoted here without proof:

Proposition 1. *If the plaintext sequence encrypted by means of a non-expanding secret-key cipher is completely random, then the ciphertext sequence is also completely random, and is also statistically independent of the secret key.*

Corollary 2. *If the plaintext sequence encrypted by a non-expanding secret-key cipher is completely random, then the cipher is strongly ideal (regardless of the probability distribution for the secret key). Moreover, the conditional entropy of the plaintext sequence, given the ciphertext sequence, satisfies*

$$H(X^n|Y^n) \approx H(Z)$$

for all n sufficiently large.

Therefore, in a ciphertext-only attack, the cryptanalyst can do no better than guessing at random from among as many possibilities as there are possible values of the secret key Z. This illustrates the fact that virtually any secret key cipher can be used as the cipher in a strongly ideal cipher system, *provided that the plaintext source emits a completely random sequence*. The goal of homophonic substitution is precisely to convert a non-uniformly distributed source into a completely random one.

Definition 3. A Homophonic coding scheme is called *perfect* if the resulting encoded binary sequence $X_1, X_2, X_3, \ldots$ is completely random.

Proposition 4. *For the homophonic coder shown in Fig. 1 the expression*

$$H(U) \leq H(V) \leq E[W] = \sum_i P(v_i) l_i$$

holds with equality on the left if and only if the homophonic channel is deterministic, and with equality on the right if and only if the homophonic coder is perfect. Moreover, there exists a binary arithmetic source coding of V such that the scheme is **perfect** *if only if $P(v_i)$ is dyadic, i.e. if and only if $P(V = v_i) = 2^{-l_i}$ holds for all values v_i of V where l_i is the length of the binary codeword assigned to v_i.*

For a proof, see [5].

Definition 5. A homophonic coding scheme is called *optimum* if it is perfect and minimizes the expected length $E[W]$ of the binary codeword assigned to the homophonic channel output V, when the input is the message source output U.

Definition 6. The associated probability distribution $P(v_1), P(v_2), \ldots, P(v_L)$ of a message source V is called *dyadic* if $-\log_2 P(v_i)$ is an integer for all values of i.

Proposition 7. *A homophonic channel is optimum if and only if, for every $u \in U$ its homophones equal (in some order) the terms in the unique dyadic decomposition*

$$P(U = u_i) = \sum_{j \geq 1} P(v_i)^{(j)}$$

where the dyadic decomposition is either a finite or an infinite sum.

For a proof, see [5].

Proposition 4 shows that the task of designing an optimum homophonic coder requires a homophonic channel that minimizes the entropy V. According to Proposition 7, this is equivalent to the requirement that the homophones are to be associated according to the *dyadic* decomposition of the source probabilities $P(u_i)$.

As an example, consider again the homophonic channel shown in Fig. 1. The source symbols occur with probabilities $P(U = u_1) = 1/4$ and $P(U = u_2) = 3/4$. Since $P(U = u_1) = 1/4 = 2^{-2}$ is already in dyadic form, no further decomposition of u_1 is required, and v_1 may be assigned directly to it, i.e. $P(v_1) = 1/4$. For the assignment of homophones to u_2, several possibilities exist. One possibility would be the following decomposition: $P(u_2) = 3/4 = 1/4 + 1/4 + 1/4$, with the associated three homophones each occurring with probability 1/4. According to Proposition 4 this decomposition is not optimal, since it is not dyadic. Another possibility is $P(U = u_2) = 3/4 = 1/2 + 1/8 + 1/16 + 1/32 + \ldots$, which is also not optimal.

Finally, $P(u_2) = 1/2 + 1/4$ gives the desired optimal (dyadic) decomposition. The homophones v_2 and v_3 are assigned to u_2 with probability 1/2 and 1/4 respectively to form a dyadic homophonic channel. Hence, this is an example of a *perfect* homophonic channel. Jendahl et al [5] have also derived the following useful upper bound on $H(V)$ for an optimum homophonic coder.

Proposition 8. *For an optimum binary homophonic coder*

$$H(U) \leq H(V) = E[W] < H(U) + 2$$

In the remainder of this article we will show how arithmetic coding may be applied in optimum homophonic coding.

4   Arithmetic Coding

Arithmetic source coding was first introduced by Rissanen and Langdon [7, 8, 11], and may viewed as a generalisation of Shannon-Fano-Elias coding [3]. In spite of the fact that arithmetic coding is more complex than Huffman coding, it offers some important advantages:

1. As a source coding scheme, arithmetic coding is able to approach the entropy of the source arbitrarily close (for details see [3, 7, 11]).
2. A clear separation exists between modelling of source statistics and encoding of source symbols, which has distinct practical advantages.
3. The algorithm is easily adaptable to varying source statistics.
4. It is not necessary to arrange the symbol probabilities in any particular order, as is required for Huffman coding.

The main idea behind arithmetic source coding is best illustrated by means of an example. We will find it beneficial to take a slight detour, and to consider Huffman source coding first. Thereafter, arithmetic will readily follow as a generalization of Huffman coding.

Table 1. Sample source statistics.

Symbol	Huffman Codewords	$P(v_i)$ (in binary)	Cumulative probability $\sum P(v_i)$	Associated subinterval
v_1	11	.01	.00	$[.0 - .01)$
v_2	0	.1	.01	$[.01 - .11)$
v_3	10	.01	.11	$[.11 - 1)$

4.1 Review of Huffman Source Coding

Consider a source V, producing symbols from the alphabet $V = \{v_1, v_2, v_3\}$ with associated probabilities 1/4, 1/2 and 1/4 respectively, as shown in Table 1. Note that the probabilities are expressed as binary fractions.

The Huffman codewords shown in column two are *prefix-free*, i.e. no codeword is the prefix of another. Suppose the following string of source symbols is to be encoded: v_2, v_2, v_3, v_1. The encoding process is illustrated below by means of Table 2.

Table 2. Illustration of Huffman encoding.

Symbol no.	Symbol	Codewords					
1	v_2	0					
2	v_2		0				
3	v_3			1	0		
4	v_1					1	1
Codeword		0	0	1	0	1	1

The resulting codeword is 001011, obtained by *non-overlapping concatenation* of the individual codewords for each symbol in the sequence. Decoding amounts to a comparison process, starting with the first bit of the received code string. Note that this code is an example of a First-In-First-Out code (FIFO). This is a highly desirable attribute, which implies that the first received symbol can be decoded before the last symbol has been received.

4.2 Review of Arithmetic Source Coding

Next, consider the encoding of the same sequence by means of arithmetic coding. In arithmetic coding, the codewords representing the source symbols can be viewed as *code points* on the half-open unit interval $[0, 1)$, which divide the unit

interval into non-overlapping subintervals. Each codeword (code point) is equal to the *cumulative sum* of the probabilities of the preceding symbols, $\sum P(v_i)$, as shown in column four of Table 1. Note that the probabilities are expressed as *binary fractions*. Each symbol is identified with the *lower* point of the corresponding subinterval, and customarily this is referred to as the *code point C*. For example, the code point for symbol v_1 is .0 and its associated subinterval is $[.0 - .01)$, and for symbol v_3 the code point is .11 and the corresponding subinterval is $[.11 - 1)$. The notation "$[.11, 1)$" means that .11 is included in the interval, as well as all fractions equal to or greater than .11 but less than 1. Symbol v_2 is assigned $1/2$ of the unit interval, and symbols v_1 and v_3 each $1/4$, as illustrated in Fig. 3. The size of the interval above each code point corresponds to the probability of that symbol, and is referred to as the *interval width A*. It is important to note that the symbols in Table 1 appear with their probabilities $P(v_i)$ arranged in *arbitrary* order.

In arithmetic coding, encoding essentially amounts to repeated division of subintervals. The encoding of the given symbol string v_2, v_2, v_3, v_1 is done as follows: For the encoding of the first symbol, v_2, the interval $[.01, .11)$, is selected. To encode the second symbol, v_2, the subinterval $[.01, .11)$ is divided into the same proportions as the original unit interval. The subinterval assigned to the second symbol, v_2, is then $[.001, .101)$. For the third symbol this subinterval is again divided, and the corresponding subinterval for v_3 is $[.1001, .1010)$. The encoding process is graphically illustrated in Fig. 3. This repeated subdivision is continued until all symbols in the sequence have been encoded. Decoding amounts to magnitude comparison, essentially following the inverse of this procedure.

This description of arithmetic coding leads tot he conclusion that two recursions are needed, one for the *code point C*, and one for the *interval width A*. This leads to the following two encoding steps.

Step 1: New code point
The first recursion determines the new code point as the sum of the current code point C, and the product of the width A of the current interval and the cumulative probability $\sum P(v_i)$ of the symbol v_i:

$$C_k = C_{k-1} + A_{k-1} \times \sum P(v_i) \quad ; k = 1, 2, 3 \cdots \qquad \text{with } C_0 = 0 \text{ and } A_0 = 1$$

Step 2: New interval width
The second recursion determines the width A of the new interval, which is the product of the probabilities of the data symbols encoded so far. Thus, the new interval width for the symbol v_i is:

$$A_k = A_{k-1} \times P(v_i) \quad ; k = 1, 2, 3 \cdots \qquad \text{with } A_0 = 1$$

In this way the next code point and its interval width are recursively calculated from the current code point C and interval width A. This double recursion is central to arithmetic coding, and facilitates FIFO-encoding, which is a highly desirable feature for any practical source coding scheme, as noted in the previous section.

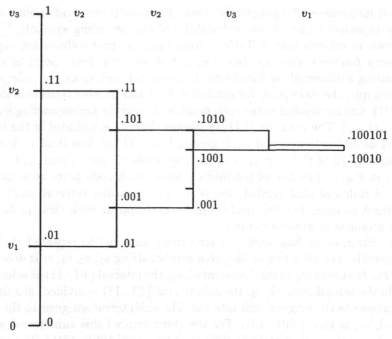

Fig. 3. Illustration of arithmetic coding.

In the arithmetic coding literature the expression $A_{k-1} \times \sum P(v_i)$, which is added to the previous code point C_{k-1}, is often referred to as the *augend* [7]. In Table 3 it is illustrated that the encoding process amounts to the addition of properly scaled cumulative probabilities $\sum P(v_i)$ (augends) to the code string.

Table 3. Arithmetic coding as sums of augends.

Symbol no.	Symbol	Augends				
1	v_2	.0	1			
2	v_2		0	1		
3	v_3			1	1	
4	v_1					0
Codeword		.1	0	0	1	0

The bit string output by the encoder is .10010. Note that 5 bits are needed to represent the given source string, compared to the 6 bits required for Huffman

coding. Comparison with the Huffman coding shown in Table 2 shows that arithmetic coding determines the code string by *overlapped addition* of the individual codewords, instead of *non-overlapped* concatenation. This is the primary reason why arithmetic coding is able to approach the source's entropy arbitrarily close.

The encoding of the third source symbol exemplifies a small problem, called the *carry-over problem*. After encoding the first two symbols, we find that when encoding the third symbol, v_3, all the bits in the codeword are being changed due to the carry-bit, which has a ripple-effect on the already calculated codeword bits. This problem is easily alleviated by means of *bit-stuffing* [8], whereby an additional 0 is introduced into the codeword bit string if a certain number of consecutive 1s have occurred.

4.3   Arithmetic Coding: Decoder Operation

The receiver obtains the code string .10010, which tells the decoder what the encoder did. In essence, the decoder recursively "undoes" the encoder's recursion. This is done in three steps:

Step 1: Decoder comparison
The decoder compares the magnitude of the received binary code string with the cumulative probabilities $\sum P(v_i)$ in Table 1. This shows that the magnitude of the binary string .10010 is equal to, or greater than .01 but less than .11. Hence the first received symbol is decoded as v_2, since the received code string value lies in the sub-interval $[.01, .11)$. Had the first symbol been, for example, v_1, then the code string magnitude would have been less than .01. Once the symbol is decoded it is possible to use the same recursion as the encoder to calculate the interval width:

$$A_k = A_{k-1} \times P(v_i) \quad ; k = 1, 2, 3 \cdots \quad \text{with } A_0 = 1 .$$

For the first decoded symbol this gives: $A_1 = 1 \times .1 = .1$

Step 2: Decoder readjust
Next, the decoder subtracts from the received code string the value of the cumulative probability $\sum P(v_i)$ which corresponds to the decoded symbol v_i. For the first decoded symbol the value $\sum P(v_2) = .01$ is subtracted: $.10010 - .01 = .01010$.

Step 3: Decoder scaling
During encoding the new code point C_k is determined by multiplying $\sum P(v_i)$ with the current interval width A_k. The effect of this multiplication can be "undone" by division with the interval width, or alternatively, by multiplication with the inverse of the interval width. This gives: $.01010 \times 2 = .1010$.

4.4 Discussion

A salient feature of arithmetic coding is the fact that *modelling* and *coding* are separable by a clear-cut line. This separation enables the encoder to adapt to changing source statistics "on the fly". Furthermore, the encoder can be combined with any suitable source modelling algorithm, which has distinct practical advantages when encountering widely varying source statistics.

It should be noted that there does not exist just one *single* arithmetic coding algorithm. Rather, several classes of arithmetic coding can be identified [1, 7, 8, 11]. Although the underlying idea of arithmetic coding is simple, in practice various ad hoc tricks are needed to cater for some awkward practical problems. The specific version of arithmetic coding proposed in this paper, which we shall refer to as the *shift-and-add algorithm*, is much simpler than any of the other cited implementations.

This is a direct consequence of the fact that the symbol probabilities were chosen to be dyadic, i.e. negative powers of two. As shown in Table 3, encoding amounts to the adding of correctly shifted augends. At first glance the requirement that symbol probabilities are constrained to negative powers of two may appear severely restrictive. However, in the next section it will become clear that this requirement can be utilized to great advantage in the case of homophonic coding.

5   A Homophonic Coding Algorithm Based on Arithmetic Coding

We are now in a position to introduce the algorithm for homophonic coding based on arithmetic coding to homophonic coding. In the sequel each of these steps will be discussed briefly.

Step 1: Source modelling
The first step is to estimate the statistics of the source, which requires source modelling. It is beyond the scope of this paper to discuss this important topic, but the interested reader is referred to the book by Bell et al [1], and the articles by Langdon and Rissanen [7][11]. At the end of this section we present some simulation results of the homophonic coding of an ASCII text file. Although no sophisticated source modelling was employed, the simulations still yielded good, useful results.

Step 2: Design of the homophonic channel
The next step is to design the homophonic channel. As mentioned earlier, this requires a dyadic decomposition of the source symbol probabilities. This step is best illustrated by means of an example. Consider a message source which generates symbols from the 5-letter alphabet $U = \{u_1, u_2, u_3, u_4, u_5\}$. Suppose the following sequence of 15 symbols is to be encoded:

$$u_1\ u_1\ u_2\ u_3\ u_3\ u_5\ u_2\ u_1\ u_4\ u_4\ u_5\ u_1\ u_2\ u_1\ u_4 \ .$$

Since the purpose is to illustrate the design of the homophonic channel, it suffices to obtain a crude estimate of symbol probabilities by counting the frequency of occurrence of each symbol. The resulting estimated source probabilities are shown in Table 4, expressed as a decimal fraction in column 2, and as a binary fraction in column 3.

Table 4. Example of designing a homophonic channel.

Source Symbols				Homophones		
Symbol	$P(u_i)$ (decimal)	$P(u_i)$ (binary)	$P(u_i)$ (truncated)	Symbol	$P(v_{ij})$ (binary)	$\sum P(v_{ij]'})$ (binary)
u_1	.3333	.01010101	.0101	v_{11}	.01	.0000
				v_{12}	.0001	.0100
u_2	.2	.00110011	.0011	v_{21}	.001	.0101
				v_{22}	.0001	.0111
u_3	.1333	.00100010	.0010	v_{31}	.001	.1000
u_4	.2	.00110011	.0011	v_{41}	.001	.1010
				v_{42}	.0001	.1100
u_5	.1333	.00100010	.0010	v_{51}	.001	.1101

One important aspect of the shift-and-add arithmetic coding algorithm is the need to represent source probabilities with finite precision arithmetic, which depends on the register-size of the arithmetic processor being used. For purposes of illustration the precision has been limited to 4 bits, as shown in column 4 of Table 4. The dyadic decomposition of the truncated source probabilities readily follows, and each symbol is mapped to a finite number of homophones. This is illustrated below for u_1:

$$P(u_1) = 0.0101\ldots \qquad \text{(truncated)}$$
$$= 1/4 + 1/16 + \epsilon$$
$$= P(v_{11}) + P(v_{12}) + \epsilon$$

As a result of the truncation of the probabilities to a finite precision there will always occur a small error ϵ in the dyadic approximation. However, the magnitude of this error depends on the choice of register-size can be made arbitrarily small.

The right-most column of Table 4 contains the accumulated symbol probabilities $\sum P(v_{ij})$, which represent the codewords for the homophones, as previously discussed. Once the $P(v_{ij})$ and $\sum P(v_{ij})$ have been calculated, the design of the homophonic channel is completed.

In Fig. 4 the allocation of homophones to the source symbols is illustrated. Note that the source symbols do not fill the entire interval $[0, 1)$, but that a small error occurs as a result of the truncation of symbol probabilities to a resolution of 4 bits.

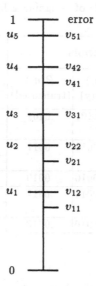

Fig. 4. Illustration of the allocation of homophones.

Step 3: Random selection of homophones

In the third step of the homophonic coding algorithm, each source symbol to be encoded is mapped into one of its associated homophones, chosen at random. The homophones are selected by means of an external randomiser. This introduces additional randomness into the message, which accounts for the increase in entropy of maximally 2 bits/symbol (see Proposition 8). In [5] a novel method is introduced for the generation of the required "awkward" random numbers by means of a binary symmetric source (BSS).

It is interesting to note that the sequence of random binary digits produced by the BSS could be used to transmit useful information. This provides a low-speed subliminal channel, which could be used very effectively in a practical cryptosystem.

Step 4: Arithmetic coding of the homophones

The final step is to encode each randomly selected homophone by means of the shift-and-add arithmetic coding algorithm introduced in the previous section. The output of the arithmetic coder is a string of binary digits. The homophonic

coder thus maps a given (non-uniformly distributed) sequence of source symbols into a uniformly-distributed bit string. It should be pointed out that any implementation of arithmetic coding could be used in this step, once the probabilities $P(v_{ij})$ are known. However, experiments have shown that the shift-add-add algorithm introduced here is about 25% faster than the algorithm given in [1].

In summary, the homophonic coding algorithm consists of the following steps:

1. Modelling: Estimate source statistics.
2. Design of homophonic channel: Decompose symbol probabilities dyadically to form the homophones.
3. Homophonic coding: Map each symbol to be encoded into one of its associated homophones, chosen at random.
4. Arithmetic coding: Encode each selected homophone with the shift-and-add arithmetic coding algorithm into a sequence of bits.

To demonstrate the operation of the homophonic coding algorithm, a file consisting of about 225 000 ASCII characters (7-bit) was encoded. No source modelling was employed, and source statistics were determined by simply counting the frequency of occurrence of each character. The validity of this approach is based on the assumption that the source statistics remain fairly constant throughout the entire file. The results in Fig. 5 show that this is approximately true. For comparison the estimated source entropy of is also shown.

Table 5. Results of statistical tests.

Probabilities	$P(0)$	$P(1)$	$P(00)$	$P(01)$	$P(10)$	$P(11)$
Uncoded Data	0.562	0.438	0.2780	0.3150	0.2519	0.1551
Homophonic Coding	0.499	0.501	0.2495	0.2500	0.2498	0.2507

Note that the per-symbol entropy after homophonic coding is increased by less than 2 bits, according to theory. The resulting encoded file is smaller than the original file, which implies that source coding (data compression) has taken place. It is instructive to ascertain whether the binary output of the arithmetic coder is indeed uniformly distributed. Table 5 shows the probability of occurrence of single bits and pairs of bits. The improvement in distribution of the coded data over the uncoded data is immediately evident.

6   Homophonic Coding Versus Source Coding

In [2] Boyd investigated three source coding (data compression) schemes, viz. Huffman coding, Lempel-Ziv coding and arithmetic coding, and calculated the

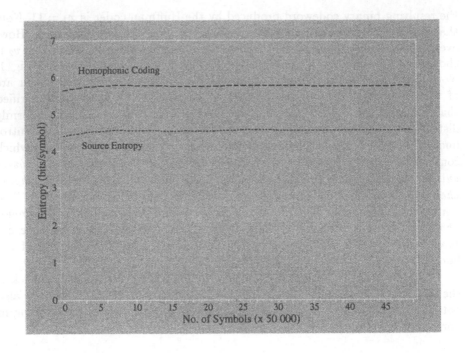

Fig. 5. Simulation results for homophonic coding of ASCII text file.

enhancement in security provided by these three techniques. On the basis of theoretical results and published information he estimated how far these source coding schemes can increase the unicity distance of a cryptosystem. Furthermore, he also posed the question whether homophonic coding offers any advantage above source coding, since it does not increase the value of the unicity distance beyond that provided by source coding.

To answer this question, consider again Fig. 2, and suppose that one of the data compression schemes investigated by Boyd is used in place of the Homophonic Coder. Depending on the accuracy of source modelling, the sequence $X_1, X_2, \ldots$ which is input to the Encrypter will resemble a completely random sequence, and therefore the resulting ciphertext $Y_1, Y_2, \ldots$ will also be random. Essentially the same argument applies when the source coder is replaced by a homophonic coder. Hence, source compression and homophonic coding both provide protection against a *ciphertext-only* attack.

However, in the case of a *known plaintext attack*, or a *chosen plaintext attack*, source coding provides no protection at all, since there exists a *deterministic mapping* between the message source symbols $U_1, U_2, \ldots$ and the output $X_1, X_2, \ldots$ of the source coder. Homophonic coding, on the other hand, produces a *probabilistic mapping* between the source symbols and the output $X_1, X_2, \ldots$.

Therefore, if an encryption system merely needs to be protected against ciphertext-only attacks, it is sufficient to apply source coding. However, if the system is to be protected against known/chosen plaintext attacks as well, additional protection is required which is provided by homophonic coding.

7   Conclusion

This paper introduces a practical homophonic coding algorithm based on arithmetic coding. Homophonic coding provides protection against ciphertext-only attacks as well as chosen/known-plaintext attacks, whereas source coding merely protects against ciphertext-only attacks. It is shown that an optimum homophonic coder can be designed for any given message source by dyadic decomposition of the source probabilities into distinct negative powers of two. The dyadic decomposition leads to a particularly simple implementation of arithmetic coding, termed the *shift-and-add algorithm*, which requires no multiplications, only shifts and additions. The operation of this algorithm was demonstrated by the encoding of an ASCII text file, making use of a very simple form of source estimation. Simulation results show that the resulting binary output is indeed uniformly distributed. Furthermore, it was found that the shift-and-add algorithm is about 25% faster than the arithmetic coding algorithm given in [1].

Acknowledgement
The author would like to thank Professor G J Kühn from the University of Pretoria for helpful discussions and comments in the preparation of this paper.

References

1. T C Bell, J G Cleary and I H Witten, *Text Compression*. Prentice Hall, 1990.
2. C Boyd, "Enhancing secrecy by data compression: theoretical and practical aspects", *Advances in Cryptology – Eurocrypt '91*, LNCS no. 547, Springer-Verlag, pp. 266-280, 1991.
3. T M Cover and J A Thomas, *Elements of Information Theory*. Wiley, New York, 1991.
4. C Günther, "A universal algorithm for homophonic coding", *Advances in Cryptology — Eurocrypt '88*, LNCS no. 330, Springer-Verlag, pp. 405-41, 1988.
5. H N Jendal, Y J B Kuhn and J L Massey, "An information-theoretic treatment of homophonic substitution", *Advances in Cryptology — Eurocrypt '89*, LNCS no. 434, Springer-Verlag, pp. 382-394, 1990.
6. D Kahn, *The Codebreakers: The Story of Secret Writing*. Weidenfeld and Nicolson, London, 1967.
7. G Langdon, "An introduction to arithmetic coding", *IBM J. Res. Develop.*, vol. 28, no. 2, pp. 135-149, March 1984.
8. G Langdon and J Rissanen, "Compression of black-white images with arithmetic coding", *IEEE Trans. Commun.*, vol. COM-29, pp. 858-867, June 1981.
9. J L Massey, "On probabilistic encipherment", *1987 IEEE Information Theory Workshop*, Bellagio, Italy.

10. J L Massey, "Some applications of source coding in cryptography", *European Transactions on Telecommunications*, vol. 5, no. 4, pp. 7/421-15/429, July-August 1994.
11. J Rissanen and G Langdon, "Arithmetic coding", *IBM J. Res. Develop.*, vol. 23, no. 2, pp. 149-162, March 1979.
12. C E Shannon, "Communication theory of secrecy systems", *Bell Syst. Tech. J.*, vol. 28, pp. 656-715, Oct. 1949.

On Fibonacci Keystream Generators

Ross Anderson

Computer Laboratory, Pembroke Street, Cambridge CB2 3QG
Email: rja14@cl.cam.ac.uk

Abstract. A number of keystream generators have been proposed which are based on Fibonacci sequences, and at least one has been fielded. They are attractive in that they can use some of the security results from the theory of shift register based keystream generators, while running much more quickly in software. However, new designs bring new risks, and we show how a system proposed at last year's workshop, the Fibonacci Shrinking Genertor (FISH), can be broken by an opponent who knows a few thousand words of keystream. We then discuss how such attacks can be avoided, and present a new algorithm, PIKE, which is based on the A5 algorithm used in GSM telephones.

1 Introduction

For many years, cryptologists have studied keystream generators based on linear feedback shift registers [1]. When implemented in hardware, such systems can use a relatively small number of gates for a given level of security; they were very popular in the days before very large scale integration, and are still used in applications such as mobile communications where low power consumption, and thus low gate count, are a priority.

However, most cryptographic algorithms are now implemented in software, and shift register generators tend to be slower in software than many alternatives. One problem is that many of them are easier to attack if their feedback polynomials are sparse (or have sparse multiples). This leads a prudent designer to specify a large number of feedback taps, or even to make the feedback depend on the key. While simple to implement in hardware, such schemes are tricky to program.

Consider for example the shrinking generator [2]. Although this is a modern algorithm, designed for efficient hardware implementation, it is scarcely faster in software than DES. The best way its implementers could find to update its key-dependent shift registers was to multiply the current state vectors by suitable binary matrices [3].

2   Generators based on Fibonacci sequences

The performance problem has led some designers of fast software encryption algorithms to abandon the shift register tradition in favour of nonlinear finite state machines [4] [5]. However, we have relatively little theory on these comparable to the cycle length and linear complexity results which can often be obtained for shift register systems, and this has led some designers to use shift register ideas to design generators based on generalised Fibonacci sequences [6] [7].

A generalised Fibonacci sequence is the sequence generated by a monic recurrence relation. More specifically, we will consider $s_i = a_{n-1}s_{i-1} + a_{n-2}s_{i-2} + \ldots + a_1 s_{i-n+1} + a_0 s_{i-n}$ (mod m), where m is a convenient power of 2.

The characteristic polynomial of such a sequence is $X^n + \sum a_i X^i$, and there are conditions on this polynomial which are necessary and sufficient for the sequence it generates to have maximal length [8]. These conditions are not quite the same as those on shift register polynomials, but they coincide for trinomials of degree greater than two. In passing we should note that the least significant bits of a Fibonacci sequence form a linear feedback shift register sequence with characteristic polynomial $X^n + \sum a'_i X^i$, where $a'_i \equiv a_i$ (mod 2).

3   The Fibonacci Shrinking Generator

We will now show how to break Siemens' Fibonacci shrinking generator (FISH), which was presented at the 1993 Cambridge Algorithms workshop [6], and is based on the shrinking generator.

It is driven by two Fibonacci sequences, which are called A and S. These start off with key material and thereafter satisfy the following recurrence relations:

$$a_i = a_{i-55} + a_{i-24} \quad (\text{mod } 2^{32}) \tag{1}$$

$$s_i = s_{i-52} + s_{i-19} \quad (\text{mod } 2^{32}) \tag{2}$$

The least significant bits of s_i are now used to shrink a_i to z_i and s_i to h_i. We will write the j-th bit of a_i as $a_{i,j}$, so that our shrinking rule is the following: if $s_{i,0} = 1$, then we append a_i to the sequence z_k, and s_i to h_k.

Next, writing $\oplus$ for bitwise xor and $\wedge$ for bitwise and, c_i is given by:

$$c_{2i} = z_{2i} \oplus (h_{2i} \wedge h_{2i+1}) \tag{3}$$

$$c_{2i+1} = z_{2i+1} \tag{4}$$

Finally, the output keystream r_i is derived from c_i by swapping the bits $c_{(2i),j}$ and $c_{(2i+1),j}$ whenever $h_{(2i+1),j} = 1$.

Now the first observation to be made about this generator is that the least significant bits of A make up a standard shrinking generator sequence; $\{z_{j,0}\}$ is just $\{a_{i,0}\}$ shrunk by $\{s_{i,0}\}$. By guessing $\{s_{i,0}\}$, we can break this sequence by brute force with an average of 2^{51} trials.

However, we can do significantly better than this. Our fish recipe will have two steps: we will first find ways to speed up the keysearch, and secondly show how to reconstruct the whole key given its least significant bits.

3.1   Sparsity and the shrinking generator

As mentioned in [2], the use of sparse shift registers in the shrinking generator can be dangerous. In fact, since A enjoys the trinomial relation (1), we will find that $1/8$ of these triples $\{a_{i,0}, a_{(i+31),0}, a_{(i+55),0}\}$ will show up in $z_{i,0}$. Their separation is a random quantity, determined by S, but if we look near the likely separations, say $z_{(i-12),0} + z_{i,0} + z_{(i+15),0}$, we will expect to see this relation holding more often than random. In fact one can look at all the likely separations, $z_{(i-x),0} + z_{i,0} + z_{(i+y),0}$, where x is between (say) 12 and 19, and y is between 9 and 15.

To estimate the work factor of an attack based on this, note that every bit in $z_{i,0}$ will be the middle bit in one relation in A. Thus there is a probability of one quarter than the other two bits are in $z_{i,j}$, and about 12% that both of them will be within the ranges mentioned. If $z_{i,0} = 1$, and we find that $z_{(i-19),0}$ ... $z_{(i-12),0} = 11110111$ while $z_{(i+9),0}$ ... $z_{(i+15),0} = 111111$, we can try the six relations $(0,1,1)$ in the knowledge that we have a 12% chance that one of them is actually a relation in A. So whenever we have enough keystream to look for a pattern as advantageous as this, we can expect to perform about 25 keysearches to recover the state of S.

Once we succeed in guessing a relation, we have a divide and conquer attack on S. If, for example, we find one at $z_{(i-12),0} + z_{i,0} + z_{(i+15),0}$, then precisely 15 of $s_{(i-31),0}, ..., s_{i,0}$ are one, together with precisely 12 of $s_{i,0}, ..., s_{(i+24),0}$; and we can derive 3 of these last 25 bits from the others using the relation (2). This reduces the attack complexity significantly (but it is still well over 2^{40} trials).

The next point is that we can cut the complexity of each key trial from $O(|A|^3)$ to $O(|A|)$ by using a trick from [9]; rather than recovering $|A|$ bits from the keystream, solving for $a_{i,0}$ using linear algebra and testing this prediction against the rest of the keystream, we simply insert bits from the keystream into $a_{i,0}$ one at a time until the recurrence relation gives us a clash. Absent an implementation, we expect that about $2|A|$ bits would suffice to detect almost all bad choices of S, it is also significant that this kind of keysearch can be implemented fairly efficiently in hardware.

The most important point, however, is that we get side information from the higher order bits of r_i about which of the possible triples $z_{(i-x),0} + z_{i,0} + z_{(i+y),0}$ represents an actual relation in the A generator. To see this, consider a series of output words r_i. For brevity, we show only the 16 rightmost (least significant) bits:

$$r_0 = 0110101011010110$$
$$r_1 = 1010001101011010$$
$$r_2 = 1001011010110111$$
$$r_3 = 0011100110011011$$

$$\ldots \tag{5}$$

Now when $h_{(2i+1),j} = 0$, $z_{(2i+1),j} = r_{(2i+1),j}$; else $z_{(2i+1),j} = r_{(2i),j}$. Thus, whenever $r_{(2i+1),j} = r_{(2i),j}$, this is equal to $z_{(2i+1),j}$. So the above table of $r_{i,j}$ gives us a partial table for $z_{i,j}$:

$$z_0 = ???????????????0$$
$$z_1 = ??10?01??101??10$$
$$z_2 = ??????????????1$$
$$z_3 = 00?1????10?1??11$$

$$\ldots \tag{6}$$

In other words, we get about half the bits of the $z_{(2i+1)}$, which is a quarter of z_i or an eighth of a_i. From here things are straightforward; given about 2^{12} words of keystream, for example, we would expect to find about one real and three false hits of the form

$$z_{i-x} = ????????????0110$$
$$\ldots$$
$$z_i = ????????????1011$$
$$\ldots$$
$$z_{i+y} = ????????????0001 \tag{7}$$

Finding such patterns reduces the attack complexity from about 25 keysearches to about two of them.

As we get more keystream, we will expect to find relations with x and y significantly more or less than the mean. For example, with 20 million words of keystream we would expect to find a relation which was four standard deviations better than (7), such as $x = 6$ and $y = 8$. In this case, a keysearch would involve about 2^{39} trials.

3.2   Reconstructing the rest of the key

Given $s_{i,0}$ and $a_{i,0}$, we next reconstruct the higher order bits of a_i; we know about every eighth bit of $a_{i,j}$ for each j. The tables in [10] show that with a correlation of 0.125 and a trinomial recurrence relation, a reconstruction will succeed when the length of available keystream is somewhere between 550 and 5500 bits long (and nearer the former). Of course, this recurrence relation operates modulo 2^k for $1 \leq k \leq 31$ rather than over $GF(2)$, so we have to deal with carry bits as they arise; to do this, we use an algorithm like that in [11] to reconstruct first $a_{i,1}$, then $a_{i,2}$ and so on up to $a_{i,31}$.

Once we have a_i, we can get the higher order bits of s_i using those bits of r_i where $r_{(2i),j} \neq r_{(2i+1),j}$. Comparing these with our reconstructed values of z_i gives us about half of the values of $h_{(2i+1),j}$, and thus an eighth of the values of $s_{i,j}$. From this, a correlation attack can proceed as before from $s_{i,0}$ to $s_{i,1}$, $s_{i,2}$, and so on up to $s_{i,31}$.

4   An improved Fibonacci generator

The above break, together with Cain and Sherman's recent break of the Gifford cipher [12], inspires us to ask whether we can find a Fibonacci generator which is both fast in software and reasonably strong. We find some ideas in A5.

This is the algorithm used in many GSM telephones to encrypt voice traffic. It consists of three shift registers of lengths 19, 22 and 23, with sparse feedback taps, which are interlocked in the sense that a threshold function of the middle bits of each register is used to decide which registers are clocked in any given cycle (usually two of them are) [13].

The best known attack on A5 consists of guessing the state of two of the registers and then working back from the keystream to get the state of the third. There has been controversy about the work factor involved in each key trial, and at least one telecom engineer has argued that this is about 2^{12} operations giving a real attack complexity on A5 of 2^{52} rather than the 2^{40} which one might naïvely expect. As we understand that a hardware keysearch unit is under design, we expect that this debate will be settled shortly.

Nonetheless, A5 passes all the standard series randomness tests [14]; its only known weaknesses are that its registers are too short, and that it has a minimum cycle length of $\frac{4}{3}2^k$, where k is the length of the longest shift register [15] (the cycle length is not a concern in the GSM protocol, as the generator is re-keyed after each packet). A5 is also efficient; the main engineering constraint on GSM equipment is battery life, and so one may surmise that its developers sought to produce an algorithm of adequate strength, but with the smallest possible gate count. In this, they appear to have done a competent job; and this motivates us to look for a fast software algorithm which is uses the underlying ideas of A5.

Our proposal is therefore as follows. We start off with the three Fibonacci generators whose relations are:

$$a_i = a_{i-55} + a_{i-24} \pmod{2^{32}} \tag{8}$$

$$a_i = a_{i-57} + a_{i-7} \pmod{2^{32}} \tag{9}$$

$$a_i = a_{i-58} + a_{i-19} \pmod{2^{32}} \tag{10}$$

We next observe that in FISH, had the control bits been the carry bits rather than the least significant bits, then our attack would have been much harder; so we will use the carry bits as controls. If all three of them are the same, then we will step all three generators; if not, we will step those two generators whose carry bits agree. This control will be delayed eight cycles; after we update the state, we inspect the control bits and write one control nybble to a register which is shifted four bits with the next update. With some processors, it may be more convenient to use the parity bits as a control; this appears to be an acceptable variant.

The next keystream word is the xor of the least significant words of all three generators. Note that this algorithm should be slightly faster than FISH, as each keystream word will take on average 2.75 generator updates to compute rather than 3; and finally, in order to ensure that we use only a small fraction of the minimum sequence length, we specify that the generator should be re-keyed after 2^{32} words have been generated. The keying method is not part of this description, but clearly a short user supplied key can be expanded using a hash function such as SHA to provide the 700 bytes of initial state.

We call this algorithm PIKE; the pike is at the top of the food chain of our local waters, being longer, leaner and meaner than the other fish. Fishermen are of course invited to try their arm.

Acknowledgements: David Wheeler first expressed doubt about the non-linear combining operations in FISH; Don Coppersmith pointed out the vulnerability of the sparse shrinking generator; David Wheeler pointed out the efficacy of using carry bits as controls; and Gideon Yuval remarked at the workshop that parity might be more convenient on some processors.

References

[1]   RA Rueppel, *'Analysis and Design of Stream Ciphers'*, Springer Verlag Communications and Control Engineering Series (1986)

[2]   D Coppersmith, H Krawczyk, Y Mansour, "The Shrinking Generator", in *Advances in Cryptology - CRYPTO '93*, Springer LNCS v **773** pp 22–39

[3]   H Krawczyk, "The Shrinking Generator: some practical considerations", in *Fast Software Encryption*, Springer LNCS v **809** pp 45–46

[4]   DJ Wheeler, "A Bulk Data Encryption Algorithm", in *Fast Software Encryption*, Springer LNCS v **809** pp 126–134

[5]   P Rogaway, D Coppersmith, "A Software-Optimised Encryption Algorithm", in *Fast Software Encryption*, Springer LNCS v **809** pp 56–63

[6]   U Blöcher, M Dichtl, "Fish: a fast software stream cipher", in *Fast Software Encryption*, Springer LNCS v 809 pp 41–44

[7]   JD Golić, "Linear Cryptanalysis of Stream Ciphers", *this volume*

[8]   RP Brent, "On the periods of generalised Fibonacci sequences", in *Mathematics of Computation* v **63** no 207 (July 1994) pp 389–401

[9]   RJ Anderson, "Solving a Class of Stream Ciphers", in *Cryptologia* v **XIV** no 3 (July 1990) pp 285–288

[10]   W Meier, O Staffelbach, "Fast Correlation Attacks on Certain Stream Ciphers", in *Journal of Cryptology* v **1** (1989) pp 159–176

[11]   DJC MacKay, 'A Free Energy Minimization Framework for Inference Problems in Modulo 2 Arithmetic" *in this volume* pp 179–195

[12]   TR Cain, AT Sherman, "How to break Gifford's Cipher", in *Proceedings of the 2nd ACM Conference on Computer and Communications Security* (Fairfax, 1994) pp 198–209

[13]   RJ Anderson, "A5 (Was: HACKING DIGITAL PHONES)", message number <2ts9a0$95r@lyra.csx.cam.ac.uk> posted to usenet newsgroup sci.crypt, 17 Jun 1994 13:43:28 GMT.

[14]   M Roe, *private communication*

[15]   WG Chambers, "On Random Mappings and Random Permutations", *this volume* pp 22–28

Cryptanalysis of McGuffin

Vincent Rijmen*   Bart Preneel**

Katholieke Universiteit Leuven
ESAT-COSIC
K. Mercierlaan 94, B-3001 Heverlee, Belgium

{bart.preneel,vincent.rijmen}@esat.kuleuven.ac.be

Abstract. This paper shows that the actual proposal for an unbalanced
Feistel network by Schneier and Blaze is as vulnerable to differential
cryptanalysis as the DES.

1   McGuffin

Schneier and Blaze introduce in [SB95] a new kind of block ciphers: the *Generalized Unbalanced Feistel Network*. Together with the general architecture they give a complete specification of an example. The basic idea is to split the input of each round into unequal parts. In the example, the 64-bit input is split into a 48-bit input of the F-function, and a 16-bit part that is exored with the output of the F-function. The F-function consists of the 8 S-boxes of the DES, but the two middle output bits of each S-box are neglected in order to obtain a 16-bit output.

2   Differential Characteristics

In [Ma94] Matsui demonstrated that one can find the best differential characteristics and linear relations for the DES with a clever search algorithm. This encouraged us to try the same for McGuffin. For the DES it is very important to depart from very good starting values in order to obtain the characteristics in relatively short time (a few hours). For McGuffin, we had no good guesses for the starting values and became the best characteristics for two to 32 rounds in about the same time. This indicates that McGuffin is very vulnerable for differential cryptanalysis. Table 1 gives the probabilities of the best differential characteristics of McGuffin. It turns out that the probability of the best $2n$-round characteristic of McGuffin is significantly larger than the probability of

---

* N.F.W.O. research assistant, sponsored by the National Fund for Scientific Research (Belgium).

** N.F.W.O. postdoctoral researcher, sponsored by the National Fund for Scientific Research (Belgium).

the best n-round characteristic of the DES. From this viewpoint 32 rounds of McGuffin is weaker than 16 rounds of the DES. Figure 1 shows the four-round iterative building block of the best differential characteristic for McGuffin. It has a probability of $\frac{1}{149}$, which should be compared with $\frac{1}{234}$ for the best two-round iterative building block for a DES-characteristic. Since every input bit enters three S-boxes before it is modified, these probabilities are key dependent. This implies that the attack can probably be improved.

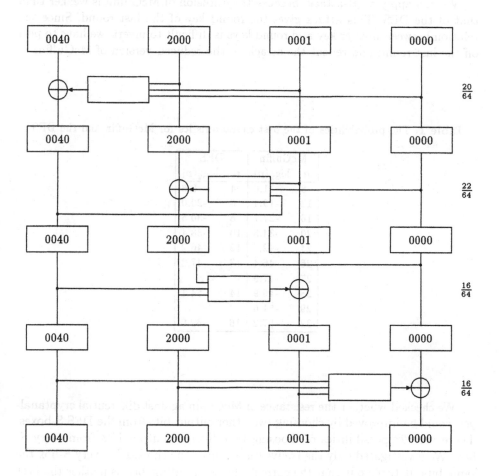

Fig. 1. The 4-round iterative building block of the best differential characteristic for McGuffin

Biham and Shamir [BS91] used a 13-round characteristic for their attack on the full DES. The first round is passed with probability one by enciphering large structures of plaintexts. The last two rounds are treated by the 2R-attack.

In our attack on McGuffin, we use a 27-round characteristic from the second to the 28th round. Extended to the first round, this characteristic has an input exor that is different from zero for S-box eight only. Since each S-box has only two output bits, only four different output exors are possible. Therefore we can pass the first round with probability one by enciphering structures of only eight plaintexts (in comparison to 8192 for the DES). Such a structure consists of the messages $P \oplus (v, 0000, 0000, 0000)$, $P \oplus \alpha \oplus (v, 0000, 0000, 0000)$, where $\alpha = (4040, 2000, 0001, 0000)$ and v takes the values from the set $\{0000, 0001, 0002, 0003\}$.

We can apply a 4R-attack, because the diffusion of McGuffin is weaker than that of the DES. This attack gives the round key of the last round. Since the relation between master key and round keys is difficult to invert, we have to peel off the last round and repeat the attack on the reduced version of McGuffin.

Table 1. The probabilities of the best characteristics for McGuffin and the DES

McGuffin		DES	
n	$\log_2(p)$	n	$\log_2(p)$
8	-11.6	4	-9.6
12	-19.4	6	-20.0
16	-27.7	8	-30.5
20	-34.8	10	-38.4
24	-42.7	12	-46.2
26	-46.4	13	-47.2
27	-47.9		
28	-49.9	14	-54.1
29	-51.6		
32	-57.2	16	-62.0

We checked whether the resistance of McGuffin against differential cryptanalysis could be improved by choosing two other output bits from the DES S-boxes. There are 12^8 possibilities of choosing two different output bits from every S-box. We investigated only the twelve cases where one chooses in every S-box the same bits. It turns out that the current choice of output bits is neither the best nor the worst possible choice. Selecting the second and the first output bit from the S-boxes of the DES to be the first and second bit from the McGuffin S-boxes seems the best alternative: the probability for the best 27-round characteristic becomes $2^{-50.8}$.

The results we presented at the workshop contained a small error. After correction, we see that McGuffin is slightly more resistant against differential cryptanalysis than we first thought. The improvement is however not enough to make the cipher resistant against differential cryptanalysis.

3   Linear Relations

We searched for linear relations for up to 32 rounds of McGuffin. When searching for linear relations over several rounds, every 'forked branch' in the algorithm gives the cryptanalyst a choice. A linear relation can be viewed as the tracing of bits through the different rounds of the algorithm. Every 'forked branch' is then a 'crossroad' where the cryptanalyst can choose either way to follow the bits. As a consequence of the inbalance of McGuffin, there are 50 % more forked branches in each round than for the DES (48 'forked' bits instead of 32). On the other hand, the reduction of the output of the S-boxes reduces the number of possible linear approximations of each S-box with 80 % (only 3 possible output masks instead of 15). The effect of reducing the number of output bits on the expected value of the probability of the best linear relation is discussed in [Ny95].

Table 2. The probabilities of the best linear relations for McGuffin and the DES

McGuffin		DES	
n	$\log_2(|p - 0.5|)$	n	$\log_2(|p - 0.5|)$
4	-2.0	2	-1.7
8	-5.0	4	-4.0
12	-9.7	6	-8.0
16	-13.7	8	-10.7
20	-18.4	10	-14.4
24	-21.9	12	-16.8
28	-26.6	14	-20.8
30	-28.6	15	-21.8
32	-30.1	16	-23.4

The probabilities of $2n$-round relations for McGuffin are lower than the probabilities of n-round DES-relations (cf. Table 2). This means that a straightforward application of Matsui's algorithm 2 would need $2^{2 \times (28.6 - 20.8)} \times 2^{43} = 2^{58.6}$ plaintexts in order to find 12 bits of the round keys of the first and the last round. This is still faster than exhaustive key search. In order to determine the remaining part of these two round keys, other linear relations should be used.

The structure of the best 30-round linear relation is shown in Figure 2.

In Sect. 2 we showed that McGuffin could be strengthened against differential cryptanalysis by selecting other output bits from the DES S-boxes. Selecting the second and the first output bit produced the strongest cipher. We searched also for the linear relations of this adapted version. It has approximately the same resistance against linear cryptanalysis as the original version.

structure: - - E - - - A - B C D - - - A - B C D - - - A - B C D - - -

box (1-8)	α	β	$2|p - 0.5|$
E	4	3_x	2_x	0.1250
A	4	38_x	1_x	0.3125
B	2	26_x	1_x	0.1875
C	6	15_x	1_x	0.1875
D	4	$2F_x$	1_x	0.3125

Fig. 2. Structure of the optimal 30-round linear relation

4 Conclusion

McGuffin is not more resistant against a differential attack than the DES. Selecting other output bits from the DES S-boxes strenghtens the algorithm. Modifying a scheme with only existing attacks in mind however is not a good design principle. The extension of the linear attack to McGuffin is not as straightforward as the extension of the differential attack, but we feel that the increase in security against this attack is marginal. A more detailed study will probably reveal a linear-like attack with about the same probability for success as the linear attack on the DES (e.g., by using multiple linear relations [KR95]).

Acknowledgements

We would like to thank David Wagner for helpful comments.

References

[BS90]   E. Biham and A. Shamir, "Differential cryptanalysis of DES-like cryptosystems," *Journal of Cryptology*, Vol. 4, No. 1, 1991, pp. 3–72.

[BS91]   E. Biham and A. Shamir, "Differential cryptanalysis of the full 16-round DES," *Technion Technical Report # 708*, December 1991. (See also *Proceedings of Crypto'92, LNCS 740*, E.F. Brickell, Ed., Springer Verlag, pp. 487–496.)

[FI46-77] FIPS 46, *"Data Encryption Standard,"* National Bureau of Standards, 1977.

[KR95]   B. Kaliski and M. Robshaw, "Linear cryptanalysis using multiple approximations and FEAL," *Fast Software Encryption*, these proceedings, pp. 249–264.

[Ma93a] M. Matsui, "Linear cryptanalysis method for DES cipher," *Advances in Cryptology, Proc. Eurocrypt'93, LNCS 765*, T. Helleseth, Ed., Springer-Verlag, 1994, pp. 386–397.

[Ma93b] M. Matsui, "Cryptanalysis of DES cipher (I)," December 1993, preprint.

[Ma94]   M. Matsui, "On correlation between the order of S-boxes and the strength of DES", *Advances in Cryptology, Proc. Eurocrypt'94, LNCS*, A. De Santis, Ed., Springer-Verlag, to appear.

[Ny95]   K. Nyberg, "S-boxes and round functions with controllable linearity and differential uniformity," *Fast Software Encryption*, these proceedings, pp. 111–130.

[SB95]   B. Schneier and M. Blaze, "The MacGuffin block cipher algorithm," *Fast Software Encryption*, these proceedings, pp. 97–110.

Performance of Block Ciphers and Hash Functions — One Year Later

Michael Roe

Cambridge University Computer Laboratory,
Pembroke Street, Cambridge CB2 3QG, UK
Email: mrr@cl.cam.ac.uk

1   Introduction

This paper extends the algorithm performance measurements which were presented at the 1993 workshop on fast software encryption [15]. The measurement techniques which were used are described in the original paper [15].

The main changes from the original paper are as follows:

- The NIST Secure Hash Algorithm (SHA) has been replaced with a new algorithm, SHA-1 [10]. The reason for this change is that NIST (or NSA) discovered an attack against the original SHA algorithm [11].
- This year's measurements are based on a faster implementation of GOST 28147.
- This year's measurements were made with a different Sun workstation. The new machine is significantly slower; as a result, all the figures in the "Sparc" column of the tables have changed.
- Some stream ciphers have been included. Many of the most interesting new algorithms in 1994 were stream ciphers. In particular, 1994 saw the publication of what were alleged to be the specifications of two proprietary stream ciphers, RC4 [1] and A5.

2   Apparatus

These measurements were carried out on Unix workstations from two different manufacturers:

- A DEC 3000/400 "Sandpiper". This uses an Alpha CPU clocked at 133 MHz. It has two times 8 KB of primary cache memory and 512 KB of secondary cache.
- A Sparc Station SLC. This uses a Sparc CPU (clock speed unknown).

---

[1] RC4 is a registered trademark of RSA Data Security Inc.

3   Experimental Approach

The performance test for hash algorithms measures the time taken to hash a message containing 6,400,000 octets. This is implemented by clearing a 64 octet buffer, calling the routine to initialise hashing, calling the routine to hash a buffer 100,000 times, and then calling the routine to finish hashing.

The performance test for symmetric key algorithms measures the time taken to encrypt a message of 6,400,000 bits with a fixed key. This is implemented by clearing a 64-bit buffer, and then calling the routine to encrypt the buffer 100,000 times.

The performance test for stream ciphers measures the time taken to generate 6,400,000 bits of keystream. The different stream cipher implementations generate different amounts of keystream per subroutine call. The algorithms from Bill Chambers generate 32 bits of keystream at a time, while SEAL generates 32768 bits, and WAKE generates 4096 bits. This is shown in the "b/call" column of figure 2. Implementations which only generate a few bits per subroutine call are clearly at a disadvantage in this test; more subroutine calls will be needed to generate the same amount of keystream, and so they will be slower. It should be possible to re-implement Bill Chambers' stream ciphers in such a way that more keystream is generated with each call. This ought to improve their performance.

4   Algorithm Parameters

Some of the algorithms which were tested provide a variable level of security. That is, they have a parameter which the user can set so as to select an appropriate compromise between security and execution speed. For performance measurements of such algorithms to be meaningful, the settings of the parameters must be given. In these experiments, the following parameter settings were used:

RC5	32 bit word size, 32 rounds, 160 bit key
SAFER-K64	6 rounds
Blowfish	64 bit key

Fig. 1. Algorithm Parameters

5 Results

Figure 2 shows the speeds (in Mbits/second) of the algorithms on the two test machines.

Stream Ciphers				
Proposer	Name	b/call	Sparc Mb/s	Alpha Mb/s
David Wheeler	WAKE [20]	4096	14.12	117.0
Phil Rogaway	SEAL [16]	32768	16.64	114.8
Bill Chambers	Clock Controlled [2]	32	2.40	19.2
Ron Rivest	RC4	1024	3.06	15.4
Bill Chambers	Linear Congruential [2]	32	0.031	0.738

Block Ciphers				
Proposer	Name	Block Size	Sparc Mb/s	Alpha Mb/s
Burt Kaliski	[6]	8192	2.87	26.8
Bruce Schneier	Blowfish [18]	64	1.62	11.63
Ron Rivest	RC5 [14]	64	1.42	7.68
Jim Massey	SAFER-K64 [7]	64	1.31	7.68
GOST	GOST 28147 [17, 19]	64	0.75	7.25
Joan Daemen	3WAY [4]	96	0.56	5.24
Meyer,Tuchman	DEA-1 [9]	64	0.294	1.855
Meyer	DEA-1, EDE mode [8]	64	0.168	1.200

Hash Functions				
Proposer	Name	Hash Size	Sparc Mb/s	Alpha Mb/s
Ron Rivest	MD4 [12]	128	9.45	78.77
Ron Rivest	MD5 [13]	128	7.28	60.02
RIPE project	RIPE-MD [3]	128	5.76	48.00
NIST	SHA-1 [10]	160	4.23	41.51
Burt Kaliski	MD2 [5]	128	0.137	0.755

Fig. 2. Algorithm Speeds — Long Messages

References

1. R. J. Anderson, editor. *Fast Software Encryption*, number 809 in Lecture Notes in Computer Science. Springer-Verlag, December 1993.
2. B. Chambers. Two stream ciphers. In Anderson [1], pages 51 – 55.
3. CWI, Amsterdam. *RIPE Integrity Primitives — Final Report of RACE Integrity Primitives Evaluation (R1040)*, June 1992.
4. J. Daemen, R. Govaerts, and J. Vandewalle. A new approach to block cipher design. In Anderson [1], pages 18 – 32.
5. B. Kaliski. *RFC 1319 : The MD2 Message-Digest Algorithm*, April 1992.
6. B. Kaliski and M. Robshaw. Fast block cipher proposal. In Anderson [1], pages 33 – 40.
7. J. L. Massey. SAFER K-64: A byte-oriented block-ciphering algorithm. In Anderson [1], pages 1 – 17.
8. C. H. Meyer and S. M. Matyas. *Cryptography: a new dimension in computer data security*. John Wiley and Sons, 1982.
9. National Bureau of Standards. *Federal Information Processing Standard — Publication 46: Data Encryption Standard*, 1977.
10. National Institute of Standards and Technology. *Federal Information Processing Standard — Publication 180-1: Secure Hash Standard*, May 1994.
11. National Institute of Standards and Technology. NIST announces technical correction to secure hash standard. Press release, April 1994.
12. R. L. Rivest. *RFC 1320 : The MD4 Message-Digest Algorithm*, April 1992.
13. R. L. Rivest. *RFC 1321 : The MD5 Message-Digest Algorithm*, April 1992.
14. R. L. Rivest. The RC5 encryption algorithm. In *Fast Software Encryption*, Lecture Notes in Computer Science. Springer-Verlag, pages 86–96 (these proceedings).
15. M. Roe. Performance of symmetric ciphers and one-way hash functions. In Anderson [1].
16. P. Rogaway and D. Coppersmith. A software optimised encryption algorithm. In Anderson [1], pages 56 – 63.
17. [Russian] State Committee for Standardization, Metrology and Certification. *GOST 28147 : Cryptographic Protection for Data Processing Systems — Cryptographic Transformation Algorithm*, 1990.
18. B. Schneier. Description of a new variable-length key, 64-bit block cipher (Blowfish). In Anderson [1], pages 191 – 204.
19. B. Schneier. The GOST encryption algorithm. *Dr. Dobb's Journal*, pages 143 – 144, January 1995.
20. D. Wheeler. A bulk data encryption algorithm. In Anderson [1], pages 127 – 134.

TEA, a Tiny Encryption Algorithm

David J. Wheeler   Roger M. Needham

Computer Laboratory
Cambridge University
England

Email: {David.Wheeler,Roger.Needham}@cl.cam.ac.uk

Abstract. We give a short routine which is based on a Feistel iteration and uses a large number of rounds to get security with simplicity.

Introduction

We design a short program which will run on most machines and encypher safely. It uses a large number of iterations rather than a complicated program. It is hoped that it can easily be translated into most languages in a compatible way. The first program is given below. It uses little set up time and does a weak non linear iteration enough rounds to make it secure. There are no preset tables or long set up times. It assumes 32 bit words.

Encode Routine

Routine, written in the C language, for encoding with key k[0] - k[3]. Data in v[0] and v[1].

```
void code(long* v, long* k)  {
unsigned long y=v[0],z=v[1], sum=0,   /* set up */
 delta=0x9e3779b9,   /* a key schedule constant */
 n=32 ;
while (n-->0) {                /* basic cycle start */
  sum += delta ;
    y += ((z<<4)+k[0]) ^ (z+sum) ^ ((z>>5)+k[1]) ;
    z += ((y<<4)+k[2]) ^ (y+sum) ^ ((y>>5)+k[3]) ;
              }    /* end cycle */
v[0]=y ; v[1]=z ; }
```

Basics of the routine

It is a Feistel type routine although addition and subtraction are used as the reversible operators rather than XOR. The routine relies on the alternate use of

XOR and ADD to provide nonlinearity. A dual shift causes all bits of the key and data to be mixed repeatedly.

The number of rounds before a single bit change of the data or key has spread very close to 32 is at most six, so that sixteen cycles may suffice and we suggest 32.

The key is set at 128 bits as this is enough to prevent simple search techniques being effective.

The top 5 and bottom four bits are probably slightly weaker than the middle bits. These bits are generated from only two versions of z (or y) instead of three, plus the other y or z. Thus the convergence rate to even diffusion is slower. However the shifting evens this out with perhaps a delay of one or two extra cycles.

The key scheduling uses addition, and is applied to the unshifted z rather than the other uses of the key. In some tests k[0] etc. were changed by addition, but this version is simpler and seems as effective. The number delta, derived from the golden number is used where

$$\text{delta} = (\sqrt{5} - 1)2^{31}$$

A different multiple of delta is used in each round so that no bit of the multiple will not change frequently. We suspect the algorithm is not very sensitive to the value of delta and we merely need to avoid a bad value. It will be noted that delta turns out to be odd with truncation or nearest rounding, so no extra precautions are needed to ensure that all the digits of sum change.

The use of multiplication is an effective mixer, but needs shifts anyway. It was about twice as slow per cycle on our implementation and more complicated.

The use of a table look up in the cycle was investigated. There is the possibility of a delay ere one entry of the table is used. For example if k[z&3] is used instead of k[0], there is a chance one element may not be used of $(3/4)^{32}$, and a much higher chance that the use is delayed appreciably. The table also needed preparation from the key. Large tables were thought to be undesirable due to the set up time and complication.

The algorithm will easily translate into assembly code as long as the exclusive or is an operation. The hardware implementation is not difficult, and is of the same order of complexity as DES [1], taking into account the double length key.

Usage

This type of algorithm can replace DES in software, and is short enough to write into almost any program on any computer. Although speed is not a strong objective with 32 cycles (64 rounds), on one implementation it is three times as fast as a good software implementation of DES which has 16 rounds.

The modes of use of DES are all applicable. The cycle count can readily be varied, or even made part of the key. It is expected that security can be enhanced by increasing the number of iterations.

Selection of Algorithm

A considerable number of small algorithms were tried and the selected one is neither the fastest, nor the shortest but is thought to be the best compromise for safety, ease of implementation, lack of specialised tables, and reasonable performance. On languages which lack shifts and XOR it will be difficult to code. Standard C does makes an arithmetic right shift and overflows implementation dependent so that the right shift is logical and y and z are unsigned.

Analysis

A few tests were run to detect when a single change had propagated to 32 changes within a small margin. Also some loop tests were run including a differential loop test to determine loop closures. These tests failed to show any unexpected behaviour.

The shifts and XOR cause changes to be propagated left and right, and a single change will have propagated the full word in about 4 iterations. Measurements showed the diffusion was complete at about six iterations.

There was also a cycle test using up to 34 of the bits to find the lengths of the cycles. A more powerful version found the cycle length of the differential function.

$$d(x)=f(x \text{ XOR } 2^p) \text{ XOR } f(x)$$

which may test the resistance to some forms of differential crypto-analysis [2].

Conclusions

We present a simple algorithm which can be translated into a number of different languages and assembly languages very easily. It is short enough to be programmed from memory or a copy. It is hoped it is safe because of the number of cycles in the encoding and length of key. It uses a sequence of word operations rather than wasting the power of a computer by doing byte or 4 bit operations.

Acknowledgements

Thanks are due to Mike Roe and other colleagues who helped in discussion and tests and to the helpful improvements suggested by the reviewer.

References

1 National Institute of Standards, Data Encryption Standard, Federal Information Processing Standards Publication 46. January 1977
2 E. Biham and A. Shamir, Differential Analysis of the Data Encryption Standard, Springer-Verlag, 1993
3 B. Schneier, Applied Cryptology, John Wiley & sons, New York 1994.

Appendix

Decode Routine

```
void decode(long* v,long* k)  {
 unsigned long n=32, sum, y=v[0], z=v[1],
 delta=0x9e3779b9 ;
sum=delta<<5 ;
                               /* start cycle */
while (n-->0) {
    z-= ((y<<4)+k[2]) ^ (y+sum) ^ ((y>>5)+k[3]) ;
    y-= ((z<<4)+k[0]) ^ (z+sum) ^ ((z>>5)+k[1]) ;
    sum-=delta ;  }
                               /* end cycle */
v[0]=y ; v[1]=z ;  }
```

Implementation Notes

It can be shortened, or made faster, but we hope this version is the simplest to implement or remember.

A simple improvement is to copy k[0-3] into a,b,c,d before the iteration so that the indexing is taken out of the loop. In one implementation it reduced the time by about 1/6th.

It can be implemented as a couple of macros, which would remove the calling overheads.

Author Index

Springer-Verlag
and the Environment

We at Springer-Verlag firmly believe that an international science publisher has a special obligation to the environment, and our corporate policies consistently reflect this conviction.

We also expect our business partners – paper mills, printers, packaging manufacturers, etc. – to commit themselves to using environmentally friendly materials and production processes.

The paper in this book is made from low- or no-chlorine pulp and is acid free, in conformance with international standards for paper permanency.

Lecture Notes in Computer Science

For information about Vols. 1–926

please contact your bookseller or Springer-Verlag